Artificial Intelligence and Blockchain in Industry 4.0

The book addresses the challenges in designing blockchain-based secured solutions for Industry 4.0 applications using artificial intelligence. It further provides a comparative analysis of various advanced security approaches such as edge computing, cybersecurity, and cloud computing in the realm of information technology.

This book:

- Addresses the challenges in designing blockchain-based secured solutions for Industry 4.0 applications using artificial intelligence.
- Provides a comparative analysis of various advanced security approaches such as edge computing, cybersecurity, and cloud computing in the realm of information technology.
- Discusses the evolution of blockchain and artificial intelligence technology, from fundamental theories to practical aspects.
- Illustrates the most recent research solutions that handle the security and privacy threats while considering the resource-constrained in Industry 4.0 devices.
- Showcases the methods and tools necessary for intelligent data analysis and gives solutions to problems resulting from automated data collection.

The text aims to fill the gap between the theories of blockchain and its practical application in business, government, and defense, among other areas. It further highlights the challenges associated with the use of blockchain for various Industry 4.0 applications such as data analytics, software-defined networks, cyber-physical systems, drones, and cybersecurity. The text is primarily written for senior undergraduate, graduate students, and academic researchers in the fields of electrical engineering, electronics and communication engineering, computer engineering, manufacturing engineering, and industrial engineering.

Future Generation of Soft and Intelligent Computing
Series Editors: Raghavendra Kumar and Hoang Viet Long

Artificial Intelligence and Blockchain in Industry 4.0
Rohit Sharma, Rajendra Prasad Mahapatra and Gwanggil Jeon

Artificial Intelligence and Blockchain in Industry 4.0

Edited by
Rohit Sharma
Rajendra Prasad Mahapatra
Gwanggil Jeon

Boca Raton London New York

CRC Press is an imprint of the
Taylor & Francis Group, an **informa** business

First edition published 2024
by CRC Press
6000 Broken Sound Parkway NW, Suite 300, Boca Raton, FL 33487-2742

and by CRC Press
4 Park Square, Milton Park, Abingdon, Oxon, OX14 4RN

CRC Press is an imprint of Taylor & Francis Group, LLC

© 2024 selection and editorial matter, Rohit Sharma, Rajendra Prasad Mahapatra and Gwanggil Jeon; individual chapters, the contributors

Reasonable efforts have been made to publish reliable data and information, but the author and publisher cannot assume responsibility for the validity of all materials or the consequences of their use. The authors and publishers have attempted to trace the copyright holders of all material reproduced in this publication and apologize to copyright holders if permission to publish in this form has not been obtained. If any copyright material has not been acknowledged please write and let us know so we may rectify in any future reprint.

Except as permitted under U.S. Copyright Law, no part of this book may be reprinted, reproduced, transmitted, or utilized in any form by any electronic, mechanical, or other means, now known or hereafter invented, including photocopying, microfilming, and recording, or in any information storage or retrieval system, without written permission from the publishers.

For permission to photocopy or use material electronically from this work, access www.copyright.com or contact the Copyright Clearance Center, Inc. (CCC), 222 Rosewood Drive, Danvers, MA 01923, 978-750-8400. For works that are not available on CCC please contact mpkbookspermissions@tandf.co.uk

Trademark notice: Product or corporate names may be trademarks or registered trademarks and are used only for identification and explanation without intent to infringe.

Library of Congress Cataloging-in-Publication Data
Names: Sharma, Rohit (Assistant professor of electronics), editor. | Mahapatra, Rajendra Prasad, editor. | Jeon, Gwanggil, editor.
Title: Artificial intelligence and blockchain in Industry 4.0 / edited by Rohit Sharma, Rajendra Prasad Mahapatra, Gwanggil Jeon.
Description: Boca Raton : CRC Press, 2024. | Series: Future generation of soft and intelligent computing | Includes bibliographical references and index. | Summary: "The book addresses the challenges in designing blockchain-based secured solutions for Industry 4.0 applications using artificial intelligence. It further provides a comparative analysis of various advanced security approaches such as edge computing, cybersecurity, and cloud computing in the realm of information technology."-- Provided by publisher.
Identifiers: LCCN 2023020297 (print) | LCCN 2023020298 (ebook) | ISBN 9781032460581 (hardback) | ISBN 9781032590264 (paperback) | ISBN 9781003452591 (ebook)
Subjects: LCSH: Blockchains (Databases) | Artificial intelligence. | Industry 4.0
Classification: LCC QA76.9.B56 A78 2024 (print) | LCC QA76.9.B56 (ebook) | DDC 005.74--dc23/eng/20230807
LC record available at https://lccn.loc.gov/2023020297
LC ebook record available at https://lccn.loc.gov/2023020298

ISBN: 978-1-032-46058-1 (hbk)
ISBN: 978-1-032-59026-4 (pbk)
ISBN: 978-1-003-45259-1 (ebk)

DOI: 10.1201/9781003452591

Typeset in Sabon
by SPi Technologies India Pvt Ltd (Straive)

Contents

Preface	viii
Authors	xii
Contributors	xiv

1 CNN-based oral cancer and dental caries detection for computer-aided diagnosis   1
ABHILASHA SINGH, RITU GUPTA AND ARUN KUMAR

2 Artificial Intelligence for Healthcare 4.0   18
K. UMAPATHY, S. OMKUMAR, T. DINESHKUMAR, M. A. ARCHANA, S. PRABAKARAN AND ARWA N. ALEDAILY

3 Artificial Intelligence's foresight in cybersecurity   38
RADHIKA NAUTIYAL, RADHEY SHYAM JHA, SAMTA KATHURIA, SHWETA PANDEY, RAJESH SINGH, ANITA GEHLOT, PRAVEEN KUMAR MALIK AND AHMED ALKHAYYAT

4 Theories of blockchain and distributed systems   52
K. UMAPATHY, S. PRABAKARAN, T. DINESHKUMAR, M. A. ARCHANA, D. KHYATHI SRI AND ARWA N. ALEDAILY

5 Analysis of critically polluted locations using the IoT and AI infrastructure   69
LULWAH M. ALKWAI

6 Analysis of deep learning techniques in biomedical images   78
TINA AND RITU GUPTA

7 Time-frequency representations of one-dimensional signals using Wigner-Ville Distribution 95

FATIMA ZAHRA LAMZOURI, BOUTAINA BENHMIMOU, NIAMAT HUSSAIN, SANDEEP KUMAR ARORA, RACHID AHL LAAMARA, ALEXANDER KOGUT, BASEEM KHAN AND MOHAMED EL BAKKALI

8 An efficient 3U CubeSat downlink transmission based on an S-band lightweight CPW-fed slot antenna 108

BOUTAINA BENHMIMOU, FATIMA ZAHRA LAMZOURI, NIAMAT HUSSAIN, NANCY GUPTA, RACHID AHL LAAMARA, SANDEEP KUMAR ARORA, JOSEP M. GUERRERO AND MOHAMED EL BAKKALI

9 WGMs diffractive emission for Mm-wave all-round antennas with Internet of Things 120

ALEXANDER KOGUT, NIAMAT HUSSAIN, IGOR KUZMICHEV, BOUTAINA BENHMIMOU, RACHID AHL LAAMARA, FATIMA ZAHRA LAMZOURI, SANDEEP KUMAR ARORA AND MOHAMED EL BAKKALI

10 Apache Hadoop framework for Big Data analytics using AI 130

URVASHI GUPTA AND ROHIT SHARMA

11 Envisioning the future of blockchain in SMEs: Insights from a survey 141

TA THI NGUYET TRANG, TA PHUONG THAO, PHAM CHIEN THANG AND KUSUM YADAV

12 Deep learning techniques for the prediction of traffic jam management for smart city infrastructure 163

ARWA N. ALEDAILY AND KUSUM YADAV

13 The role of ethical chatbots for enhancing customer experience: An interdisciplinary perspective 184

PRIYANKA TYAGI AND NARENDRA MOHAN MISHRA

14 An efficient gas leakage detection and smart alerting system using IoT 194

K. MUTHUMANICKAM, P. VIJAYALAKSHMI, S. KUMARGANESH, T. KUMARAVEL, K. MARTIN SAGAYAM AND LULWAH M. ALKWAI

15 Principles and goals of Industry 4.0 214

K. UMAPATHY, D. MUTHUKUMARAN, G. POOJITHA, A. SAI SAMVIDA, S. PRABAKARAN AND SAFIA YASMEEN

16 The positionality of culture in teaching EFL in
technology-supported classrooms: Teachers'
perceptions and practices   231
CHAU THI HOANG HOA AND LIEN BAO TRAN

17 Algorithm for secured energy-efficient routing in wireless
sensor networks: A review   246
JYOTI SRIVASTAVA AND JAY PRAKASH

18 Role of cloud computing and blockchain technology
in paradigm shift to modern online teaching culture
in the education sector   263
NAZREEN KHANAM AND MD SAFIKUL ISLAM

19 Artificial intelligence-based communication systems
used in Industry 4.0: For multiple input and multiple
output antenna 5G wireless devices   277
SUVERNA SENGAR, PRAVEEN KUMAR MALIK AND MONIKA AGARWAL

20 Wireless technologies and the Internet of Things   291
VIKRANT PACHOURI, SAMTA KATHURIA, SHWETA PANDEY, RAJESH SINGH,
ANITA GEHLOT, SHAIK VASEEM AKRAM, PRAVEEN KUMAR MALIK
AND SAFIA YASMEEN

21 Artificial intelligence and blockchain-based intervention
in building infrastructure   302
VIKRANT PACHOURI, SHWETA PANDEY, SAMTA KATHURIA, RAJESH SINGH,
ANITA GEHLOT, SHAIK VASEEM AKRAM, PRAVEEN KUMAR MALIK
AND AHMED ALKHAYYAT

Index   314

Preface

This book aims to address the challenges in designing blockchain-based secured solutions for Industry 4.0 applications using AI and provides a comparative analysis of various advanced security approaches in the realm of information technology.

Chapter 1 uses a Convolutional Neural Network (CNN) to classify images into different classes. The model labels images into three classes: Normal tooth, Caries, and Oral Cancer. The exactness of the classification by the proposed framework lies in the range of 85–95%. This framework helps in the timely identification of specific oral diseases. The proposed model can assist dentists in an effective and efficient manner. This finding represents the capability of AI to assist in working on oral well-being.

Chapter 2 enunciates the evolution and constituents of Healthcare 4.0, the importance of AI, and its inevitability to enhance the characteristics and functionalities of different medical devices used in health sectors such as the treatment of diabetes, heart diseases, and surgery. The impact of AI in enhancing accuracy in diagnosis, ability in decision-making, and assessment of risks are also discussed. Ultimately, the chapter presents the potential of AI by which appropriate medical systems can be designed for Healthcare 4.0.

Chapter 3 is one of the most promising approaches for combating cybersecurity risks and providing security. Here, we examine the potential of AI in enhancing cybersecurity solutions by highlighting both its advantages and disadvantages. We also talk about the potential for future research in the realm of cybersecurity related to the development of AI approaches across many application domains.

Chapter 4 focuses on theories of blockchain technology by paying special attention to its evolution, applications, benefits, and challenges in distributed transaction ledgers and financial transaction systems. It also provides a complete overview of blockchain technology, significant issues, and applicability in several sectors. The role of blockchain is discussed elaborately along with a description of several cryptocurrencies with respect to different transaction systems.

Preface ix

Chapter 5 compares the CEPI score for some developing cities over 2009, 2011, 2013, and 2018. The research highlights the truth that how the pollution in some developing cities is increasing day by day. This chapter evaluates the pollution situation in various locations and provides a comparative analysis using a comprehensive environmental pollution index.

Chapter 6 covers the basic deep learning techniques applied to three dimensions. The second is the analysis of different deep learning models applied to medical images, along with advantages, disadvantages, and applications with respect to the summarization of the technique applied.

Chapter 7 presents and applies the principles of time-frequency analysis of one-dimensional signals using proprieties of Wigner–Ville distribution. We will study the main methods of time-frequency representations while presenting the continuous and discrete versions for each technique as well as the computer implementation.

Chapter 8 introduces a graceful mechanism of integrating planar antennas on the CubeSat's box, which minimizes the antenna transmission power consumption in the satellite. This approach presents a new slot antenna for ISM-band CubeSat communications that is excited by a coplanar waveguide feed line. The suggested CPW-fed slot antenna (CPW-SA) reveals good impedance at 2450 MHz; it's also small and has a large impedance bandwidth of 2360 MHz.

Chapter 9 presents a segmented dielectric resonator antenna proposed for mm-wave all-around antenna uses. The developed WGM configuration allows therefore introducing a very effective solution for diffractive emissions at mm-wave frequencies using low cost and simple configurations of all-round antennas.

Chapter 10 introduces a brief theory about the Apache Hadoop framework. The Apache Hadoop framework is a popular distributed computing platform used for big data analytics, particularly in AI and machine learning applications. Hadoop's ability to handle large datasets is made possible through its distributed file system, HDFS, and Map Reduce programming model, allowing for efficient parallel processing across multiple nodes. Hadoop also offers a range of tools such as Apache Spark, Hive, and HBase for batch processing, real-time data access, and querying data stored in Hadoop. Overall, Hadoop provides a reliable and scalable platform for organizations to gain valuable insights and drive innovation in their AI applications.

Chapter 11 analyzes and identifies the factors influencing blockchain application in small- and medium-sized enterprises (SMEs) in Vietnam. The results show that blockchain cost has the most significant relationship with blockchain adoption with a negative influence, followed by technology readiness with a positive effect on blockchain adoption in SMEs.

Chapter 12 focuses on how the detection of the type of vehicles is possible with deep learning algorithms like YOLOv3 and how the results can be reported in the form of visualizations to clearly analyze the data patterns with respect to time.

Chapter 13 critically reviews communication strategies in artificial intelligence and its applications. Artificial intelligence is a broad category of technologies with several benefits for businesses. The ethical and social implications of artificial intelligence in the existing literature are controversially and complicatedly addressed. The current study aims to assess how customer experiences are affected by ethical chatbots. The results of structural equation modeling demonstrate the strong impact of ethical chatbots (relationship commitment, perceived benefits, and safety) on consumer experiences. The study's findings will help marketers to get a competitive edge in an ever-changing business environment.

In **Chapter 14**, we categorize and analyze the research on ML-reliant IoT from several angles. Next, we use the IoT real-time application as a case study to illustrate how it may be used to address social issues. We also examine IoT's difficulties from many angles. Utilizing IoT potential and challenges can help our society become more sustainable and affluent.

Chapter 15 introduces the basic concept of Industry 4.0. Industry 4.0 has extensively anticipated effects in various fields and its implementation will have an impression on changing the working habitat. KUKA organization is a typical implementation of, for example, automated industries, brilliant machines, and robots. The above implementations aid Industry 4.0 to split up swiftly. This chapter also enunciates the overview, characteristics, and effects of Industry 4.0 in the current scenario.

Chapter 16 aims to explore how Vietnamese EFL teachers perceive and practice intercultural integration during online teaching due to the COVID-19 pandemic. Data collected from 64 teachers show that (1) teachers had a higher level of perceptions than practices of online intercultural integration; (2) illustrating intercultural content in the coursebooks was most frequent activity; and (3) online teaching made more favorable conditions to intercultural integration than offline teaching.

In **Chapter 17**, the authors look into something that happens in multi-hop clustered wireless sensor networks (MCWSN). Each node in these networks uses a method called "simultaneous wireless information and power transfer" to decode data and get energy from a radio-frequency signal (SWIPT).

Chapter 18 explores the impact of influential technologies on transforming the educational system from the conventional Ashram to a more modern, web-based system. Additionally, it sheds light on the benefits and drawbacks of these technologies for students, educators, and guardians. Meanwhile, the study reveals that cloud computing and blockchain technologies play a crucial role in inaccessible conditions, notwithstanding some pivotal limitations.

Chapter 19 discusses the Internet-of-Things (IoT) standards upgraded due to the Industry 4.0 revolution, which has had significant effects on the development of 5G wireless communication and the 5G wireless technology for antenna designing and antenna basic parameters, which are very important for designing a 5G antenna.

In **Chapter 20**, the vision of the IoT that includes the upcoming digital revolution is presented. After conducting a survey, the analytical criteria for various wireless technologies are provided in this chapter. The assessment that has been done presents a new study topic and future research direction in enabling wireless technologies, which will be useful for both industry and academia and in overcoming those challenges by offering solutions appropriately. The findings will aid and provide direction for further research on wireless communication techniques for the IoT in the era of the fifth generation (5G).

In **Chapter 21**, the use of AI and blockchain in Industry 4.0 is carefully assessed. To gain industrial sustainability, Industry 4.0 has been evaluated in order to identify the difficulties that may arise. The result of this evaluation offers a new research area and future research directions in several Industry 4.0 research domains, which will be helpful for businesses and academics in achieving sustainability and overcoming the difficulties posed by incorporating AI and blockchain into Industry 4.0.

Authors

Dr. Rohit Sharma is currently working as an Associate Professor in the Department of Electronics and Communication Engineering, SRM Institute of Science and Technology, Delhi-NCR Campus, Ghaziabad, India. He is an active member of ISTE, IEEE, ICS, IAENG, and IACSIT and a senior member of IEEE. He is an editorial board member and reviewer of more than 12 international journals and conferences, including the topmost journal *IEEE Access* and *IEEE Internet of Things Journal*. He serves as a Book Editor for seven different titles to be published by CRC Press, Taylor & Francis Group, USA, Apple Academic Press, USA, Springer, etc. He has received the Young Researcher Award in "2nd Global Outreach Research and Education Summit & Awards 2019" hosted by the Global Outreach Research & Education Association (GOREA). He is serving as a Guest Editor in the SCI journal of Elsevier, *CEE*. He has actively been at the organizing end of various reputed International conferences. He is serving as an Editor and Organizing Chair to the 3rd Springer International Conference on Microelectronics and Telecommunication (2019), and has served as the Editor and Organizing Chair of the 2nd IEEE International Conference on Microelectronics and Telecommunication (2018), Editor and Organizing Chair to IEEE International Conference on Microelectronics and Telecommunication (ICMETE-2016) held in India, and Technical Committee member of "CSMA2017, Wuhan, Hubei, China", "EEWC 2017, Tianjin, China" IWMSE2017 "Guangzhou, Guangdong, China", "ICG2016, Guangzhou, Guangdong, China", and "ICCEIS2016 Dalian Liaoning Province, China".

Prof. Rajendra Prasad Mahapatra, B. Tech, M. Tech, and Ph.D, is currently working at SRM Institute of Science & Technology, NCR Campus, as Professor and Head of the Department of the Computer Science & Engineering. He has vast experience of 17 years as an academician, researcher, and administrator. During these 17 years, he has worked in India and abroad. He has been associated with Mekelle University, Ethiopia, for more than 2 years. Two candidates have successfully completed their Ph.D.s under his supervision and seven students are currently perusing their research under his guidance. Prof. Mahapatra has authored more than 70 research papers which are published in international journals like *Inderscience, Emerald, Elsevier, IEEE,* and *Springer*. He is a fellow member of I.E. (India), a senior member of IACSIT, Singapore, a life member of ISTE, and a member of IEEE and many more reputed bodies.

Prof. Gwanggil Jeon received his B.S., M.S., and Ph.D. (summa cum laude) degrees from the Department of Electronics and Computer Engineering, Hanyang University, Seoul, Korea, in 2003, 2005, and 2008, respectively. From September 2009 to August 2011, he was with the School of Information Technology and Engineering, University of Ottawa, Ottawa, Canada, as a Post-Doctoral Fellow. From September 2011 to February 2012, he was with the Graduate School of Science and Technology, Niigata University, Niigata, Japan, as an Assistant Professor. From December 2014 to February 2015 and from June 2015 to July 2015, he was a Visiting Scholar at Centre de Mathématiques et Leurs Applications (CMLA), École Normale Supérieure Paris-Saclay (ENS-Cachan), France. From 2019 to 2020, he was a Prestigious Visiting Professor at Dipartimento di Informatica, Università degli Studi di Milano Statale, Italy. He is currently a Full Professor at Incheon National University, Incheon, Korea. He was a Visiting Professor at Sichuan University, China; Universitat Pompeu Fabra, Barcelona, Spain; Xinjiang University, China; King Mongkut's Institute of Technology Ladkrabang, Bangkok, Thailand; and University of Burgundy, Dijon, France. Dr. Jeon is an IEEE Senior Member and Associate Editor of *Sustainable Cities and Society, IEEE Access, Real-Time Image Processing, Journal of System Architecture,* and *MDPI Remote Sensing*. Dr. Jeon was a recipient of the IEEE Chester Sall Award in 2007, the ETRI Journal Paper Award in 2008, and the Industry-Academic Merit Award by the Ministry of SMEs and Startups of Korea Minister in 2020.

Contributors

Monika Agarwal
School of Commence and Management
IIMT University
Meerut, India

Shaik Vaseem Akram
Uttaranchal Institue of Technology
Uttaranchal University
Dehradun, India

Arwa N. Aledaily
College of Computer science and Engineering
University of Ha'il
Ha'il, Kingdom of Saudi Arabia

Ahmed Alkhayyat
College of Technical Engineering
The Islamic University
Najaf, Iraq

Lulwah M. Alkwai
College of Computer Science and Engineering
University of Ha'il
Ha'il, Kingdom of Saudi Arabia

M. A. Archana
SCSVMV Deemed University
Kanchipuram, India

Sandeep Kumar Arora
School of Electronics and Electrical Engineering
Lovely Professional University
Phagwara, India

Boutaina Benhmimou
LPHE-MS, Faculty of Sciences of Rabat
Mohammed Five University in Rabat (UM5R)
Agdal, Rabat, Morocco

Mohamed El Bakkali
LPHE-MS, Faculty of Sciences of Rabat, Mohammed Five University in Rabat (UM5R)
Agdal, Rabat, Morocco

T. Dineshkumar
SCSVMV Deemed University
Kanchipuram, India

Anita Gehlot
UIT, Division of Research & Innvovation
Uttaranchal University
Dehradun, India

Josep M. Guerrero
Center for Research on Microgrids (CROM)
Department of Energy Technology,
Aalborg University
Aalborg East, Denmark

Nancy Gupta
ECE, LKCE, I.K.G.
Punjab Technical University
Jalandhar, India

Ritu Gupta
Bhagwan Parshuram Institute of Technology
GGSIP University
New Delhi, India

Urvashi Gupta
Department of Electronics & Communication
SRM Institute of Science and Technology, Delhi-NCR Campus
Ghaziabad, India

Niamat Hussain
Department of Smart Device Engineering
Sejong University
Seoul, South Korea

Chau Thi Hoang Hoa
Tra Vinh University
Tra Vinh city, Vietnam

Md Safikul Islam
Dr. Ambedkar International Centre (DAIC)
Ministry of Social Justice and Empowerment
New Delhi, India

Radhey Shyam Jha
Law College of Dehradun
Uttaranchal University
Dehradun, India

Samta Kathuria
Law College of Dehradun
Uttaranchal University
Dehradun, India

Baseem Khan
School of Electrical and Computer Engineering
Hawassa Institute of Technology
Hawassa University
Awasa 05, Ethiopia

Nazreen Khanam
Department of Geography
Jamia Millia Islamia
New Delhi, India

D. Khyathi Sri
SCSVMV
Kanchipuram, India

Alexander Kogut
O.Ya.Usikov
Institute for Radiophysics and Electronics of NAS of Ukraine
Kharkov, Ukraine

Arun Kumar
SRM Institute of Science and Technology
Delhi-NCR Campus
Modinagar, Ghaziabad, India

T. Kumaravel
Department of CSE
Kongu Engineering College
Erode, India

S. Kumarganesh
Department of ECE
Knowledge Institute of Technology
Salem, India

Igor Kuzmichev
O.Ya.Usikov
Institute for Radiophysics and Electronics of NAS of Ukraine
Kharkov, Ukraine

Rachid Ahl Laamara
LPHE-MS, Faculty of Sciences of Rabat
Mohammed Five University in Rabat (UM5R)
Agdal, Rabat, Morocco

Fatima Zahra Lamzouri
LPHE-MS, Faculty of Sciences of Rabat
Mohammed Five University in Rabat (UM5R)
Agdal, Rabat, Morocco

Praveen Kumar Malik
School of Electronics and Electrical Engineering
Lovely Professional University
Phagwara, India

Narendra Mohan Mishra
Faculty of Management
SRM Institute of Science and Technology
Delhi-NCR Campus
Ghaziabad, India

D. Muthukumaran
SCSVMV
Kanchipuram, India

K. Muthumanickam
Department of IT
Kongunadu College of Engineering and Technology
Trichy, India

Radhika Nautiyal
Law College of Dehradun,
Uttaranchal University
Dehradun, India

S. Omkumar
SCSVMV Deemed University
Kanchipuram, India

Vikrant Pachouri
Uttaranchal Institue of Technology
Uttaranchal University
Dehradun, India

Shweta Pandey
Law College of Dehradun
Uttaranchal University
Dehradun, India

G. Poojitha
SCSVMV
Kanchipuram, India

S. Prabakaran
SCSVMV Deemed University
Kanchipuram, India

Jay Prakash
Department of ITCA
Madan Mohan Malviya University of Technology
Gorakhpur, India

K. Martin Sagayam
Department of ECE
Karunya Institute of Technology and Sciences
Coimbatore, India

A. Sai Samvida
SCSVMV
Kanchipuram, India

Suverna Sengar
School of Electronics and Electrical Engineering
Lovely Professional University
Jalandhar, India

Rohit Sharma
Department of Electronics & Communication
SRM Institute of Science and Technology, Delhi-NCR Campus
Ghaziabad, India

Abhilasha Singh
SRM Institute of Science and
 Technology
Delhi-NCR Campus
Modinagar, Ghaziabad, India

Rajesh Singh
UIT, Division of Research &
 Innvovation
Uttaranchal University
Dehradun, India

Jyoti Srivastava
Department of ITCA
Madan Mohan Malviya University
 of Technology
Gorakhpur, India

Pham Chien Thang
International School
Thai Nguyen University
Thai Nguyen, Vietnam

Ta Phuong Thao
International School, Thai Nguyen
 University
Thai Nguyen, Vietnam

Tina
Bhagwan Parshuram Institute of
 Technology
GGSIPU
New Delhi, India

Lien Bao Tran
Tra Vinh University
Tra Vinh city, Vietnam

Ta Thi Nguyet Trang
International School, Thai Nguyen
 University
Thai Nguyen, Vietnam

Priyanka Tyagi
Faculty of Management
SRM Institute of Science and
 Technology
Delhi-NCR Campus
Ghaziabad, India

K. Umapathy
SCSVMV Deemed University
Kanchipuram, India

P. Vijayalakshmi
Department of CSE
Knowledge Institute of Technology
Salem, India

Kusum Yadav
College of Computer science and
 Engineering
University of Ha'il
Ha'il, Kingdom of Saudi Arabia

Safia Yasmeen
Alfaisal University
Riyadh, Saudi Arabia

Chapter 1

CNN-based oral cancer and dental caries detection for computer-aided diagnosis

Abhilasha Singh
SRM Institute of Science and Technology, Ghaziabad, India

Ritu Gupta
GGSIP University, New Delhi, India

Arun Kumar
SRM Institute of Science and Technology, Ghaziabad, India

1.1 INTRODUCTION

Advances in science and technology have transformed us in modern society. The manner in which things used to work many years prior has changed dramatically as of late. Clinical imaging innovations have likewise adapted and are developing at an extremely fast rate [1, 2]. A stream, known as Image Classification, can help effectively classify images into various categories based on certain features identified in the images. These features might be unobservable by the naked human eye but can be easily noted if the software or the model is trained in such a manner. Based on numerous parameters in dental images like color, textureand shape of gums and teeth, these images can be classified accordingly and prediction of possible diseases can be done. Different cameras may have different settings based on varied resolutions and angles; also since environmental conditions such as lighting, brightness and contrast may vary, images need to be pre-processed before providing them to the model or software. Eventually, after learning from the training data, the model tries to identify the images, and usingthe test data, the final results can be observed.

CNN is a kind of Artificial Neural Network which falls under the Deep learning domain [19]. It is an exceptional sort of multi-faceted neural organization which expects practically zero image pre-handling and can straightforwardly identify patterns and features from digital images. Different such activities have recently been explored by many researchers. As medical imaging has emerged at a fast pace and many researchers have noticed the ways of leveraging advanced AI and Machine Learning models, much researchisbeing done in this field to develop newer, more efficient techniques

DOI: 10.1201/9781003452591-1

that can give more accuracy in this sector. Radhika R. et al. suggested a similar model wherein they analyzed dental caries in the images using Open CV [3]. Thamaraiet et al. classified teeth images with periodontal illness utilizing Dental Radiographs [4]. Wenzhe You et al., in their venture work, recognized dental plaque by utilizing profound learning strategies which can effectively distinguish dental plaque and assist dental specialists in diagnosing the infection [5]. Deep learning approaches were employed by Hassan Aqeel Khan et al. to look at automated feature detection, segmentation and quantification of typical findings in dental periapical radiographs [6]. Hu Chen etal. detected dental lesions to determine disease categories and predict severity by training CNNs [7]. LuyaLian et al. employed deep learning algorithms to identify caries lesions, categorize them and then aimed to compare their classification outcomes with those of expert dentists [17]. Numerous other such comparative investigations are proceeding to help dental wellbeing as it is a significant piece of our regular routines [3, 9, 10, 29, 30].

The novelty of the proposed work over other such comparable works is as follows:

1. Most of the available works are bound to diagnose just one dental disease at a time. The proposed model can recognize one or more infections and help patients find early treatment [27].
2. Oral Cancers are detected by Biopsy and Biopsy is invasive and painful. The majority of the studies have been developed based on advanced imaging tools and techniques such as fiber optics trans-illumination, laser and light fluorescence, electrical resistance, digital radiography, etc. to detect dental caries [11]. But these also many times lead to the detection of false positives and so many experts still believe that visual tactile inspection is the best practice fordiagnosing dental caries. The proposed version can assist in figuring out those diseases with extra efficiency, as it should be, and in a cost-effective manner.
3. This model can be of great use to dentists to help them in the identification of tooth diseases in terms of time-saving. Even, common people can take the help of this software to detect the disease without the help of an expert. Proper treatment can be taken later under the supervision of the dentist.

    The framework that is proposed and the specific algorithm are explained in Section 1.3, "Proposed Framework". The parameters used to evaluate the framework are covered in Section 1.4 of the chapter, "Predictions of the model". Section 1.5 displays the outputs and results of the suggested system. The suggested system is compared to a few current, related systems in Section 1.6, "State-of-the-Art Comparison". Section 1.7 concludes and goes through the parameters of the work that is proposed.

1.2 BACKGROUND AND RELATED WORK

Maintaining good oral health can help us in preventing many diseases that may critically affect our life. Overlooking dental care can cause serious dental problems as well as may lead to other serious health issues [12, 13]. People's approach to living has changed seriously with the expanding urbanization in developing areas. This has prompted obliviousness toward dental wellbeing, subsequently expanding the event of dental sicknesses [14]. A few general dental diseases are discussed further.

1.2.1 Dental caries (tooth decay)

Dental caries occur by the breakdown of the tooth enamel. The constant breakdown of food by the bacteria present in teeth results in the production of acid that eventually destroys the tooth enamel and decays the tooth [23, 24]. Figure 1.1 shows a caries tooth and a healthy tooth without dental caries.

1.2.2 Oral cancer

Oral cancer is one of the top three diseases in various Asian nations, with an estimated 4 cases per 100,000 persons worldwide, according to the WHO. Every year, more than 49,000 new cases of oral cancer are found in the United States alone. It mostly occurs in people over 40 years and is more common in men and strongly varies by socioeconomic condition [15, 16]. Figure 1.2 shows an oral cancer image.

Numerous scholars have recently studied similar activities. A lot of research is being done in this area to create newer, more effective strategies that can provide greater accuracy in the medical imaging industry because the field is developing quickly and many researchers have discovered the possibilities to use advanced AI and Machine Learning models. A similar

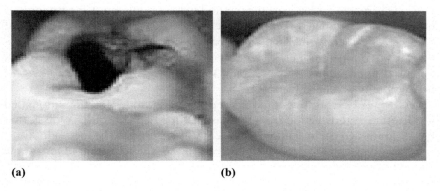

Figure 1.1 (a) Caries infected tooth and (b) normal tooth.

4 Artificial Intelligence and Blockchain in Industry 4.0

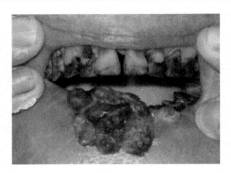

Figure 1.2 Cancer image.

model was proposed by Radhika R. et al. [3], who used Open CV to analyze dental caries in the photos. Using dental radiographs, Thamaraiet et al. identified teeth pictures with periodontal disease [4]. In their research project, Wenzhe You et al. successfully distinguished dental plaque using deep learning algorithms, which can help dental specialists identify the illness [5].

Convolutional neural networks (CNNs) were used by Sharma D. et al. [18] to differentiate oral pre-cancerous and malignant lesions from normal mucosa using a clinically annotated image dataset. They got accuracy in the range of 36–76%.

García-Pola M. etal. [20] examined the data about various tools based on Machine Supporting non-harmful investigative techniques such as telemedicine, clinical images, fluorescence images, exfoliative cytology and indicators of oral disease are leaned toward. The outcomes recommend that such devices can givean unproblematicassurance to the early determination of oral disease. According to Khanagar S. et al. [21], AI is more accurate and precise than current clinical systems, standard measurements and human judgment when it comes to making decisions and anticipating events.

By analyzing patient hyperspectral images, Jeyaraj P. R. and Samuel Nadar E. R. [22] created a deep learning method for a computer-aided autonomous oral cancer-detecting system and achieved a classification accuracy of 91.4%. A deep convolutional CNN-based computer-assisted detection system was developed by Lee J. H. et al. [25], who also evaluated the system's potential applicability and accuracy for the detection and forecasting of periodontally damaged teeth (PCT).

Casalegno F. et al. presented a deep learning model for the automated localization and detection of dental lesions in TI images [26]. A CNN that has been trained to do semantic segmentation is used in the method.

Rimi I. F. et al. [28] used machine learning to predict dental illness in the context of people's daily activities. With the concerned doctors and dentists, the primary features of dental disease were discussed. With a collection of dental caries images, Talpur S. et al. [31] examined various machine learning algorithms, with deep learning being one of the most popular.

1.3 PROPOSED FRAMEWORK

The framework being put forward comprises image classification algorithm named CNN [6]. The dataset is the algorithm's foundation because it contains the primary values that the algorithm uses to extract features from the images.

The image size in the dataset is 500 × 500. Images are pre-processed and resized to 128 × 128. Images are isolated into groups of 32 arbitrarily. There are 598 photos in total in the database, of which 402 are used for training and 196 for testing. The model classifies images as "Normal tooth", "Caries" or "Oral Cancer".

Steps of the framework are discussed further.

1.3.1 Input image

The first step is to provide a color (RGB) or grayscale image to the system.

1.3.2 Pre-processing

In this step, images are scaled and resized. Generally, image scaling is the process that is widely used in almost all image processing techniques. They are also converted into gray level from RGB format.

1.3.3 Creation of model

In the proposed model, 70% of the information is utilized for training and 30% is for testing. The model utilizes numerous functions: "to_categorical()"– it changes over a vector (integer) to a logical matrix which assists in changing the training data prior to passing it to the model. Classes are represented as numbers in the proposed model's training set; however, to_categorical() converts these numbers into vectors before using them in the model. Epochs of 50 are employed with a batch size of 32 images to improve accuracy on training and testing images.

The model comprises three convolutional blocks with a MaxPooling layer and a kernel of size 2 × 2. A kernel of size 3 × 3 and 32 and 64 filters on top are totally related tolayers. "relu" is the activation function employed. The model with a 128-unit size also uses a thick layer and a "softmax" activation function to predict a multinomial probability distribution. After two MaxPool layers, a dropout layer has been added to prevent overfitting. It sets input units to 0 with a recurrence of the rate at each progression during training time randomly. Figure 1.3 shows the methodology and CNN architecture for the proposed framework. The input image to the system is first pre-processed (Section 1.3.2) and then the model is created (Section 1.3.3) and executed on the image to identify its class.

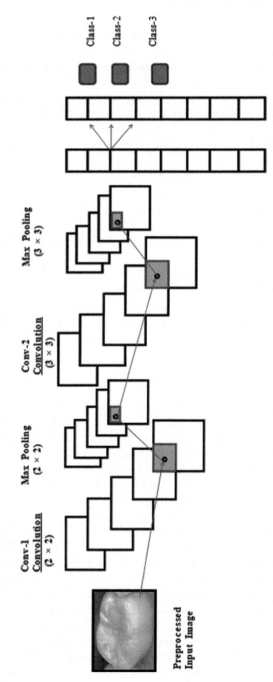

Figure 1.3 CNN architecture for proposed model.

1.4 PREDICTIONS OF THE MODEL

The proposed model is executed on dataset images to assess itsworking efficiency.

1.4.1 Confusion matrix for the proposed model

Figure 1.4, confusion matrix, reveals that 188 predictions were correct out of the total 198 predictions.

The following factors are utilized to assess the performance of the proposed model:

- **Accuracy** – It tells the portion of the total image sperfectly classified by the proposed model.

$$\text{Accuracy} = \frac{TP + TN}{TP + TN = FP + FN} \quad (1.1)$$

- **Precision** – Positive images identified as positive by the model.

$$\text{Precision} = \frac{TP}{TP + FP} \quad (1.2)$$

- **Sensitivity/Recall** – Proportion of all positive (diseased) images that the model accurately predicted.

$$\text{Recall} = \frac{TP}{TP + FN} \quad (1.3)$$

- **F1-score** – This parameter is used to combine precision and recall into a single value.

$$f1 - \text{score} = \frac{2TP}{2TP + FP + FN} \quad (1.4)$$

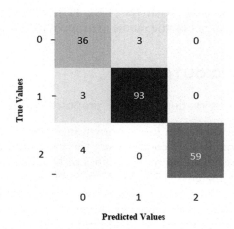

Figure 1.4 Generated confusion matrix.

Table 1.1 Proposed model's confusion matrix

Output class	TP	TN	FP	FN
Caries (0)	36	152	3	7
Cancer (1)	93	99	3	3
Normal Tooth (2)	59	135	4	0

- **Macro Avg** – Unweighted mean of the metrics measured for each class individually.

Macro avg =
$$\frac{f1-\text{score Class}-1 + f1-\text{score Class}-2 + \ldots + f1-\text{score Class}-n}{n} \quad (1.5)$$

- **Weighted Avg** – It takes a weighted mean of f1 scores for different classes.

Weighted avg =
$$\frac{\begin{array}{c}f1-\text{score Class}-1 * \text{No. of images of Class}-1 + f1-\text{score Class}- \\ 2 * \text{No. of images of Class}-2 + \ldots + f1-\text{score Class}- \\ n * \text{No. of images of Class}-n\end{array}}{\begin{array}{c}\text{No. of images of Class}-1 + \text{No. of images of Class}-2 + \ldots + \\ \text{No. of images of Class}-n\end{array}} \quad (1.6)$$

Output from the proposed model is as follows:

- The first class corresponds to caries images and is labeled with (0),
- The second class corresponds to cancer images and is labeled with (1)
- The third class corresponds to normal tooth images and is labeled with (2)

True Positives (TP), True Negatives (TN), False Positives (FP) and False Negatives (FN) identified by the model are shown in Table 1.1.

1.5 RESULTS AND OUTCOMES

Following are the observations from the proposed model:

- Training and validation loss is shown in Figure 1.5. It demonstrates that as the number of epochs rises, the validation and training loss both decrease considerably. The values of training and validation loss reach approximately 0.15 and 0.23, respectively.
- Training and validation accuracy of the model is displayed in Figure 1.6. It illustrates that as the number of epochs increases, validation and training accuracy increases significantly. The values of training and validation accuracy are approximately 0.96 and 0.90, respectively.

CNN-based oral cancer and dental caries detection 9

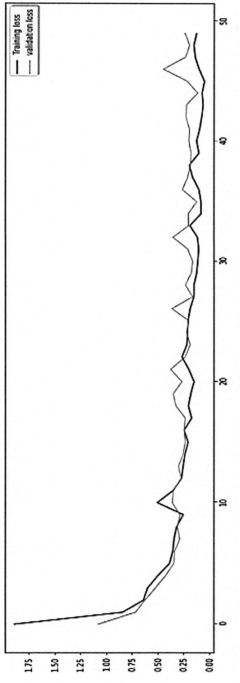

Figure 1.5 Plot for training and validation loss.

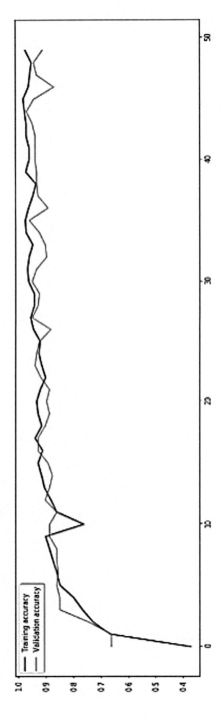

Figure 1.6 Plot for training and validation accuracy.

	precision	recall	f1-score	support
0	0.92	0.84	0.88	43
1	0.97	0.97	0.97	96
2	0.94	1.00	0.97	59
accuracy			0.95	198
macro avg	0.94	0.94	0.94	198
weighted avg	0.95	0.95	0.95	198

Figure 1.7 Values of assessment parameters.

- Values of precision, recall, f1-score, accuracy, macro avg and weighted avg are shown in Figure 1.7. Overcall accuracy of the proposed model comes out to be in the range of 85–95%. The average value of precision for all classes is 0.94, recall is 0.93and f1-score is 0.94. Values of macro avg and weighted avg are also more than 0.9. All the results show that the overall performance of the system is up to the mark and can be helpful in the prediction of healthy or diseased teeth.
- Figure 1.8 demonstrates the outcome of the proposed model after classifying the input image as a "caries" image. It also shows the possible treatments for the cure of disease.
- Figure 1.9 demonstrates the outcome of the proposed model after classifying the input image as a "Cancer" image. It also shows the possible treatments for the cure of disease.
- Figure 1.10 demonstrates the outcome of the proposed model after classifying the input image as a "Healthy tooth" image.

1.6 STATE-OF-THE-ART COMPARISON

Image Processing and Classification is a budding strategy and as of now, hasbegun to be widely utilized in the clinical area in diagnosing different diseases that can be distinguished byprocessing and classification of clinical imagery. In the proposed model of recognizing specific dental issues and healthy teeth through feature extraction, it reasonably contends with other such methods utilized in identifying dental infections. In any case, the proposed model can help in diagnosing oral issues for dental specialists as well as patients and commoners.

Table 1.2 shows the state-of-the-art comparison of the proposed framework.

12 Artificial Intelligence and Blockchain in Industry 4.0

The following image is a caries image. So for the same one of the following treatments should be followed.

1. Fluoride Treatment. If your cavity just started, a fluoride treatment may help restore your tooth's enamel and can sometimes reverse a cavity in the very early stage.
2. Fillings. Fillings, also called restorations, are the main treatment option when decay has progressed beyond the early stage. Fillings are made of various materials such as tooth-colored composite resins, porcelain or dental amalgam that is a combination of several materials.
3. Crowns. For extensive decay or weakened teeth, you may need a crown – a custom fitted covering that replaces your tooth's entire natural crown.
4. Root canals. When decay reaches your inner material of your tooth (pulp), you may need a root canal. This is a treatment to repair and save a badly damaged or infected tooth instead of removing it.
5. Tooth extractions. Some teeth become so severely decayed that they can't be restored and must be removed. Having a tooth pulled can leave a gap that allows other teeth to shift. If possible, consider getting a bridge or dental implant to replace the missing tooth.

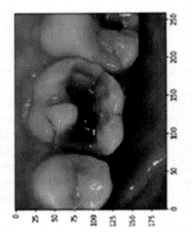

Figure 1.8 Prediction of "caries" image with a possible treatment.

CNN-based oral cancer and dental caries detection 13

The following image is a cancer image. So for the same one of the following treatments should be followed.

1. Surgery to remove the tumor. Your surgeon may cut away the tumor and a margin of healthy tissue that surrounds it to ensure all of the cancer cells have been removed.
2. Surgery to remove cancer that has spread to the neck. If cancer cells spread to the lymph nodes in your neck of if there's a high risk that this has happened based on the size or depth of your cancer, your surgeon may recommend a procedure to remove lymph nodes and related tissue in your neck (neck dissection).
3. Radiation therapy uses high energy beams, such as x-rays or protons, to kill cancer cells. Radiation therapy is most often delivered from a machine outside your body (external beam radiation). Though it can also come from radioactive seeds and wires placed near your cancer (brachytherapy).
4. Chemotherapy is a treatment that uses chemicals to kill cancer cells. Chemotherapy drugs can be given alone, in combination with other chemotherapy drugs or in combination with other cancer cells. Chemotherapy may increase the effectiveness of radiation therapy, so the two are of the combined.

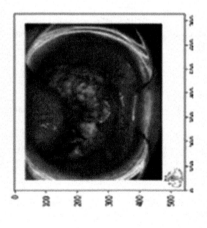

Figure 1.9 Prediction of "Cancer" image with a possible treatment.

The Following image is a normal tooth image. So for the same no treatment is needed.

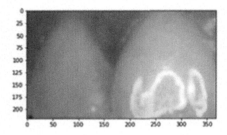

Figure 1.10 Identification of normal tooth.

Table 1.2 State-of-the-art comparison

Papers	Features	Technique employed	No. of diseases considered	Diseases identified	Accuracy
Pearce A. et al. [9]	Superpixel segmentation, Shape detection	Open CV	2	Chipped tooth, caries	Not mentioned
M. Thamarai et al. [7]	Bifurcation, Contour detection, Dental Radiography	Digital Subtraction Radiography	1	Periodontal disease	Not mentioned
You W. et al. [14]	LabelMe software used to detect features	Convolutional NN	1	Plaque	Around 75%
U. R. Acharya et al. [12]	Higher-order features	Linear Discriminant analysis	1	Thyroid tumour	87.50%
Morikawa T et al. [8]	Fluorescence visualization	Optical instruments	1	Oral cancer	98.0% sensitivity and 43.2% specificity on subjective evaluation
Proposed system	**Global features**	**Convolutional NN**	**2**	**Healthy tooth, dental caries, oral cancer detection**	**95%**

1.7 CONCLUSION

Artificial Intelligence is an emerging sector thathas helped the medical sector exponentially. This chapter proposed a model based on CNN which identifies a tooth as healthy, caries or cancer-infected. The outcomes of the chapter exhibit that the accuracy of the model in its purpose is 95%. Apart from it, average values of other parameters considered are: precision – 0.94, recall – 0.93, f1-score – 0.94, macro average – 0.94 and weighted average – 0.95. All the results show that the overall performance of the system is up to the mark and can be helpful in the prediction of healthy or diseased teeth. The output of the model also includes the possible treatments of the predicted disease, however, proper treatment can only be taken later under the supervision of the dentist. The state-of-the-art comparison also shows that the proposed model outperforms the parallel comparable models. Therefore, the proposed model demonstrates the capability of AI to assist in working on oral wellbeing. The future scope of the presented work can be to consider more than two dental diseases, work on an enhanced dataset and increase the value of accuracy and other parameters considered in the work.

REFERENCES

1. Bhan, A., Vyas, G., Mishra, S., & Pandey, P. (2016). "Detection and Grading Severity of Caries in Dental X-Ray Images", *International Conference on Micro-Electronics and Telecommunication Engineering (ICMETE)*, pp. 375–378.
2. Zandona, A. F., & Zero, T. (2006). Diagnostic tools for early caries detection. *The Journal of the American Dental Association*, 137, 1675–1684.
3. Lu, D., & Weng, Q. (2007). A survey of image classification methods and techniques for improving classification performance, *International Journal of Remote Sensing*, 28(5), 823–870, https://doi.org/10.1080/01431160600746456
4. Olsen, G. F., Brilliant, S. S., Primeaux, D., & Najarian, K. (2009). An image-processing enabled dental caries detection system, *2009 ICME International Conference on Complex Medical Engineering*, Tempe, AZ, pp. 1–8. https://doi.org/10.1109/ICCME.2009.4906674
5. https://www.who.int/news-room/fact-sheets/detail/oral-health [Accessed on 17 February 2021].
6. Kumar, A., & Mahapatra, R. P. (2022). Detection and diagnosis of COVID-19 infection in lungs images using deep learning techniques. *The International Journal of Imaging Systems and Technology*, 1–14. https://doi.org/10.1002/ima.22697
7. Thamarai, M., & Kalpa, M. (2014). Automated diagnosis of periodontal diseases using image processing techniques. *IJIRSET*, 3(1), 8466–8472.
8. Morikawa, T., Kozakai, A., Kosugi, A., Bessho, H., & Shibahara, T. (2020). Image processing analysis of oral cancer, oral potentially malignant disorders, and other oral diseases using optical instruments. *International Journal of Oral and Maxillofacial Surgery*, 49(4), 515–521. https://doi.org/10.1016/j.ijom.2019.08.016. Epub 2019 Sep 7. PMID:31500953.

9. Pearce, A., & Radhika, R. (2018). Computer aided diagnosis of oral disease indental images using open CV. *International Journal of Pure and Applied Mathematics*, 118, 389–396. https://doi.org/10.12732/ijpam.v118i10.79
10. Khatamino, P., Cantürk, İ., & Özyılmaz, L. (2018). "A Deep Learning-CNN Based Systemfor Medical Diagnosis: An Application on Parkinson's Disease Handwriting Drawings," *6th International Conference on Control Engineering & Information Technology (CEIT)*, Istanbul, Turkey, 2018, pp. 1–6. https://doi.org/10.1109/CEIT.2018.8751879
11. Yamany, S. M., & Farag, A. A. (2011). A system for human jaw modeling using intra-oral images. *Proceedings of the 20th Annual International Conference of the IEEE Engineering in Medicine and Biology Society*, 2, 562–566.
12. Acharya, U. R., Sree, S. V., Swapna, G., Gupta, S., Molinari, F., Garberoglio, R., Witkowska, A., & Suri, J. S. (2013). Effect of complex wavelet transform filter on thyroid tumor classification in three-dimensional ultrasound. *The Proceedings of the Institution of Mechanical Engineers, Part H: Journal of Engineering in Medicine*, 227(3), 284–292.
13. Wang, L., Zhang, K., Liu, X. etal. (2017). Comparative analysis of image classification methods for automatic diagnosis of ophthalmic images. *Scientific Reports*, 7, 41545.
14. You, W., Hao, A., Li, S. et al. (2020). Deep learning-based dental plaque detection on primary teeth: A comparison with clinical assessments. *BMC Oral Health*, 20, 141. https://doi.org/10.1186/s12903-020-01114-6
15. Khan, H. A., Haider, M. A., Ansari, H. A., Ishaq, H., Kiyani, A., Sohail, K., Muhammad, M., Khurram, S. A. (2021). Automated feature detection in dental periapical radiographs by using deep learning. *Oral Surgery, Oral Medicine, Oral Pathology and Oral Radiology*, 131(6), 711–720, ISSN 2212-4403, https://doi.org/10.1016/j.oooo.2020.08.024
16. Chen, H., Li, H., Zhao, Y. et al. (2021). Dental disease detection on periapical radiographs based on deep convolutional neural networks. *International Journal of Computer Assisted Radiology and Surgery*, 16, 649–661 https://doi.org/10.1007/s11548-021-02319-y
17. Lian, L., Zhu, T., Zhu, F., & Zhu, H. (2021). Deep learning for caries detection and classification. *Diagnostics*, 11(9), 1672. https://doi.org/10.3390/diagnostics 11091672
18. Sharma, D., Kudva, V., Patil, V., Kudva, A., & Bhat, R. S. (2022). A convolutional neural network based deep learning algorithm for identification of oral precancerous and cancerous lesion and differentiation from normal Mucosa: A retrospective study. *Engineered Science*, 18, 278–287.
19. Warin, K., Limprasert, W., Suebnukarn, S., Jinaporntham, S., & Jantana, P. (2021). Automatic classification and detection of oral cancer in photographic images using deep learning algorithms. *Journal of Oral Pathology & Medicine*, 50(9), 911–918.
20. García-Pola, M., Pons-Fuster, E., Suárez-Fernández, C., Seoane-Romero, J., Romero-Méndez, A., & López-Jornet, P. (2021). Role of artificial intelligence in the early diagnosis of oral cancer. A scoping review. *Cancers*, 13(18), 4600.
21. Khanagar, S. B., Naik, S., Al Kheraif, A. A., Vishwanathaiah, S., Maganur, P. C., Alhazmi, Y., ... Patil, S. (2021). Application and performance of artificial intelligence technology in oral cancer diagnosis and prediction of prognosis: A systematic review. *Diagnostics*, 11(6), 1004.

22. Jeyaraj, P. R., & Samuel Nadar, E. R. (2019). Computer-assisted medical image classification for early diagnosis of oral cancer employing deep learning algorithm. *Journal of Cancer Research and Clinical Oncology*, 145(4), 829–837.
23. Hung, M., Voss, M. W., Rosales, M. N., Li, W., Su, W., Xu, J., ... Licari, F. W. (2019). Application of machine learning for diagnostic prediction of root caries. *Gerodontology*, 36(4), 395–404.
24. Hwang, J. J., Jung, Y. H., Cho, B. H., & Heo, M. S. (2019). An overview of deep learning in the field of dentistry. *Imaging Science in Dentistry*, 49(1), 1–7.
25. Lee, J. H., Kim, D. H., Jeong, S. N., & Choi, S. H. (2018). Diagnosis and prediction of periodontally compromised teeth using a deep learning-based convolutional neural network algorithm. *Journal of Periodontal & Implant Science*, 48(2), 114–123.
26. Casalegno, F., Newton, T., Daher, R., Abdelaziz, M., Lodi-Rizzini, A., Schürmann, F., ... Markram, H. (2019). Caries detection with near-infrared transillumination using deep learning. *Journal of Dental Research*, 98(11), 1227–1233.
27. Shan, T., Tay, F. R., & Gu, L. (2021). Application of artificial intelligence in dentistry. *Journal of Dental Research*, 100(3), 232–244.
28. Rimi, I. F., Arif, M., Islam, A., Akter, S., Rahman, M., Islam, A. H. M., & Habib, M. (2022). Machine learning techniques for dental disease prediction. *Iran Journal of Computer Science*, 5, 1–9.
29. Reyes, L. T., Knorst, J. K., Ortiz, F. R., & Ardenghi, T. M. (2022). Machine learning in the diagnosis and prognostic prediction of dental caries: A systematic review. *Caries Research*, 56, 161–170.
30. Lee, J. H., Kim, D. H., Jeong, S. N., & Choi, S. H. (2018). Detection and diagnosis of dental caries using a deep learning-based convolutional neural network algorithm. *Journal of Dentistry*, 77, 106–111.
31. Talpur, S., Azim, F., Rashid, M., Syed, S. A., Talpur, B. A., & Khan, S. J. (2022). Uses of different machine learning algorithms for diagnosis of dental caries. *Journal of Healthcare Engineering*, 2022.

Chapter 2

Artificial intelligence for Healthcare 4.0

K. Umapathy, S. Omkumar, T. Dineshkumar, M. A. Archana and S. Prabakaran
SCSVMV Deemed University, Kanchipuram, India

Arwa N. Aledaily
University of Ha'il, Ha'il, Kingdom of Saudi Arabia

2.1 INTRODUCTION

Health 4.0 is a derivative of the principles of Industry 4.0. Industry 4.0 is a means of creating revolution or innovation by which companies produce, enhance and market their products using the concept of digitization. The relevant technologies connected with Industry 4.0 are illustrated in Figure 2.1. Health 4.0 promotes a new perception for the medical industry. It concatenates various techniques such as the Internet of Things (IoT), health cloud, machine learning, big data and blockchain. The objective is to provide enhanced, quality and economical healthcare services to needy people, thereby improving the efficiency of the medical industry. It promotes a model of business in order to improve the connectivity among the patients, stakeholders, hospital infrastructures and value factors [1]. By this model, various factors such as low cost, production frequency and reliability of healthcare services can be enhanced for the satisfaction of the patients. However, employing those healthcare applications with respect to the concept of Health 4.0 is totally a complex process. Moreover, in this context, advanced Health-4.0-based applications are yet to be explored.

Machine learning is a technique corresponding to AI by which the system will learn from the environment for executing smart and intelligent decisions. Machine learning plays a vital role in the field of Healthcare 4.0 [2]. It is a derivative of artificial intelligence. It has an inevitable growth of the global market until the year 2025. Similarly, the size of the market of the IoT for healthcare also progresses at the same frequency [3]. Figure 2.2 shows IoT applications in various sectors of healthcare [4]. The merits of employing the IoT in healthcare are the reduction of waiting periods during urgency and easy identification of inventory, personnel and patients. The other advantages are cost reduction, easy reporting, management of drugs, emergency alerts to physicians, speedy diagnosis of health complications, etc. Nowadays, due to busy schedules, it is radically difficult to monitor the status of patients all over the day thereby providing medicines to them

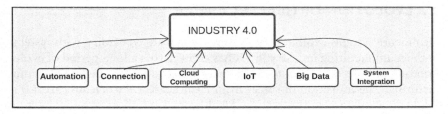

Figure 2.1 Technologies connected with Industry 4.0.

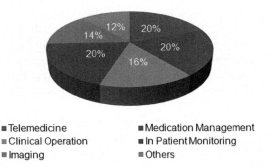

Figure 2.2 Impact of the IoT on healthcare sectors.

appropriately. Hence, the usage of devices connected with the IoT will help us to track the status of patients and give reminder messages on time regarding medicines.

Generally, people suffer from various health complications because of their improper habits and lifestyle. It is difficult to predict the health complication at the beginning stage because of the time consumed for data analysis of the patients. AI-based approaches will simplify the prediction of health complications in an accurate and timely manner. In this context, the IoT will collect the relevant data from respective sensors thereby reducing the time for data collection from various sources [5].

There are various constraints such as cost factors, shortage of suitable services, skilled physicians and appropriate equipment in the field of healthcare. The key challenges faced by this industry are heavy competition and the need for collaboration among the service providers. The above challenges provide a necessity for exploring new and innovative models of healthcare with the inclusion of the latest technologies. The technologies include the IoT, Artificial Intelligence, Data Analytics, Blockchain and Cyber Security. The aim of Industry 4.0 with respect to Healthcare 4.0 is to improve the mode of operation, reduce cost factors and enhance performance by means of automation, optimization and skilled intelligence. The background and technologies of Healthcare 4.0 and the impact and application of AI in the medical industry will be discussed in the rest of the chapter.

2.2 EVOLUTION OF HEALTHCARE 4.0

Healthcare systems exhibit many features which are coinciding with systems in the manufacturing industry [6]. It has come across a long period of evolution on par with other sectors of industry, especially with the line of manufacturing. The transition of stages from Healthcare 1.0 to Healthcare 4.0 is gradually explained in this section. Healthcare 1.0 indicates the interaction of a patient with members of the clinic such as doctors, receptionists and pharmacists. Generally, a doctor at the clinic after testing the patient will prescribe the medicines and give instructions regarding the treatment of the disease. This is the methodology which is quite common in the society for a number of years. This situation is called Healthcare 1.0. But due to progressive developments in the fields of medicine, science and technology, a number of new and innovative equipment have been developed for the purpose of monitoring and tracking in the healthcare industry. Typical examples of those medical equipment are pulse oximeters, CT scans, etc. This development is called Healthcare 2.0.

From the beginning of the last decade, medical records in electronic format have become popular thanks to the evolution of information technology. This is called an electronic medical record (EMR). The impact of these electronic records seemed to be good on the operations and processes connected with the clinic. A number of actions are registered in the medical records with appropriate time stamping. Some of the actions are digitized too. One typical example of this digitization is the Summary Visit of the patient. Moreover, remote patient monitoring and virtual visits are carried out with the help of computer networks. The virtual visit is the interaction between the patient and the doctor through online means. The impact of the Corona pandemic has increased the necessity for remote patient monitoring and virtual visits. This sort of transition is called Healthcare 3.0.

Healthcare 4.0 is currently booming on par with Industry 4.0. In this regard, the total process of healthcare has emerged as a system supported by various types of medical equipment and blended with different techniques such as big data, artificial intelligence and cloud computing. In this proposed system, health sectors, hospitals, all sorts of facilities, patients and all types of medical devices are totally integrated with each other. Data related to patients, medical expenses billing and claims connected with insurance are generally distributed by means of appropriate protocols. In addition, prevention and monitoring of diseases, patient care, etc., can be achieved through AI methodologies. Figure 2.2 shows the evolution of Healthcare 4.0. From Figure 2.3, it is clear that healthcare has emerged from a simple process to a complicated and intelligent process for the treatment of diseases. The care of patients is given utmost importance and was delivered through means of multiple numbers of clinics, teams of people and societies. In the beginning, it was concentrated on one facility and then multiplexed by involving multiple facilities and entities.

Artificial intelligence for Healthcare 4.0   21

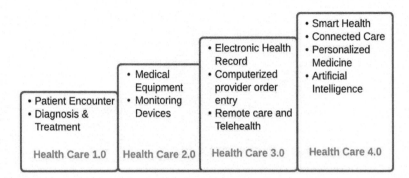

Figure 2.3 Evolutions of Healthcare 4.0.

Originally, simple data were employed for the transition of healthcare. But currently, streams of big data with greater changes in size, format and quality are presently used with Healthcare 4.0. Hence, the factor of automation is included in this Healthcare 4.0 which is very much similar to that of Industry 4.0. The vital discrimination between Healthcare 4.0 and Industry 4.0 is the engagement of people. Here, the involvement of patients and clinicians play a vital role in the purpose of health tracking, identification of symptoms and making appropriate decisions with respect to the treatment of diseases and patient care.

2.3 CONSTITUENTS OF HEALTHCARE 4.0

The following are the various constituents of Healthcare 4.0. They are:

1. Intelligence: Tracking and monitoring of diseases, treatment of diseases and cooperation between patients and clinic personnel are the various factors to be considered in order to concentrate on specialized care for the patients. This is implemented with the aid of AI methods and techniques. Hence, to achieve special care for patients, the requirements and traits of the patients have to be understood appropriately and classified into various groups based on the type of risk factor. Typical examples of risk factors among patients are cancer, paralysis, heart attack, etc. [7]. Understanding the traits and requirements of the patients is very much required in order to provide specialized care. In this context, the patients can be grouped into various classes of health complications such as cancer, heart attack and stroke. The diseases can be traced based on the information of the above classification. By creating an association between the patients and health complications, the clinic personnel will be able to provide the support of decisions for appropriate treatment of those health complications or diseases.

22 Artificial Intelligence and Blockchain in Industry 4.0

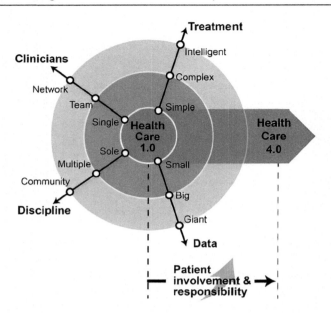

Figure 2.4 Traits of Healthcare 1.0 to 4.0 [1].

2. Accurate Analysis: Observations of health complications and aftereffects are predicted and collected based on the information regarding the grouping and classification of diseases. These observations will be much useful for the diagnosis of various diseases like heart-related problems and Corona infections. This sort of analysis will help to provide the values of risk factors and the parameters connected with the analysis [8]. The above factors and parameters are correlated as a set of instructions or guidelines for making decisions in the medical industry. The traits of Healthcare 4.0 are shown in Figure 2.4.
3. Preventive Action: The outcomes from the accurate analysis can be exploited to promote the preventive methodology for various types of diseases [9]. It is always wise to prevent the occurrence of diseases rather than concentrating on curing them [10]. This will enhance the safety of the concerned patients. Tracking the aftereffects of health complications and response to those complications with respect to treatment will help to go for various care methodologies.
4. Observation and Treatment: The patients must be monitored for important symptoms and other related crucial factors in order to enhance the outcomes of the patients. The status of the patient and accurate analysis are very much required to go for respective and appropriate caretake methodologies. For monitoring and tracking of patients in the remote place, devices such as sensors, wearable components and websites are essentially required [11]. The caretake methodologies can be improved by promoting analysis and models at advanced levels [12].

Artificial intelligence for Healthcare 4.0   23

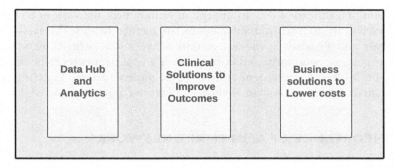

Figure 2.5 Layout of a smart healthcare system.

Figure 2.6 Healthcare system in a connected approach [1].

Figure 2.5 shows a smart healthcare system which has the following vital components:

- Suitable and appropriate interaction between the patients and caretakers
- Healthy communication among the members of the care team
- Employment and utilization of respective equipment and devices
- Billing and insurance activities

Promoting Healthcare 4.0 is an important requirement for various technical fields such as the IoT, artificial intelligence, Augmented Reality, Virtual Reality, Medicine and Pharmacy. The association between the patients, caretakers, clinic personnel, equipment and instruments is a vital parameter to be concentrated in this healthcare system. Above all, the patients are the key parameter of the smart healthcare system as shown in Figure 2.6 in a connected way.

2.4 IMPORTANCE OF AI IN SMART NETWORKS

Artificial Intelligence (AI) is the most wanted topic in medical industries because of its special features such as processing of large data, production of precious results and generation of optimized results [13]. AI is not a new technique because algorithms and respective machines have been in use for a long period for the purpose of making decisions and sorting out the aftereffects of health complications. Various parameters such as clarity, reliability and standardization can be obtained by means of coordination between the algorithms and respective machines [14]. Generally, AI means the capability of the computer for reproducing the intelligence of human beings with respect to coding and respective algorithms. The processes such as discovery of drugs, projected surgery and imaging can be performed by AI in an intellectual manner. AI may be booming recently, but this idea was identified by scientists in the year 1940 itself [15]. The suggestions provided by these scientists formed a basis for many investigations based on the concept of AI. A typical example is a robot named WABOT-1 constructed by Japanese scientists in the year 1972. But, further investigations started on AI models only in the late 1990s [16]. From these models of AI, social media, service providers of email and various industries obtained many more advantages in the 20th century. The need for large computational power and handling of big data has increased the demand for AI-based systems.

Machine Learning is the commonly used technique for generating observations from the given patterns under the methods of AI. This is illustrated in Figure 2.7. Machine learning can be categorized into different types as shown in Figure 2.8. The methods of learning can be categorized into three types – reinforced, supervised and unsupervised methods – based on the structure of algorithms.

In the case of supervised learning, the input data are subject to training by the algorithm. This type of learning is meant for applications where the processing of historical data is carried out and employed to measure the possible events connected with the future. The historical information will be used by these algorithms for the purpose of training and they possess absolute accuracy. There are two types of these algorithms – one is regression and the other one is classification. In the case of regression algorithms, there will be a relation between the input and output variables, for example,

Artificial intelligence for Healthcare 4.0 25

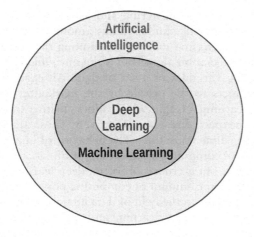

Figure 2.7 Relation between AI and ML.

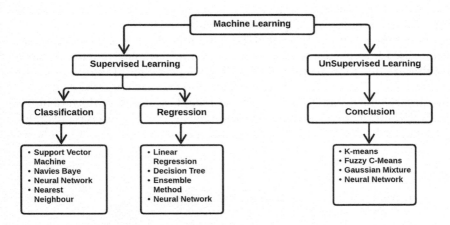

Figure 2.8 Classification of ML.

forecasting climate. The output variables and input variables are related to each other based on the categorization of classes. A typical example for the classification of classes may be "True/False" or "Yes/No". Hence, because of the above specialties, supervised learning is used to sort out problems in real time and will be able to determine the expected outputs with respect to available data. Contrary to this, unsupervised learning methods can be used for the identification of patterns from the given set of data even if they are grouped appropriately.

The above algorithms can be categorized into a set of groups with respect to similarities and differences in the collected information. They can be classified into two groups – clustering and association. A clustering algorithm is supposed to include similar things among components of data. A typical example is purchasing mannerisms among the customers in a marketing group. The association algorithm is the one which identifies the relationships among the variables in a set of data. An example of this is the suggestion of products to the people by the marketing group. The third type – reinforced learning – is that which allots positive and negative values to expected and unexpected results respectively. These algorithms are difficult to train and consume more time. The behavior of human beings will be reflected in the case of computers by a set of algorithms called Deep Learning (DL). In order to sort out a crucial problem, deep learning employs neural networks by which a large amount of computing power can be applied. Due to the latest developments in the field of data analytics, DL is able to handle crucial situations by monitoring, learning and appropriately responding. DL can apply all three types of approaches – supervised, unsupervised and reinforced on any type of expected application. One typical example is electronic email where AI is employed to segregate spam emails from regular emails with a sufficient percentage of accuracy. Nowadays, the forecasting of weather is completely dependent on models of AI.

As per the review of Jiang et al. [17], among all methods of learning, the supervised learning approach seemed to be the best one for applications of healthcare due to its feature of providing relevant clinical information and results. They also emphasized that support vector machines and neural networks are widely employed for almost all applications in healthcare industries. The impact of AI methods on various medical applications is illustrated in Figure 2.9. Table 2.1 shows the merits and demerits of algorithms based

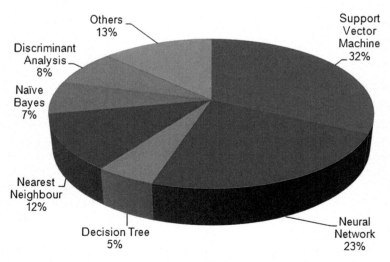

Figure 2.9 Impact of AI on medical applications.

Table 2.1 Comparison of specific AI algorithms for medical applications

S. No	Name of AI algorithm	Applications	Merits	Demerits
1	Support Vector Machine (SVM)	Screening of cancer, determining infection at surgery, tracking of glucose levels, management of resources for pandemic	Great accuracy, Speedy and sorts out complicated problems, better scaling capacity	Longer time for training, Expensive computation, Model difficult to understand
2	Neural Networks (NN)	Screening of cancer, cardiac screening, prediction of diabetes, surgery, etc.	Speedy and adaptive algorithm, generates output without specific rules and consistent learning, handles multiple tasks and crucial databases	Longer time for training, expensive and complicated programs, hard to interpret and modify
3	Decision Tree (DT)	Tracking of glucose levels, surgery, general health check-up, systems maintenance, etc.	Speedy, adaptive and efficient, simplicity, easy for handling, good learning	Complicated manipulations, expensive and lengthy computations, less efficient to handle noise and over-fitting, poor accuracy
4	Random Forest (RF)	Screening of diseases, screening of heart complications, general health monitoring, etc.	Very fast, can handle noisy and voluminous data, automation	Hard to operate and implement, Poor accuracy, subject to over-fitting, depth and tree number to be defined

on AI. Recently, another approach called Natural Language Processing (NLP) is integrated with AI and implemented for real-time applications. By using NLP, computers are able to analyze, apply, compute and provide human language. The input information shall be in written or spoken format. Hence to solve complex processes, both NLP and ML are blended. Suitable examples of this integration of NLP and ML are Google Assistant,

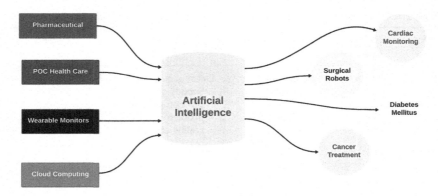

Figure 2.10 Role of AI in different sectors of healthcare.

Alexa, etc. This NLP can be also employed to encode medical documents in an automated manner.

A new technology for transition called "PathAI" has been promoted by means of the latest systems developed by AI. This technology created a record to enhance the health status and outcomes of patients with respect to various methods of diagnosis. In the same context, a health application named "PAGER" will assist the treatment of patients by providing suitable suggestions and ideas. It also helps to identify appropriate and suitable medicines for the treatment of diseases by training relevant data. Moreover, AI aids in promoting techniques for interfacing human beings with machines. This sort of human-to-machine interface will be much beneficial in the medical industry. Artificial limbs and appropriate sensors are employed to gather real-time information regarding patients [18].

2.5 AI FOR HEART MONITORING

Heart complication is the main reason for mortality across the globe. This number reached 18 million in the year 2000. Cardiac diseases are of primary concern for developing countries due to their complication toward the survival of life. The generation traits, style of life and depression are the various reasons for cardiac diseases. Heart complications can be categorized into different types – heart attack, failure of heart, pericarditis, etc. Various approaches such as CT scans, electrocardiograms (ECGs) and stress tests can be used as a means of observing the abnormalities in the function of the heart [19]. From the above tools, ECGs and stress tests can be fast tools for easy and quick evaluation of the function of the heart. A stress biomarker is employed by bio-sensing instruments to determine the level of depression [20].

Atrial Fibrillation (AF) is a status of a patient not having regular heartbeats. These patients will develop clotting of blood in their bodies which may lead to abnormal function of the heart and may also lead to stroke attack. But the challenge in this diagnosis is the inability to detect the irregular heartbeat from the beginning itself. Hence, consistent clinical practices are required to detect atrial fibrillation within a period of 30 days. However, there are some hindrances in this diagnosis process due to some technical reasons. So a reliable, consistent and efficient methodology is very essential for the detection of atrial fibrillation on a long-term basis.

ECG is a usual methodology used for the observation of heart function on a real-time basis. Since it can give clear details of the heart function, it can be employed for determining different types of cardiac complications. An ECG recorder having "12" numbers of leads is used to monitor the irregular rhythm of the heart. In this regard, the manual recording of ECG will absorb a lot of time and generate errors for a large collection of data. But in practice, the usual method of measuring ECG is totally uncomfortable to the patient wearing it and it is subject to disturbance or noise during the recording. Scientists used to employ algorithms based on machine learning to identify the rhythm of heartbeats and determine the complications in an accurate manner. From the recording of ECG, certain features are extracted and they are used to determine the heart complications like sleep apnea and irregular heartbeats [21]. With the latest developments in the field of cloud computing, AI can be employed to handle large collections of data. The systems designed based on AI can be employed for different sorts of applications which include the classification of noise in ECG-type signals, identification of irregular heartbeats and determination of atrial fibrillation. The impact of AI on the field of cardiology is shown in Figure 2.11. The data obtained from various IoT-connected devices are handled appropriately for the treatment of health complications and to get exposure to new types of diseases.

A new framework was presented by Kachnee et al. for the analysis of the data of ECG in order to execute various tasks such as the identification of heartbeats and representation of signals [22]. The proposed framework is acquainted with the Physikalisch-Technische Bundesanstalt (PTB) database. The pre-processing of ECG signals is done by AI before going for the classification of heartbeats. This framework will aid AI in determining the type of heartbeats and the type of training to be executed. The method of evaluation and type of training will be done by using a tensor flow library. The trained network was employed to determine the sufficient number of heartbeats required for arrhythmia.

2.6 AI FOR SURGERY

The impact of the IoT keeps on rising because of the integrated development of AI with machine learning, computer vision and other techniques. The above features of AI promote a basis for all respective activities connected with surgeries supported by AI [23]. The enhanced and booming features of AI can be exploited in the field of surgery. Hence, AI is very much

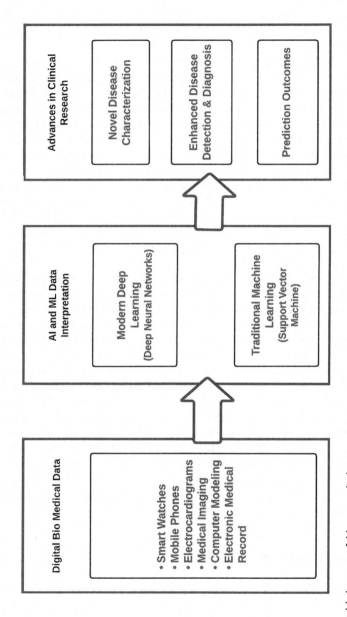

Figure 2.11 Impact of AI on cardiology.

suitable for different levels of surgery. In a conventional method of surgery, all actions are performed by human beings. But in surgeries based on AI, the total process has become independent right from imaging to operating without the intervention of human beings [24]. The system of surgery suggested by Da Vinci Systems is a famous type that is used to permit physicians to operate surgery remotely using the required techniques [25]. This is the general and usual method followed by most doctors because of its efficiency and accuracy. The implementation of mobile platforms by AI is exploited to give surgical instructions at remote places. This approach can be also used for the guidance of a robot for performing surgeries at places where sufficient and suitable resources are not available. Typical applications are aircraft in space or places of war [26].

Generally, the surgeries are of various types – one category is based on scheduling and the other is based on the condition of emergency with threatening of life. The following are four important objectives in the care of surgery.

- Improving the quality of research activities
- Identification of patients and result prediction with respect to the complexity of diseases
- Collection of data and consistent monitoring
- Performance in surgery

The analysis of data can be done in a precise and accurate manner by the surgery people with the integration of AI and ML [27]. Their impact on the domain of surgery is shown in Figure 2.12.

2.7 AI FOR DIABETES

Diabetes is one of the growing health complications that requires consistent monitoring of levels of glucose in the blood. Patients suffering from hypoglycemia and hyperglycemia will need to check their glucose levels in the blood every day. Millions of people across the globe suffer from diabetes and a lot of money is spent on the treatment of diabetes. This expenditure on the management of diabetes is likely to increase in the coming years. The side effects of improper management of diabetes include retinopathy which leads to loss of sight partially or completely [28]. Precise and consistent measurements of glucose levels are essential in order to prevent and handle the side effects caused by diabetes. The traditional method of measuring glucose levels needs puncturing the surface of the skin for taking the blood. Hence, a methodology is required for glucose measurement without puncturing the skin surface. But current glucose measuring systems depend on electrochemical methods as far as Point of Care (POC) is concerned. However, a consistent glucose measuring system provides data to the concerned on the

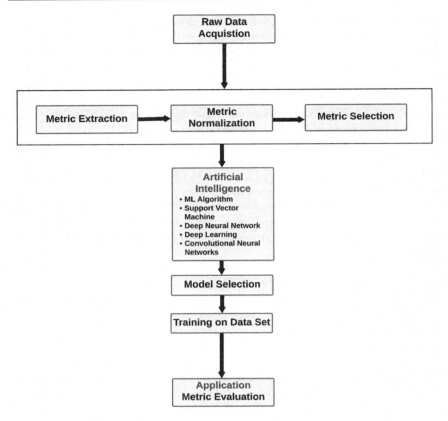

Figure 2.12 Impact of AI on surgery.

basis of real time. The approaches based on the integration of AI with ML play a significant role in managing diabetes as they determine the related patterns and process of screening in an easy manner [29, 30]. Hamdi et al. proposed an integrated system for the measurement of glucose with models of compartments. The ANN algorithm was implemented for the measurement. The measuring equipment consists of a sensor kept under the surface of the skin which is interfaced to a transmitter. This transmitter is then connected to a receiver for displaying in wireless mode [31, 32]. The algorithm is illustrated in Figure 2.13.

An optimistic and precise nutrition is a method of nutrition that depends on various factors such as age, rate of metabolism and treatment of diseases. This method of nutrition can be implemented by combining different biosensor devices together. Wang et al enunciated the idea of precision nutrition for individual persons by blending biosensor devices with that of systems based on cloud [33, 34]. The proposed system monitors the intake of food and uses sensors for the measurement of nutrients in fluids generated in the human body such as urine, blood and saliva. This is illustrated in Figure 2.14.

Artificial intelligence for Healthcare 4.0 33

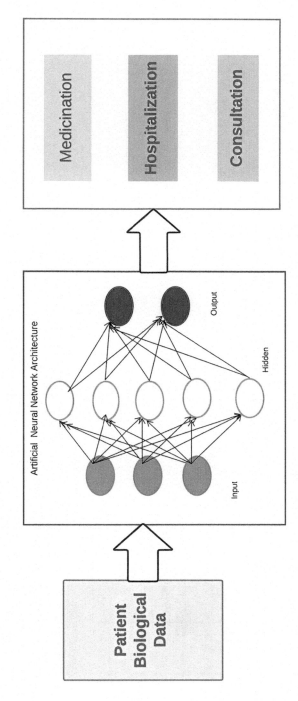

Figure 2.13 ANN model for treatment of diabetes.

34 Artificial Intelligence and Blockchain in Industry 4.0

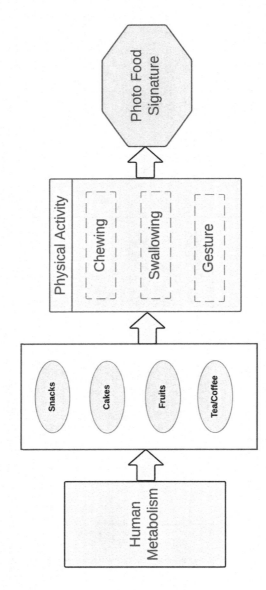

Figure 2.14 Measurement system for nutrition.

2.8 CONCLUSION

Artificial Intelligence is the latest technique by which programs are designed for performing specific activities that require optimum intelligence. Additionally, other techniques such as machine learning, speech recognition and the IoT are blended with AI in order to improve the functionality of systems in the field of health sciences. This chapter initially enunciated the evolution of Healthcare 4.0 which is a derivative of Industry 4.0. It described how Healthcare 4.0 is used to promote the model of business in order to improve the connectivity among the patients, stakeholders, hospital infrastructures and value factors. The impact of artificial intelligence on various characteristics and functions of different medical systems used in the health sector was also discussed in detail with relevant examples. Ultimately, it focused on the integration of AI and ML for the treatment of health complications such as diabetes, cardiac disease and surgery in the health sector.

REFERENCES

1. Li, J., and Carayon, P. "Health Care 4.0: A Vision for Smart and Connected Health Care", *IISE Transactions on Healthcare Systems Engineering*, 2021, Volume 11, Issue 3, pp. 171–180.
2. Kishor, Amit and Chakraborty, Chinmay, "Artificial Intelligence and Internet of Things Based Healthcare 4.0 Monitoring System", *Wireless Personal Communication*, 2021, https://doi.org/10.1007/s11277-021-08708-5
3. Market research report, Retrieved March 2021 from https://www.grandviewresearch.com/pressrelease/global-artificial-intelligence-healthcare-market.
4. Market research report, Retrieved March 2021 from https://www.grandviewresearch.com/industryanalysis/internet-of-things-iot-healthcare-market
5. Market research report, Retrieved March 2021 from https://www.appventurez.com/blog/iot-healthcare-future-scope
6. Zhong, X., Lee, H. K., and Li, J, "From Production Systems to Health Care Delivery Systems: A Retrospective Look on Similarities, Difficulties And Opportunities", *International Journal of Production Research*, 2017, Volume 55, Issue 14, pp. 4212–4227, https://doi.org/10.1080/00207543.2016
7. Rodrigues, G., Warde, P., Pickles, T., Crook, J., Brundage, M., Souhami, L., and Lukka, H., Genitourinary Radiation Oncologists of Canada, "Pre-Treatment Risk Stratification of Prostate Cancer Patients: A Critical Review", *Canadian Urological Association Journal, Journal de l'Association des urologues du Canada*, 2012, Volume 6, Issue 2, pp. 121–127, https://doi.org/10.5489/cuaj.11085
8. Kansagara, D., Englander, H., Salanitro, A., Kagen, D., Theobald, C., Freeman, M., and Kripalani, S., "Risk Prediction Models for Hospital Readmission: A Systematic Review", *JAMA*, 2011, Volume 306, Issue 15, pp. 1688–1698, https://doi.org/10.1001/jama.2011.1515
9. Vlaeyen, E., Stas, J., Leysens, G., Van der Elst, E., Janssens, E., Dejaeger, E., Dobbels, F., and Milisen, K. "Implementation of Fall Prevention in Residential Care Facilities: A Systematic Review of Barriers and Facilitators", *International*

Journal of Nursing Studies, 2017, Volume 70, pp. 110–121. https://doi.org/10.1016/j.ijnurstu.2017.02.002

10. Struckmann, V. et al. "Relevant Models and Elements of Integrated Care for Multi-morbidity: Results of a Scoping Review", *Health Policy (Amsterdam, Netherlands)*, 2018, Volume 122, Issue 1, pp. 23–35, https://doi.org/10.1016/j.healthpol.2017.08.008

11. Begam, S. S., Vimala, J., Selvachandran, G., Ngan, T. T., and Sharma, R. "Similarity Measure of Lattice Ordered Multi-Fuzzy Soft Sets Based on Set Theoretic Approach and Its Application in Decision Making", *Mathematics*, 2020, Volume 8, p. 1255.

12. Thanh, V., Rohit, S., Raghvendra, K., Le Hoang, S., Thai, P. B., Dieu, T. B., Ishaani, P., Manash, S., and Tuong, L. "Crime Rate Detection Using Social Media of Different Crime Locations and Twitter Part-of-speech Tagger with Brown Clustering", 1 Jan. 2020: 4287–4299.

13. Nguyen, P. T., Ha, D. H., Avand, M., Jaafari, A., Nguyen, H. D., Al-Ansari, N., Van Phong, T., Sharma, R., Kumar, R., Le, H. V., Ho, L. S., Prakash, I., and Pham, B. T. "Soft Computing Ensemble Models Based on Logistic Regression for Groundwater Potential Mapping", *Applied Sciences*, 2020, Volume 10, p. 2469.

14. Jha, S. et al. "Deep Learning Approach for Software Maintainability Metrics Prediction", *IEEE Access*, 2019, Volume 7, pp. 61840–61855.

15. Sharma, R., Kumar, R., Sharma, D. K., Son, L. H., Priyadarshini, I., Pham, B. T., Bui, D. T., and Rai, S. "Inferring Air Pollution from Air Quality Index by Different Geographical Areas: Case Study in India", *Air Quality, Atmosphere, and Health*, 2019, Volume 12, pp. 1347–1357.

16. Sharma, R., Kumar, R., Singh, P. K., Raboaca, M. S., and Felseghi, R.-A. "A Systematic Study on the Analysis of the Emission of CO, CO2 and HC for Four-Wheelers and Its Impact on the Sustainable Ecosystem", *Sustainability*, 2020, Volume 12, p. 6707.

17. Dansana, D., Kumar, R., Das Adhikari, J., Mohapatra, M., Sharma, R., Priyadarshini, I., and Le, D.-N. "Global Forecasting Confirmed and Fatal Cases of COVID-19 Outbreak Using Autoregressive Integrated Moving Average Model", *Frontiers in Public Health*, 2020, Volume 8, p. 580327. https://doi.org/10.3389/fpubh.2020.580327

18. Malik, P. K., Sharma, R., Singh, R., Gehlot, A., Satapathy, S. C., Alnumay, W. S., Pelusi, D., Ghosh, U., and Nayak, J. "Industrial Internet of Things and Its Applications in Industry 4.0: State of The Art", *Computer Communications*, 2021, Volume 166, pp. 125–139. https://doi.org/10.1016/j.comcom.2020.11.016. ISSN 0140-3664

19. Sharma, R., Kumar, R., Satapathy, S. C., Al-Ansari, N., Singh, K. K., Mahapatra, R. P., Agarwal, A. K., Le, H. V., and Pham, B. T. "Analysis of Water Pollution Using Different Physicochemical Parameters: A Study of Yamuna River", *Frontiers in Environmental Science*, 2020, Volume 8, p. 581591. https://doi.org/10.3389/fenvs.2020.581591

20. Dansana, D., Kumar, R., Parida, A., Sharma, R., Adhikari, J. D. et al., "Using susceptible-exposed-infectious-recovered model to forecast coronavirus outbreak", *Computers, Materials & Continua*, 2021, Volume 67, Issue 2, pp. 1595–1612.

21. Vo, M. T., Vo, A. H., Nguyen, T., Sharma, R., and Le, T. "Dealing with the class imbalance problem in the detection of fake job descriptions", *Computers, Materials & Continua*, 2021, Volume 68, Issue 1, pp. 521–535.

22. Sachan, S., Sharma, R., and Sehgal, A., "Energy Efficient Scheme for Better Connectivity in Sustainable Mobile Wireless Sensor Networks", *Sustainable Computing: Informatics and Systems*, 2021, Volume 30, p. 100504.
23. Ghanem, S., Kanungo, P., Panda, G. et al. "Lane Detection under Artificial Colored Light in Tunnels and on Highways: An IoT-based Framework for Smart City Infrastructure", *Complex & Intelligent Systems*, 2021. https://doi.org/10.1007/s40747-021-00381-2
24. Sachan, S., Sharma, R., and Sehgal, A. SINR Based Energy Optimization Schemes for 5G Vehicular Sensor Networks. *Wireless Personal Communications*, 2021. https://doi.org/10.1007/s11277-021-08561-6
25. Priyadarshini, I., Mohanty, P., Kumar, R. et al. A Study on the Sentiments and Psychology of Twitter Users during COVID-19 Lockdown Period. *Multimedia Tools and Applications*, 2021. https://doi.org/10.1007/s11042-021-11004-w
26. Panesar, S. S., and Ashkan, K. "Surgery in Space", *British Journal of Surgery*, 2018, Volume 105, pp. 1234–1243.
27. Galbusera, F., Casaroli, G., and Bassani, T. "Artificial intelligence and machine learning in spine research", *JOR Spine*, 2019, Volume 2, e1044.
28. Grzybowski, A. et al. "Artificial Intelligence for Diabetic Retinopathy Screening: A Review", *Eye*, 2020, Volume 34, pp. 451–460.
29. Hasanzad, M., Aghaei Meybodi, H. R., Sarhangi, N., and Larijani, B. "Artificial Intelligence Perspective in the Future of Endocrine Diseases", *Journal of Diabetes and Metabolic Disorders*, 2022, Volume 21, pp. 971–978.
30. Umapathy, K. "Wireless Technique Based Vehicle Speed Control System", *AIP Conference Proceedings*, 2022, Volume 2519, pp. 050022-1–050022-5, https://doi.org/10.1063/5.0109726. ISSN: 1551 7616
31. Hamdi, T. et al. "Artificial neural network for blood glucose level prediction", *Proceedings of the 2017 International Conference on Smart, Monitored and Controlled Cities (SM2C)*, Sfax, Tunisia, February 2017, pp. 91–95.
32. Umapathy, K., and Rajaveerappa, D. "Implementation of Low Power and Small Area 128-Point Mixed Radix 4-2 FFT Processor for OFDM Applications", *European Journal of Scientific Research (EJSR)*, 2012, Vol. 84, Issue 1, pp. 112–118, ISSN 1450-216X
33. Sempionatto, J. R., Montiel, V. R., Vargas, E., Teymourian, H., and Wang, J. "Wearable and Mobile Sensors for Personalized Nutrition", *ACS Sensors*, 2021, Volume 6, pp. 1745–1760.
34. Kumar, S. S., and K. Umapathy "Efficient Implementation of Downlink Resource Allocation in Wireless Networks with Inter-cell Interference", *Journal of Advanced Research in Dynamical and Control Systems (JARDCS)*, Volume 9, Issue 14, 2017, pp. 2297–2312. ISSN: 1943-023X

Chapter 3

Artificial intelligence's foresight in cybersecurity

Radhika Nautiyal, Radhey Shyam Jha, Samta Kathuria, Shweta Pandey, Rajesh Singh and Anita Gehlot
Uttaranchal University, Dehradun, India

Praveen Kumar Malik
Lovely Professional University, Phagwara, India

Ahmed Alkhayyat
The Islamic University, Najaf, Iraq

3.1 INTRODUCTION

Industry 4.0 has significantly exploded as a result of using the Internet of Things' (IoT) recent large-scale growth, which executes the intricate data analysis of confidential material that must be safeguarded against cybersecurity attacks [1]. In a number of industries, including home automation, healthcare, electricity, agriculture, automation, and industrial processes, cybersecurity threats have rapidly increased. IoT device sensors produce a lot of data due to their broad range of services, which necessitates authorisation, security, and privacy. In the past, conventional frameworks and procedures were employed to guarantee IoT security. But over time, the use of various artificial intelligence (AI) techniques for spotting cybersecurity breaches has grown in favour [2, 3].

Depending on the component of the system being attacked and what the attacker aims to achieve from the attack, there is a wide range of cyberattacks that can be used against IoTs. As a result, a lot of study has been done on IoT cybersecurity. This covers AI methods of defending IoT systems from intruders, typically by the detection of odd activity that may be an indication of an assault [4]. However, mostly in instances of IoT, cybercriminals hold the upper hand because they only need to find vulnerabilities whereas cybersecurity specialists are required to secure a number of targets. Due to the complexity of the algorithms used to detect anomalous activity, this has also led to an increase in the usage of AI by cyber-attackers in an effort to avoid detection [5]. As IoT technologies have developed, AI has drawn a lot of interest. In order to identify dangers and potential attacks, IoT cybersecurity apps have been using AI technologies including linear regression, decision trees, deep training, support vector machines, and neural networks.

Security solutions and mitigation techniques will play a bigger role as our society gets more technologically connected and advanced. Protecting our systems and society is a difficult task (that relies on these systems). As a result, developing more efficient and effective solutions for cybersecurity is a continuing research area [6]. In order to secure systems from vulnerabilities and threats and to effectively deliver the right services to users, numerous techniques, methods, and means are used in the field of cybersecurity. As a result, both external and internal system threats are included in the cybersecurity discussed in this study. Cybersecurity aims to safeguard threats thoroughly as possible while also swiftly and effectively complying with the objectives of identification before the damage, management in the accident or recovery after the accident. These risks will have a significant influence on the routine operation of the systems [7]. There have been instances of developing AI-based solutions for a range of cybersecurity applications in recent years, in part because enterprises are becoming more aware of the value of AI in reducing cyberthreats [8].

We list the following as the significant contributions of this study:

- We go over the shortcomings of conventional cybersecurity solutions and explain how cutting-edge AI technologies might enhance cybersecurity.
- We examine and summarise the main AI-based methodologies for applications in network situation awareness, monitoring harmful behaviour, and anomalous traffic identification.
- We outline several significant issues that the cybersecurity community is currently facing and that need to be resolved in the future.

The following parts of the chapter are structured as follows.

The traditional (non-AI) cybersecurity methods are covered under Section 3.2, a brief introduction of AI is covered under Section 3.3, artificial intelligence used for cybersecurity is covered under Section 3.4, and AI applications to improve cybersecurity for different application domains are covered under Section 3.5. Lastly, Section 3.6 provides a brief summary and a suggestion for further research and Section 3.7 covers the conclusion.

3.2 TRADITIONAL (NON-AI) CYBERSECURITY METHODS

We provide a brief overview of traditional (non-AI) cybersecurity methods for spotting cyber-attacks:

- Rate control: DDoS and DoS attacks target the availability of systems [9]. By decreasing the amount of receiving network traffic via simple traffic throttling and revising permission lists, rate-control approaches might reduce the consequence on such systems that reflect the operations after DoS when they undergo assaults [10].

- Intrusion detection using signatures: A signature-based intrusion detection system (IDS) uses a set of databases to record either attack signatures that correspond to malicious traffic or legitimate signatures that correspond to regular traffic [11]. In real-time, the IDS system compares information in arriving network packets with the signatures that have been previously saved. The disadvantage of this approach is that IDS's ability to precisely identify malicious traffic entering a network is constrained in the absence of pertinent signatures.
- Autonomous systems: These can assure dependability and availability, self-protect and self-heal, as in the example of the BANS [12]. The four different modules that make up this system are called the Cyber Neuron, Axon, Central Nerve, and Peripheral Nerve. The utilisation of Cyber Neuron serves in guarding against malware and spyware. A sophisticated tool to repair spyware and malware damage is called Cyber Axon [13]. Similarly, Peripheral Nerve establishes a communication link between numerous cyber neurons placed on various devices to provide strong protection against DoS/DDoS attacks. The last function of the Central Nerve is to provide facts to other security systems and act as a body of knowledge for potential attacks. It is suggested that peripheral nerves can work together to defend against DoS and DDoS attacks.
- Game theory has been used in cybersecurity before. The victim's computer and the malicious actor are both players in the same game. Each person makes an effort to maximise their incentive by strategic movement, arguing logically that the move will achieve the desired result. The actions of each participant can either be predicted in advance or remain a secret. An illustration for setting a smart grid of a game wherein the defender process to maintain interconnection between various entities while the attacker tries to interfere with communication between a power source and a residence [14]. Both the attacker as well as the defender would use techniques to accomplish their respective aims at each stage of the game.
- Heuristics: In order to determine the best rule for categorising network data as genuine or anomalous, firewalls and IDSs frequently use heuristics. One such method identifies suspect website addresses by performing a series of procedures that include substring matching. VirusTotal is a desktop application where one can enter a website address and receive the scored analysis in reference to how malicious the input website is, which is used in the second phase for the proposed scheme to scan the web addresses. To that, the lowest score of the two scans is taken into consideration when determining whether to allow the information packets to enter the network or not.
- Security controls for end users: Current end-user gadgets like mobile phones, iPads, and smart portable devices (PCs) need built-in security rather than add-ons. While some suppliers try to push automatic

upgrades, it's possible that end customers would not upload their devices with the most recent of all security patches. The difficulty would be for software vendors to make sure that security updates protect against fresh attacks, or so-called "zero-day attacks," and smoothly integrate with all pre-existing software for the end-user device [15].

3.3 ARTIFICIAL INTELLIGENCE

There are numerous methods for constructing AI. People first tried to formalise their information by using a knowledge base. However, this method requires too many manual steps to accurately explain a world with intricate rules. AI is focused on how machines may appropriately reason or respond given their knowledge. This all-encompassing definition takes into account how well machines may mimic human thought or behaviour (Figure 3.1). Machines are considered intelligent on one extreme of the spectrum if they can maximise the result at every stage of the process. The Turing Test, on the other hand, establishes the bar for artificial intelligence.

3.4 ARTIFICIAL INTELLIGENCE USED FOR CYBERSECURITY

3.4.1 AI's potential use in cybersecurity applications

Before summarising, the issue in Figure 3.2 will address AI-based approaches to user access authentication, abnormal traffic identification, harmful behaviour monitoring, and network situation awareness.

3.4.1.1 User access authentication

- Requirements for user access authentication: The system must improve the administration for user access authentication precisely identify all types of unwanted behaviours and enable the detection of illicit or it can be harmful items in its role as the defence for first-line cybersecurity [16]. The apparatus should be tested before use. It guarantees user authentication. The user information should be private while also

Figure 3.1 Range of intelligent measures (from thinking like human to maximising the outcomes).

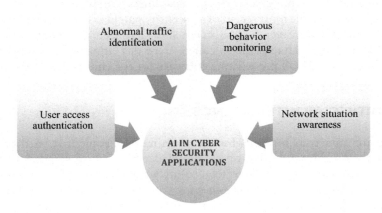

Figure 3.2 AI's potential use in cybersecurity applications.

being secure time to avoid other risk situations like harmful user information collecting.
- Cases of mode authentication: The problem in mode authentication is to figure out exactly how passwords can be compared and add additional user attributes to assure the reliability of dual authentication. For instance, modern ATMs exclusively require PIN codes to verify users' identities [17]. The privacy of authentication cannot be guaranteed by using this single method. Due to the drawbacks of one-time authentication, multiple authentication technology has been studied.
- Cases of biometric authentication: Biometric authentication has drawn more attention than mode authentication due to its own uniqueness, non-reliability, heredity, and invariance. Currently, identification is primarily reliant on biological traits (like fingerprints, irises, etc.) and behavioural traits (like voice, movement, etc.), [18] as well as the potent self-learning capabilities of AI which can effectively utilise them.

3.4.1.2 Abnormal traffic identification

Creating a solid foundation enabling network situation awareness via network traffic analysis, prompt identification of high-risk online behaviours, and effective countermeasures is crucial for improving network response and preserving overall cybersecurity.

3.4.1.3 Dangerous behaviour monitoring

The offensive techniques used by hackers are continually evolving, along with new technologies like big data and cloud computing [19]. People indulged in offensive technique of hacking cybersecurity are committed to locating "lethal points" of the existing network and launching assaults on the same at any moment of time as a result of the exponential development in data volume and the expansion of Internet access. The initial IDSs were unable to adjust to the network's features. High-speed data flow also makes it easier to detect hacking activity traces, which has become crucial support for proactive security measures. Monitoring risky actions and their types in real-time is required to achieve cybersecurity with precise procedures. In order to achieve this goal, researchers have started to innovate and enhance existing IDS systems in order to come up with their present network requirements as scalable as possible.

3.4.1.4 Network situation awareness

- Requirements for network situation awareness: The network designers might not have seen the stability and liability in the topology of the network during network building. The non-uniformly flow of the database during network use reveals the network's position, detects the weak link in advance, and provides the foundation for network fixations. This necessitates the employment of network situation awareness. The situation awareness model must have a solid ground knowledge base to enable it to swiftly identify and replicate the network situation in order to accomplish this operation. The model must simultaneously be able to extract characteristics with the goal of never appearing in a network setting. Additionally, reasoning has the potential to produce accurate perceptual findings.
- Situational awareness on a network paired with AI: By standardising the synchronous operation mechanism for AI in several data security canalisations, enhancing the information security situation inference algorithm, developing the system software structure, and adding comparative restoration steps centred on and behest the concept of security featurism parameters, [20] sought to optimisation of the designing of the data security situation awareness system.

3.4.2 Artificial Intelligence in cybersecurity

Numerous cybersecurity specialists are looking towards artificial intelligence to dynamically safeguard the systems against cyber assaults. AI is most frequently used in cybersecurity for intrusion detection by examining traffic patterns and searching for any activity that is indicative towards the proclamation of the assault [21].

- Machine learning: Unsupervised learning and supervised learning are the two main categories of machine learning. In formative supervised learning, training data are manually classified as malicious or valid before being fed into the algorithms to develop a model with "classes" of data against which it may compare the traffic it is analysing. Instead of using training data and manual labelling, unsupervised learning gathers together related pieces of information in classes and then categorises them based on the coherence of the set of data within each class and the modularity of the data existing between the classes. The other models covered in this section can also be made using machine learning algorithms.

- Decision trees: An AI system known as a decision tree developed a bunch of guidelines relying on outlining training data samples. It employs iterative division to find the description that best classifies the traffic it is studying (typically just "attack" or "normal") [22]. This method is used in cybersecurity, for instance, to identify attacks by examining the frequency, volume, and duration of traffic. For instance, a traffic jam that lasts for a long time yet having a low frequency is probably a cyber-attack and will be classified as such [23]. By categorisation of variables from CPU use, network reveals, and volume of data written, decision trees may also be used to identify command injection; this method is well-liked because the developer is aware of exactly how the AI does and doesn't regard the traffic as aberrant. The AI can also examine traffic in real-time if a successful set of criteria is identified, sending out an almost immediate notice if any strange behaviours is observed. The Rule-Learning technique is an alternative method for using decision trees. It seeks out a set of attack features while maximising a score that indicates the accuracy of the classification (i.e., the amount of mistakenly classified data samples).
- K-nearest neighbours (k-NN): By computing the overall Euclidean distance between new pieces of data with previously classified items of data, the (k-NN) technique, to put it simply, identifies which class a fresh piece of data must be placed in. The k-NN technique is useful for IDS systems because it can swiftly pick up on unique traffic patterns to detect previously unidentified, even zero-day threats [24]. K-NN applications are being studied by cybersecurity experts for real-time cyber-attack detection. The method has been used to identify attacks like fraudulent data injection attacks and is effective whenever data can indeed be described by a model that allows for the evaluation of its distance from those other data, such as by employing the Gaussian distribution or possibly using vectors.
- Support vector machines: Support vector machines (SVMs), a development of linear regression models, find a plane that divides data into two classes. Depending on the function employed in the method, this plane could very well be non-linear, linear, polynomial, sigmoid, and so on. Additionally, SVMs can divide data into even more than two categories by employing more than a single plane. Using this method, Internet traffic patterns can be broken down into classes like HTTP, FTP, SMTP, and others in cybersecurity [25]. Due to the fact that SVM

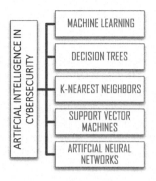

Figure 3.3 Artificial intelligence used for cybersecurity its characteristics.

is a guided machine-learning technique, it is frequently utilised in applications wherein attacks can indeed be simulated, such as when using network traffic produced by penetration testing for training data.
- Artificial neural networks (ANNs): ANNs are indeed a method based on how neurons communicate and interpret information with one another in the brain. A neuron in an ANN seems to be a mathematical equation which takes input data, produces a target value, and then relays information to the following neuron based on that value [26]. The neurons can then learn and adjust their weights by simply measuring the error between both the predicted value and the prior output value as the ANN algorithm iterates until the output value becomes acceptably near to the target value. The programme then gives the mathematical equation that generates a value that may be used to categorise the data once this process is complete.

The huge systems found in a smart city would benefit from AI cybersecurity as well, as the AI would have been able to provide extremely quick response times, which are crucial in systems like traffic control. In the future, smaller systems like self-driving cars and even smart houses might also incorporate AI cybersecurity. Other preventative security measures should be in place because many AI cybersecurity techniques identify or stop assaults already in progress instead of preventing attacks from occurring at all (Figure 3.3).

3.5 AI APPLICATIONS TO IMPROVE CYBERSECURITY FOR DIFFERENT APPLICATION DOMAINS

The Internet has now established itself as a crucial utility in people's daily lives all around the world, just like gas, water, and energy have done in the past. Devices that are connected to the Internet run the danger of becoming targets of various cyber-attacks as more and more of them do. Cybersecurity is now crucial to safeguarding both these Internet-connected products and their users.

- The Internet: Cyber-attacks are hostile patterns that are distinct from legitimate Internet traffic, according to AI. Because AI algorithms can review a significant quantity of data and adapt to the shifting nature of Internet traffic, intrusion-detection systems have been developed to discern between malicious and valid traffic. Recent cyber-attacks have targeted people, business logic, and network infrastructure [27].
- Application-layer attack: The focus of modern application-layer attacks has switched from blocking information flow to changing the meaning of information. A new type of cyber-attack has evolved with the emergence of digital social networks that tries to spread misleading information to influence the behaviours or decisions of the target audience [28].

As servers operate the essential business applications for an organisation, targeting servers is a desirable way to attack the company providing the services or the consumers of those services. Application-layer attacks have previously mostly targeted protocols like HTTP, DNS, or Session Initiation Protocol (SIP) [29]. A bigger percentage of detection and false were displayed after a proposed HTTP/2 flooding traffic has been launched against such an HTTP/2 service. Early detection of fake news is essential [30]. As a result, a strategy for early fake news identification using a set of ANNs was proposed in a work. The research examined the speed and complexity of the news-propagation path. It uses two ANN derivatives: convolutional neural networks (CNNs), a variation of directed neural networks (DNNs), and recurrent neural networks (RNNs), which resemble directed graphs and have more hidden layers [31]. Detecting fake information also uses linguistics-related knowledge to categorise documents. Here, the text categorisation methods broaden the observations and characteristics needed in cybersecurity to put automatic detection methods into practice. These studies demonstrated how false information automatic detection systems enhance human welfare and indicate how AI may leverage new features.

Despite recent improvements in text categorisation, semantic cyber-attack detection remains in its early stages. Studies that used TF-IDF (Term Frequency – Inverse Document Frequency) needed human assistance to provide pertinent terms [32]. This demonstrates that despite the usage of AI, human intelligence is still needed for cyberthreats identification at the existing application layer. Additionally, some investigations use characteristics other than verbal cues. These characteristics are exclusive to social media and not language cues.

- The Internet of Things: Many modern products come with networking capabilities and Internet connectivity, making the IoT conceivable. Research developed fog computing services by bringing the base and application nearer to the client because the demand for better data

speeds is growing. To reduce network round-trip delays, fog computing distributes servers, notably with Content Delivery Network [33]. Fog computing hence enhances website speed and offers real-time management of energy and carbon footprint. Additionally, the emergence of vehicular networking apps, which allow for quick data transfers across mobile devices, was made possible by advancements in telecommunications technology.
- An end user of the Internet, who is a human, is likely the weakest player in cybersecurity [34]. People are more concerned with their daily duties than they are with the ever-growing cyber-attack surface. While some of the well-known cyberthreats can be reduced by reengineering machines, humans need on-going training based on current and prior problems. One of the key factors influencing the effectiveness of malware propagating through contemporary phishing techniques is this necessity [35]. Software with malevolent intentions is called malware. Phishing is a technique used to get unsuspecting users to do what the attacker wants them to, like click a link or open an executable file. Such behaviours either encourage the spreading of malware or persuade the victims to divulge their private data. The most advanced phishing methods used nowadays make use of the human capacity for limited omniscience. Users must evaluate the target's credibility to prevent stumbling for phishing hooks, and frequently this may be done by looking at the code hidden behind the links, which might also call for some specialist knowledge. In this field, artificial intelligence can help to improve human intelligence.
- Privacy-Internet-connected devices are getting smaller and more prevalent. Their capacity to collect data outperforms that of people who become conscious of what they're doing (in capturing data). Devices gather information about the environment, including speech, geolocation temperature, and surrounding lighting, to enhance user experience [36]. However, research demonstrates that gathering such information may have ulterior motives. Intelligent virtual assistants (like Apple's Siri, Amazon Alexa, and Google Home) can indeed be utilised to forcibly open a smart lock. One study showed that gadgets could be used to locate locations to smuggle, cyber bully, foment panic, and distract in an airport using a user's browser history to show ads. Devices as well can be used to associate a place or person with criminal activity occurrences [37]. Legal, regulatory, and ethical frameworks will be incorporated into AI privacy responsibilities since sharing personal information can improve human wellness.

3.6 FUTURE OBJECTIVES

- The competition between white hat (defenders) and black hat (offenders) hackers has been stoked by recent developments in AI research in cybersecurity. Attackers can use AI to simulate human behaviours in order to gain personal satisfaction, dominance, or financial gain. Intelligent agents that autonomously click advertising play online games, and purchase and resale concert tickets have been developed thanks to AI. How dividing lines can be made between advancements and fundamental demands determines future research potential in cybersecurity. To manage the deployment of technology, a line must be drawn between the two groups, but this is challenging because each group's development is mirrored by the other. Therefore, it is crucial to look at how AI might be applied to meeting basic human needs and creating cybersecurity measures.

- Access to cutting-edge computer infrastructure will make it easier to effectively and efficiently solve AI problems. The need for speedy data analysis will arise as the quantity of computing devices grows and the amount of traffic correspondingly grows. Consequently, advanced computing systems are needed in order to analyse data utilising AI approaches. Cluster computing tools are used to analyse cyber traffic in order to meet this challenge. At the top end, quantum computing will be the game-changing technology that assists in resolving challenging computing issues. NASA's quantum computer, which itself runs 100 million times quicker than conventional computers, has proven able to handle complicated tasks in a fraction of the time.

3.7 CONCLUSION

AI has emerged as a critical tool in the field of cybersecurity as the pace and sophistication of assaults rise. This chapter demonstrated how cyberthreats have grown, become more complicated, and expanded in scope. We stress how current hazards are still affected by historical cyberthreats. We provided a thorough analysis of cyberthreats and available countermeasures. In particular, we discussed the impact of cyber-attacks and how they might be launched against various network stacks and apps. While the community recognises cyberthreats and creates remedies by utilising a wide array of technology and techniques, cyberthreats shall continue to increase. Therefore, this chapter analysed studies concentrating on the applications of AI in four areas of cybersecurity: user access authentication, networking condition awareness, monitoring of harmful conduct, and anomalous traffic identification.

Although research into AI's potential to address cybersecurity challenges is on-going, there are still fundamental questions about how and where AI deployment may be governed. For instance, if machine intelligence becomes a more crucial component of humanity's future, it will gradually deplete life's essential resources. When machines and people fight for limited resources, a new type of government will emerge. This will then open up a fresh line of inquiry.

REFERENCES

1. Singh, S., Sheng, Q. Z., Benkhelifa, E., & Lloret, J. (2020). Guest editorial: Energy management, protocols, and security for the next-generation networks and Internet of Things. *The IEEE Transactions on Industrial Informatics*, 16(5), 3515–3520.
2. Calderon, R. (2019). The benefits of artificial intelligence in cybersecurity. *Economic Crime Forensics Capstones*, 36. https://digitalcommons.lasalle.edu/ecf_capstones/36
3. Turransky, A., & Amini, M. H. (2022). Artificial intelligence and cybersecurity: Tale of healthcare applications. Cyberphysical smart cities infrastructures: Optimal operation and intelligent decision making, 1–11.
4. Zhang, Z., Ning, H., Shi, F., Farha, F., Xu, Y., Xu, J., ... Choo, K. K. R. (2022). Artificial intelligence in cyber security: research advances, challenges, and opportunities. *Artificial Intelligence Review*, 55, 1–25.
5. Tetaly, M., & Kulkarni, P. (2022, October). Artificial intelligence in cyber security–A threat or a solution. In *AIP Conference Proceedings* (Vol. 2519, No. 1, p. 030036). AIP Publishing LLC.
6. Sarker, I. H., Furhad, M. H., & Nowrozy, R. (2021). Ai-driven cybersecurity: an overview, security intelligence modeling and research directions. *SN Computer Science*, 2, 1–18.
7. Chan, L., Morgan, I., Simon, H., Alshabanat, F., Ober, D., Gentry, J., ... Cao, R. (2019, June). Survey of AI in cybersecurity for information technology management. In *2019 IEEE technology & engineering management conference (TEMSCON)* (pp. 1–8). IEEE.
8. Timmers, P. (2019). Ethics of AI and cybersecurity when sovereignty is at stake. *Minds and Machines*, 29, 635–645.
9. Truong, T. C., Zelinka, I., Plucar, J., Čandík, M., & Šulc, V. (2020). Artificial intelligence and cybersecurity: Past, presence, and future. In Subhransu Sekhar Dash, Paruchuri Chandra Babu Naidu, Ramazan Bayindir, & Swagatam Das (Eds.), *Artificial intelligence and evolutionary computations in engineering systems* (pp. 351–363). Springer, Singapore.
10. Stevens, T. (2020). Knowledge in the grey zone: AI and cybersecurity. *Digital War*, 1, 164–170.
11. Azad, C., Bhushan, B., Sharma, R. et al. (2021). Prediction model using SMOTE, genetic algorithm and decision tree (PMSGD) for classification of diabetes mellitus. *Multimedia Systems*. https://doi.org/10.1007/s00530-021-00817-2
12. Priyadarshini, I., Kumar, R., Tuan, L.M. et al. (2021). A new enhanced cyber security framework for medical cyber physical systems. *SICS Software-Intensive Cyber-Physical Systems*. https://doi.org/10.1007/s00450-021-00427-3
13. Priyadarshini, I., Kumar, R., Sharma, R., Singh, P. K., & Satapathy, S. C., (2021). Identifying cyber insecurities in trustworthy space and energy sector for smart grids. *Computers & Electrical Engineering*, 93, 107204.
14. Singh, R., Sharma, R., Akram, S. V., Gehlot, A., Buddhi, D., Malik, P. K., & Arya, R. (2021). Highway 4.0: Digitalization of highways for vulnerable road safety development with intelligent IoT sensors and machine learning, *Safety Science*, 143, 105407. ISSN 0925-7535
15. Sahu, L., Sharma, R., Sahu, I., Das, M., Sahu, B., & Kumar, R. (2021). Efficient detection of Parkinson's disease using deep learning techniques over medical data. *Expert Systems*, e12787. https://doi.org/10.1111/exsy.12787

16. Sharma, R., Kumar, R., Sharma, D.K. et al. (2021). Water pollution examination through quality analysis of different rivers: A case study in India. *Environment, Development and Sustainability*. https://doi.org/10.1007/s10668-021-01777-3
17. Ha, D.H., Nguyen, P.T., Costache, R. et al. (2021). Quadratic discriminant analysis based ensemble machine learning models for groundwater potential modeling and mapping. *Water Resources Management*. https://doi.org/10.1007/s11269-021-02957-6
18. Dhiman, G., & Sharma, R. (2021). SHANN: an IoT and machine-learning-assisted edge cross-layered routing protocol using spotted hyena optimizer. *Complex & Intelligent Systems*. https://doi.org/10.1007/s40747-021-00578-5
19. Sharma, R., Gupta, D., Polkowski, Z., Peng, S.-L. (2021). Introduction to the special section on big data analytics and deep learning approaches for 5G and 6G communication networks (VSI-5g6g), *Computers & Electrical Engineering*, 95, 107507, ISSN 0045-7906, https://doi.org/10.1016/j.compeleceng.2021.107507.
20. Singh, P. D., Dhiman, G., & Sharma, R. (2022). Internet of Things for sustaining a smart and secure healthcare system. *Sustainable Computing: Informatics and Systems*, 33, 100622. ISSN 2210-5379, https://doi.org/10.1016/j.suscom.2021.100622.
21. Singh, R., Gehlot, A., Mittal, M., Samkaria, R., Singh, D., & Chandra, P. (2018). "Design and development of a cloud assisted robot." In *Smart and Innovative Trends in Next Generation Computing Technologies: Third International Conference, NGCT 2017, Dehradun, India, October 30–31, 2017, Revised Selected Papers, Part II 3*, pp. 419–429. Springer, Singapore.
22. Abdullahi, M., Baashar, Y., Alhussian, H., Alwadain, A., Aziz, N., Capretz, L. F., & Abdulkadir, S. J. (2022). Detecting cybersecurity attacks in internet of things using artificial intelligence methods: A systematic literature review. *Electronics*, 11(2), 198.
23. Khanna, K., Panigrahi, B. K., & Joshi, A. (2018). AI-based approach to identify compromised meters in data integrity attacks on smart grid. *IET Generation, Transmission & Distribution*, 12(5), 1052–1066.
24. Sadik, S., Ahmed, M., Sikos, L. F., & Islam, A. N. (2020). Toward a sustainable cybersecurity ecosystem. *Computers*, 9(3), 74.
25. Wilner, A. S. (2018). Cybersecurity and its discontents: Artificial intelligence, the Internet of Things, and digital misinformation. *International Journal*, 73(2), 308–316.
26. Zhao, L., Zhu, D., Shafik, W., Matinkhah, S. M., Ahmad, Z., Sharif, L., & Craig, A. (2022). Artificial intelligence analysis in cyber domain: A review. *International Journal of Distributed Sensor Networks*, 18(4), 15501329221084882.
27. Sarker, I. H., Furhad, M. H., & Nowrozy, R. (2021). Ai-driven cybersecurity: an overview, security intelligence modeling and research directions. *SN Computer Science*, 2, 1–18.
28. Khan, N. F., Ikram, N., Murtaza, H., & Asadi, M. A. (2021). Social media users and cybersecurity awareness: predicting self-disclosure using a hybrid artificial intelligence approach. *Kybernetes*, (ahead-of-print).
29. Raban, Y., & Hauptman, A. (2018). Foresight of cyber security threat drivers and affecting technologies. *foresight*, 20(4), 353–363.
30. Qumer, S. M., & Ikrama, S. (2022). Poppy Gustafsson: redefining cybersecurity through AI. *The Case for Women*, 1–38.

31. Jun, Y., Craig, A., Shafik, W., & Sharif, L. (2021). Artificial intelligence application in cybersecurity and cyberdefense. *Wireless Communications and Mobile Computing*, 2021, 1–10.
32. Mahbooba, B., Timilsina, M., Sahal, R., & Serrano, M. (2021). Explainable artificial intelligence (XAI) to enhance trust management in intrusion detection systems using decision tree model. *Complexity*, 2021, 1–11.
33. Kumar, A., Rahmath, M., Raju, Y., Reddy Vulapula, S., Prathap, B. R., Hassan, M. M., ... Asakipaam, S. A. (2022). Enhanced secure technique for detecting cyber attacks using artificial intelligence and optimal IoT. *Security and Communication Networks*, 2022, 8024518.
34. Juneja, A., Juneja, S., Bali, V., Jain, V., & Upadhyay, H. (2021). Artificial intelligence and cybersecurity: current trends and future prospects. In Pardeep Kumar, Vishal Jain, & Vasaki Ponnusamy (Eds.), *The Smart Cyber Ecosystem for Sustainable Development*, 431–441. https://doi.org/10.1002/9781119761655.ch22
35. Pethő, Z., Török, Á., & Szalay, Z. (2021). A survey of new orientations in the field of vehicular cybersecurity, applying artificial intelligence based methods. *Transactions on Emerging Telecommunications Technologies*, 32(10), e4325.
36. Kumar, R. S., Keerthana, G., Murali, L., Chidambaranathan, S., Premkumar, C. D., & Mahaveerakannan, R. (2022). Enabling artificial intelligence and cyber security in smart manufacturing. In R. Kanthavel, K. Ananthajothi, S. Balamurugan, & R. Karthik Ganesh (Eds.), *Artificial Intelligent Techniques for Wireless Communication and Networking*, 269–286. https://doi.org/10.1002/9781119821809.ch17
37. Sharma, S. (2021). Role of artificial intelligence in cyber security and security framework. In Neeraj Bhargava, Ritu Bhargava, Pramod Singh Rathore, & Rashmi Agrawal (Eds.), *Artificial Intelligence and Data Mining Approaches in Security Frameworks*, 33–63. https://doi.org/10.1002/9781119760429.ch3

Chapter 4

Theories of blockchain and distributed systems

K. Umapathy, S. Prabakaran, T. Dineshkumar, M. A. Archana and D. Khyathi Sri
SCSVMV Deemed University, Kanchipuram, India

Arwa N. Aledaily
University of Ha'il, Ha'il, Kingdom of Saudi Arabia

4.1 INTRODUCTION

Blockchain is an integrated continuous ledger which simplifies the process of recording transactions and observing assets in a corporate network. Blockchain can track and sell almost any value by lowering risk and costs for concerned persons. The blockchain technology is used to give transparency to agricultural supply chains, protect healthcare data, revolutionize games and transform the way about data and ownership in general. Cryptocurrencies like Bitcoin may be transferred among the people using blockchain technology. The transaction is approved by numerous devices and saved like a group of coding with a number of devices. The subsequent transactions are concatenated to maintain the record in the ledger thereby removing the potential for hacking. Benefiting from this safe technique of transmitting data without the involvement of mediators is the enticing feature [1]. Blockchain may be used for transactions other than financial transactions. It is adaptable for usage in areas other than its own. Blockchain technology is being used in a variety of businesses. These sectors include energy, logistics, education and others. Furthermore, present security solutions may jeopardize the transparency of blockchain and on-chain transaction control. Figure 4.1 shows the schematic for blockchain technology.

4.2 EVOLUTION OF BLOCKCHAIN

Even though blockchain seemed to be new, it has a long and intriguing background. Its evolution covers the most significant and noteworthy events in blockchain development. The notion of a secured collection of records was first proposed by Stuart Haber in 1991. This technique gained popularity and became widely utilized during the next two decades. Blockchain reached a watershed moment in the late 20th century when it was created

Figure 4.1 Blockchain technology.

as a concept and implemented in practice. Due to this, cryptocurrency and blockchain were hosted for the first time and their influence on the IT industry started to emerge. Although previous research results demonstrate tremendous progress toward securing blockchain-powered systems, various research difficulties remain.

Transaction is a parameter that indicates the movement of value or information. In a blockchain, a transaction shall execute activities like storage and retrieval of information or denoting value transfer between the parameters. Transactions can be organized in blocks having fixed sizes and added to the blockchain. Merkle is a structure of data employed to record transactions within the block in a proper manner. Merkle tree is a type belonging to the hash tree which keeps transactions in blocks. It generates hashes using a method called SHA-256. Merkle trees are used to efficiently store and verify big data collections. Leaf nodes in this data structure will hold the hash of the transaction, whereas other nodes include labels of next-level nodes. The hash of all transactions in the block is stored in the tree's root called Merkle root. Blockchain technology is used in pharmaceutical supply chains to track and trace prescription medications. This tool can simply, prevent and regulate counterfeit medicine distribution, as well as recall ineffective and hazardous pharmaceuticals. Blockchain will record and decompose relevant data without changing the contents. Hence, a blockchain creates the basis for different records of transactions without alteration or damage. So they are called distributed ledger technologies (DLT). A blockchain permits the distribution of information over various nodes of a network at different locations. With this approach, the storage of data becomes accurate and precise.

Hayes in his work proposed Bitcoin as electronic cash means of transaction against the traditional method of payment [2]. Rose C. carried out productive transactions with Bitcoin initially [3]. Dupont purchased two pizzas by paying Bitcoin [4]. Bitcoin is the world's most valuable cryptocurrency, worth one US dollar. Several organizations have begun receiving donations in the form of Bitcoins [5]. Blockchain and cryptocurrency have received extensive media coverage including a lot of programs on TV and other media. A known developer of Bitcoin named Vitalik Buterin started a magazine specifically for it [6]. A steep increase in the price of Bitcoin was then observed. A publication called *Ethereum* emphasized the merits of blockchain and Bitcoin [7]. Bitcoin was recognized as a method of payment by various companies. Typical examples are Zynga, Hotel Vegas, etc. The integration of Bitcoin was announced by PayPal then [8]. The agreement signed between a company based in the US and NASDQ planned to implement this technique in the share market [9]. A methodology based on blockchain was declared by IBM to sort out business solutions.

Various agencies belonging to governments approved blockchain and cryptocurrencies legally in Eastern countries [10]. The popularity of Bitcoin increased highly and a government has been declared for Dubai, totally driven by blockchain recently [11]. A separate group based on blockchain and the generation of appropriate cryptocurrency is being planned by Facebook. The banking platform based on this will be implemented shortly by various large banks across the globe [12, 13].

4.3 KEY ELEMENTS OF BLOCKCHAIN

Distributed Ledger Technology: The ledger and record of transactions of this technique can be used totally by all members of the concerned network. The redundancy of data is quite usual in conventional networks. Hence, one-time recording of transactions is implemented with this technique by which redundancy of information can be eliminated. The properties of blockchain technology are shown in Figure 4.2.

Immutable Records: Once a transaction is entered in this technique, there is no provision for editing or destroying further. Any abnormalities in the current transaction can be handled by entering a new one and all relevant transactions are available for viewing.

Smart Contracts: This technique with a list of norms is recorded and implemented in order to increase the processing of transactions in an automatic manner. This contract can indicate parameters for the transfer of bonds in companies, tax payments, etc. The following are important aspects of blockchain:

Consensus: The legitimacy of a transaction must be agreed upon by all network participants.

Provenance: All network participants may see the history of a property associated with blockchain.

Theories of blockchain and distributed systems 55

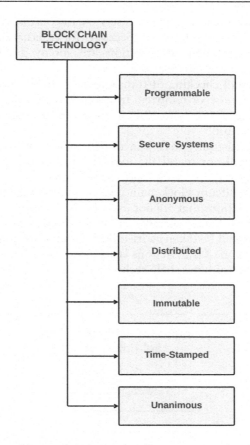

Figure 4.2 Properties of blockchain technology.

Immutability: Tapping out any particular transaction on a blockchain is highly impossible. Suppose a transaction is incorrectly performed, then it has to be repeated to correct errors of the previous transaction.

Distributed: The network of blockchain is a distributed one and based on peer-to-peer connection. Suppose any of the nodes cease to function, the network continues to operate normally. No one authority can command the whole network.

Security: Transaction security is provided by means of cryptography using public keys. Generally, members of the network will have both types of keys – public and private – for securing messages and producing signatures in digital format.

Coherence: The blockchain has the value of fact. The members of the network can see the same copy of the distributed ledger.

Decentralization: In traditional systems of transaction, actions must be recognized by relevant institutions generating value and delays in execution with respect to central servers. In contrast to the previously presented centralized form, the necessity for external persons is totally eliminated. Agreement computations are employed in blockchain networks to maintain information.

Persistence: Rapid checking of transactions is highly possible here. It is difficult to reverse or remove a transaction after it has been recorded in the blockchain. It is feasible to discover incorrect transactions in blocks in real-time.

Anonymity: A customer may join the chain and utilize existing facilities which will not reveal the identification of the concerned client. Familiarity is required with blockchain technology.

Auditability: A Bitcoin blockchain records client modifications depending on transactions that are not used. Each action affects earlier unutilized transfers. When current action is recorded in a chain, unused transactions will be taken as used during the recording. The option of easy verification of transactions is also available. Figure 4.3 shows the characteristics of blockchain technology.

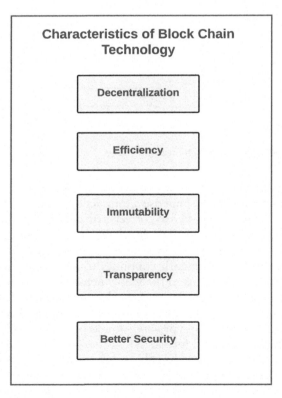

Figure 4.3 Characteristics of blockchain technology.

4.4 WORKING OF BLOCKCHAIN

Each transaction is entered as a block of information. The transition of an asset from one point to another is clearly indicated by these transactions. The block may accept any type of data such as the name of a person, type of action, place, time and date. All relevant blocks are cascaded to each other. A chain of data will be created for the movement of assets and change of ownership. The time, date and series of actions will be validated automatically by the block thereby avoiding the provision for editing and concatenation of other blocks.

4.4.1 Benefits of blockchain

Most of the activities of the chain will be concentrated on the maintenance of records in their original form and the validation of third parties. The maintenance of records will usually become prone to fraud and cyber attacks. The number of transactions has increased rapidly due to the implementation of the IoT. The merits of blockchain are illustrated in Figure 4.4.

> **Greater Trust:** Utmost confidence and trust is given to members of the relevant network with blockchain for accessing their information in a precise, accurate and periodical manner. The sharing of information is authorized and restricted only among the members.
> **Greater security:** High security and accuracy is provided for the protection of data with respect to network members with a permanent recording of their transactions. Generally, the deletion or modification of any transaction is totally prohibited.

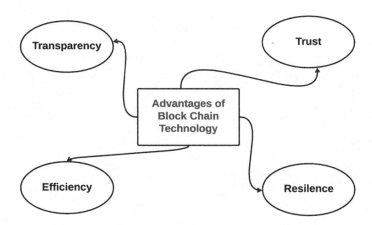

Figure 4.4 Advantages of blockchain technology.

More efficiency: Records that amount to a large time are removed, thereby distributing a common ledger among the members of the network. A list of norms is recorded and implemented in order to increase the processing of transactions in an automatic manner

4.5 TYPES OF BLOCKCHAIN NETWORKS

Networks with blockchain may be constructed in a lot of possible ways. Figure 4.5 indicates the types of networks with blockchain.

Public Blockchain Networks: Bitcoin is a typical example of this network type by which any person can represent it. Large power will be needed for processing and there is no room for safe and secured transactions. These are the demerits for implementation in the company sector.

Private Blockchain Networks: The concept of decentralization is well applicable to these networks also. However, the institution with this network can finalize the participants, methodology and maintenance of the common ledger. This approach can increase the number of members with appropriate confidence.

Permission Blockchain Networks: The creation of a blockchain of private type is very similar to the creation of a network with blockchain in a lot of commercial activities. This concept is applicable to public networks also. Hence, this will place a constraint on the joining participants and concerned transactions. But there must be a request for participation.

Consortium Blockchains: A consortium blockchain is suited for all business activities in which participants must be given absolute authorization and right for sharing within the chain.

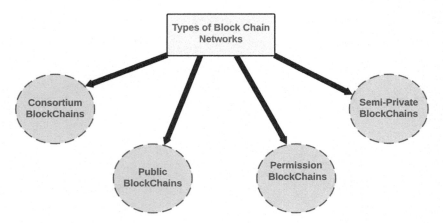

Figure 4.5 Types of blockchain networks.

4.6 VARIOUS PLATFORMS OF BLOCKCHAIN TECHNOLOGY

Blockchain has gained popularity and is widely employed across a variety of businesses. The blockchain is commonly viewed as a digital database in a distributed manner [14]. Blockchain has evolved as a disruptive invention that has the potential to transform how people can communicate, make automatic values and track the process of transactions [15]. A blockchain might eliminate the authority to observe and oversee transactions among diverse persons which could save money [16]. Other mining companies store a duplicate of the whole record including all sorts of activities and confirmation of them. Due to this method, records maintained are safe, synchronized and unmanipulated. Hence, blockchain is commonly acknowledged as a part of the IoT outside other essential domains [17]. It is an open-source process that anybody may be a part of it. Oversight does not exist and norms are totally common for all participants. The most important blockchains are enunciated further.

Bitcoin Blockchain: It is a famous blockchain belonging to public types. It was launched in 2008 as an option for existing financial processes. The primary goal of this type is the decentralization of the current banking methodology and the use of Bitcoin for transfers. It employs cryptographic approaches for cryptocurrency governance, such as transaction verification and the long-term establishment of a transaction history chain. The relevant software and concerned keys are critical components of a Bitcoin transaction address [18]. Unlike conventional currencies, it is virtual and no real money is handled. Users will have their personal keys to establish rights in the concerned network. The keys are meant for executing any sort of transaction. These keys are frequently saved in a wallet in digital format. Bitcoin is based on "mining", which is a high-powered operation based on computers for validation of all actions. Miners get positive compensations [19]. Bitcoin has exacerbated the problem of double-spent transactions when compared to that of digital money [20].

Ethereum: It is another type of platform that was created in late 2013 by Vitalik Buterin, a software developer. It contains EVM – an operating system which executes contracts in an intelligent way with nodes participating in it. Solidity – a Turing complete language will create appropriate contracts. It is of open-source type by which anybody may create and implement appropriate applications. A currency named Ether is employed for transactions among the accounts of this platform. Merkle is a structure of data employed to record transactions within a block in a proper manner. Merkle tree is a type belonging to a hash tree which keeps transactions in blocks. It generates hashes using a method called SHA-256 [21]. Merkle trees are used to efficiently store and verify big data collections. Leaf nodes in this data structure will hold the hash of the transaction, whereas other nodes include labels of next-level nodes. The hash of all transactions in a block is stored in the tree's root called the Merkle root [22]. Every sister node along the path from beginning to end provides a mean value commitment. A typical blockchain confirms the correctness and validation of data by employing the Merkle tree's proof of opening [23, 24].

Hyperledger: It is the third platform belonging to the open-source category invented in 2016 to create appropriate solutions for business. This comprises several initiatives with various sorts of demands and options. The set of initiatives includes Composer, Fabric and others. Hyperledger Fabric is a functioning platform while other types belong to the stage of incubation. Fabric is a system that includes intelligent contracts. Various methods such as Kafka and Solo are employed for implementing the mechanism of fabric. It has capabilities for building blockchains of permission types [25]. Hyperledger Fabric is a methodology given by the Hyperledger administration. It aims to establish a decentralized environment. It involves various parameters such as client, authority, instruction and end user. Furthermore, the components communicate via proper ways for secret and concealed transactions. In comparison to privacy, a component that does not have simple network access will not be able to use the chain. The concept of scalability is promoted in order to increase the quantity of information among the allocated nodes by allocating one person for each channel [26].

Multichain is another option for creating and implementing a blockchain in private mode. It is a do-it-yourself platform for banks with the goal of developing applications with decentralization. It follows in a round-robin way from a pool of approved miners. Multichain has no transaction fees or mining rewards [27]. The blockchain technology provides a secure and transparent system for documentation. The word "blockchain" may refer to a variety of things [28]. The blockchain is defined as "a distributed database or public record of all transactions or digital events conducted and shared among participating parties". The majority of system participants will provide authenticity for all relevant transactions. Once entered, it is impossible to delete the information. All recorded activities in the blockchain can be checked in an independent manner [29].

4.7 SIGNIFICANT CHALLENGES FOR BLOCKCHAIN APPLICATIONS

Blockchain technology is widely utilized on the Internet for a number of applications. It is a feasible option for various standard systems of transaction due to its decentralized nature. For scalability, interoperability and sustainability, systems with blockchain must be enhanced to an appropriate level. Bitcoin supporters, on the other hand, have various alternative possibilities [30]. Internet of Things Application (IOTA) has the most significant benefit over other cryptocurrencies in that it has no transaction fees. Unlike all other cryptocurrencies that struggle with scaling issues, IOTA's protocol offers limitless scalability [31].

4.8 CONCEPT OF CLOUD COMPUTING

Figure 4.6 shows the cloud system enabled with blockchain. In many corporate and military situations, cloud computing is employed to help with the management of information storage. Integrated environments with cloud are populated with varied system components purchased from many sources. Applications of blockchain are not restricted to the area of cryptocurrency but are also extended to company digitalization [32]. A blockchain is a networking system which employs a large number of computers known as nodes. It is critical because a large amount of data is delivered and preserved with it. This technique lowers expenses while also increasing accuracy [33].

More people will be needed to arrange and administer these servers. Data centers will safeguard every piece of information. As a result, if it is continued to focus on issues and teams, it may not be feasible to meet the commercial objectives [34]. By moving to "cloud computing", it is possible to reduce the strain of on-site maintenance. The method of preserving, maintaining and analyzing information by means of computers is referred to as cloud computing. It serves as a stand-in for a computer at its designated place [35]. Automatic updating of software will be done which enables users to simply access and edit documents saved in the cloud. Furthermore, there are various drawbacks [36]. Many diverse services are offered and divided into three primary models [37].

Software as a Service (SaaS) was the primary facility offered to clients. These services include Google E-mail and Amazon Services. Platform as a Service (PaaS) is the second accessible option.

4.8.1 Blockchain in cloud computing

The adoption of blockchain may have a significant impact on cloud computing. Enhanced reliability, better protection of data and many other benefits are a few merits. There is a significant increase in the profit of companies after the implementation of blockchain [38, 39]. The applications afforded by the technique are listed in Figure 4.7.

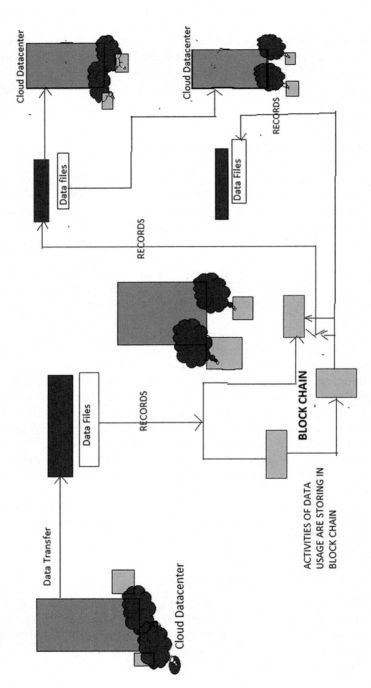

Figure 4.6 Blockchain-enabled cloud computing system.

Applications of Block Chain Technology

- Digital Records
- Security and Privacy
- Defence
- Energy Management
- Voting
- Advertising
- Assests Tracking
- Reputation System
- Law and Enforcement
- Intrusion Detection
- Agriculture Sector
- Insurance
- Internet of Things
- Internet of Senses
- Automotive
- Healthcare 4.0
- Digital Ownership
- Supply Chain
- 5G Technologies

Figure 4.7 Applications of blockchain.

The following are various uses of blockchain:

- Electronic marketing
- Management of Risks
- Management of Reputations
- Education
- Enhanced Security
- IoT
- Protection of copyrights
- Medical Electronics, etc. [40, 41]

Applications in the community correspond to marketing and defense, applications related to mobile, automation, irrigation, registration of voters, law and monitoring, digital documents and tracking of assets [42, 43].

4.9 BLOCKCHAIN SECURITY

The format of data generated by blockchain coincides with properties of security of intrinsic type. It is correlated with the concepts of cryptography and distribution in order to provide utmost confidence for transactions. Generally, blockchains organize data in terms of blocks and each block includes a list of appropriate transactions. All the blocks in the chain are interlinked with each other with no provision for tampering. Then all concerned transactions are validated in a proper manner. Decentralization is provided by blockchain so that all network members are permitted to participate. The record of the transaction cannot be tampered or changed by any member and there is no possibility for the process getting failed. Based upon the invitation given to network members, networks can be decomposed into public and private. Despite all the above advantages, the networks with blockchain are prone to various types of cyber attacks listed below:

Phishing Attempts: This is a means of obtaining the credentials of a user by sending fake emails that looks like receiving from reliable sources. Once the user data are collected, the corresponding person and the network will be in vain.

Routing Attacks: Blockchain is completely dependent on the transfer of large data for its applications on the basis of real-time. But data are tapered on the way while they are transmitted to respective service providers. This will look normal for the participants.

Sybil attacks: Here, the phony network is utilized for tapering the complete data of users. Sybil is a popular character in a novel known for identification of disorders.

51% attacks: Mining requires a lot of power of computation for processing in a public-type blockchain. By the process of mining, it is possible to obtain more than 50 percent of the computational power of the network. Hence, editing of and tampering with the ledger is done.

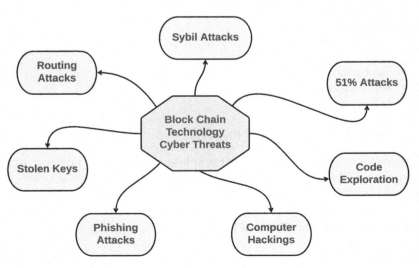

Figure 4.8 Blockchain cyber threats.

4.10 CONCLUSION

Blockchain is a novel technique of the IoT employed for providing fault-tolerant distributed systems. Blockchain has been employed in the architecture of numerous cryptocurrencies to provide and retain transactions among distributed users because of its indispensable features such as immutability and decentralization. It can be used to create decentralized systems in a variety of disciplines. This chapter enunciated the theories of blockchain technology with respect to its evolution, applications, advantages and challenges in financial transaction systems. The complete overview of blockchain technology with its significant issues and applicability in several sectors was also discussed. Finally, the transaction systems were discussed in detail along with a description of several cryptocurrencies with the impact of blockchain.

REFERENCES

1. Krdzalic Y., "Block chain explained: The complete guide", 2018 Update - Part 2, 2021.
2. Hayes A. "The socio-technological lives of bit coin", *Theory Culture and Society, SAGE Journals,* 2019, 36, pp. 49–72.
3. Rose C., "The evolution of digital currencies: Bitcoin, a crypto-currency causing a monetary revolution", *International Business and Economics Research Journal,* 2015, 14, pp. 617–622.
4. DuPont Q., *Crypto-currencies and Block chains,* John Wiley & Sons, Hoboken, NJ, 2019.
5. French L. A., "The Effects of Block chain on Supply Chain Trust", A Thesis Presented in Partial Requirements for Master of Supply Chain Management at Massey University, Palmerston North, New Zealand. Ph.D. Thesis, Massey University, 2022.
6. Mougayar W., *The Business Block chain: Promise, Practice, and Application of the Next Internet Technology,* John Wiley & Sons, Hoboken, 2016.
7. Notaro A., "All that is solid melts in the Ethereum: The brave new (art) world of NFTs", *Journal of Visual Art Practice,* 2022, 21(4), pp. 1–24.
8. Strilets B., "Current state and prospects for the legal regulation of cryptocurrencies in the European Union", *Actual Problems of State and Law,* 2022, pp. 70–76.
9. Ramadoss R., "Block chain technology: An overview", *IEEE Potentials,* 41, pp. 6–12.
10. Habbal M., "NFT-based model to manage educational assets in Metaverse", *Avrupa Bilim Teknoloji, Dergisi,* 2022, pp. 20–25, https://dergipark.org.tr/tr/download/article-file/2709574
11. Kaur H., Alam M. A., Jameel R., Mourya A. K., "A proposed solution and future direction for blockchain-based heterogeneous medicare data in cloud environment", *Journal of Medical Systems,* 2018, 42, pp. 42–48.
12. Murthy C. V. B., Shri M. L., "A survey on integrating cloud computing with blockchain", *Proceedings of 2020 "International Conference on Emerging*

Trends in Information Technology and Engineering (IC-ETITE)", 2020, Vellore, India, pp. 24–25.
13. Niranjanamurthy M., Nithya B., Jagannatha S., "Analysis of Blockchain technology: Pros, cons and SWOT", *Cluster Computing*, 2019, 22, pp. 14743–14757.
14. Baboshkin P., Mikhaylov A., Shaikh Z. A., "Sustainable Crypto-currency growth impossible? Impact of network power demand on bitcoin price", *Finansovyj žhurnal—Financial Journal*, 2022, 3, pp. 116–130.
15. Jabbar R., et al., "Blockchain technology for intelligent transportation systems: a systematic literature review". *IEEE Access*, 2022, 10, pp. 20995–21031.
16. Pagnotta E. S., "Decentralizing money: bitcoin prices and blockchain security", *The Review of Financial Studies*, 2022, 35, pp. 866–907.
17. Salem H., Mazzara M., Saleh H., Husami R., Hattab S. M., "Development of a Blockchain-Based Ad Listing Application", *Proceedings of International Conference on Advanced Information Networking and Applications"*, Sydney, Australia, April 2022, Springer, pp. 37–45.
18. Rajashekar M., Sundaram S., "Dynamic attribute tree for the data encryption and third-party auditing for cloud storage", *Indian Journal of Science and Technology*, 2022, 15, pp. 798–805.
19. Wang J., et al. "An effective blockchain-based data storage scheme in zero-trust IoT", *ACM Transactions on Internet Technology*, 2022, http://dx.doi.org/10.1145/3511902.
20. Maalla M. A., Bezzateev S. V., "Efficient incremental hash chain with probabilistic filter-based method to update blockchain light nodes", *Journal of Mechanical Science and Technology*, 2022, 22, pp. 538–546.
21. Agrawal D., Minocha S., Namasudra S., Gandomi A. H., "A robust drug recall supply chain management system using hyperledger blockchain ecosystem", *Computers in Biology and Medicines*, 2022, 140, pp. 105100.
22. Sammeta N., Parthiban L., "Hyperledger blockchain-enabled secure medical record management with deep learning-based diagnosis model", *Complex and Intelligent Systems*, 2022, 8, pp. 625–640.
23. Mirtskhulava L., Iavich M., Razmadze M., Gulua N., "Securing Medical Data in 5G and 6G via Multichain Blockchain Technology using Post-Quantum Signatures". In *Proceedings of the 2021 IEEE International Conference on Information and Telecommunication Technologies and Radio Electronics (UkrMiCo)*, Odesa, Ukraine, 29 November–3 December 2021; pp. 72–75.
24. Bouachir O., Aloqaily M., Tseng L., Boukerche A., "Blockchain and fog computing for cyber physical systems: The case of smart industry", *Computer* 2020, 53, pp. 36–45.
25. Sedlmeir J., Buhl H. U., Fridgen G., Keller R., "The energy consumption of blockchain technology: Beyond myth", *Business & Information Systems Engineering*, 2020, 62, pp. 599–608.
26. Gill S. S., Tuli S., Xu M., Singh I., Singh K. V., Lindsay D., Tuli S., Smirnova D., Singh M., Jain U., et al. "Transformative effects of IoT, Blockchain and Artificial Intelligence on cloud computing: Evolution, vision, trends and open challenges", *Internet Things* 2019, 8, pp. 100118.
27. Nguyen D. C., Pathirana P. N., Ding M., Seneviratne A., "Integration of blockchain and cloud of things: Architecture, applications and challenges", *IEEE Communications Surveys and Tutorials*, 2020, 22, pp. 2521–2549.

28. Reyna A., Martín C., Chen J., Soler E., Díaz M., "On blockchain and its integration with IoT. Challenges and opportunities," *Future Generation Computer Systems*, 2018, 88, pp. 173–190.
29. Al-Jaroodi J., Mohamed N., "Blockchain in industries: A survey", *IEEE Access*, 2019, 7, pp. 36500–36515.
30. Sharma, N., Sharma, R. Real-time monitoring of physicochemical parameters in water using big data and smart IoT sensors. *Environment, Development and Sustainability* (2022). https://doi.org/10.1007/s10668-022-02142-8.
31. Anandkumar R., Dinesh K., Obaid A. J., Malik P., Sharma R., Dumka A., Singh R., Khatak S., "Securing e-Health application of cloud computing using hyperchaotic image encryption framework", *Computers & Electrical Engineering*, 2022, 100, 107860, ISSN 0045-7906, https://doi.org/10.1016/j.compeleceng.2022.107860.
32. Sharma R., Xin Q., Siarry P. and Hong W.-C. "Guest editorial: Deep learning-based intelligent communication systems: Using big data analytics," *IET Communications*, (2022). https://doi.org/10.1049/cmu2.12374.
33. Sharma R., Arya R., "UAV based long range environment monitoring system with Industry 5.0 perspectives for smart city infrastructure", *Computers & Industrial Engineering*, 2022, 168, 108066, ISSN 0360-8352, https://doi.org/10.1016/j.cie.2022.108066.
34. Rai M., Maity T., Sharma R., et al., "Early detection of foot ulceration in type II diabetic patient using registration method in infrared images and descriptive comparison with deep learning methods", *The Journal of Supercomputing*, 2022. https://doi.org/10.1007/s11227-022-04380-z.
35. Sharma R., Gupta D., Maseleno A., Peng S.-L. "Introduction to the special issue on big data analytics with internet of things-oriented infrastructures for future smart cities", *Expert Systems*, 2022, 39, p. e12969. https://doi.org/10.1111/exsy.12969.
36. Sharma R., Gavalas D., Peng S.-L., "Smart and future applications of Internet of Multimedia Things (IoMT) using Big Data analytics", *Sensors*, 2022, 22, p. 4146. https://doi.org/10.3390/s22114146.
37. Sharma R., Arya R. "Security threats and measures in the Internet of Things for smart city infrastructure: A state of art", *Transactions on Emerging Telecommunications Technologies*, 2022, e4571. https://doi.org/10.1002/ett.4571.
38. Zheng J., Wu Z., Sharma R., Haibin L. V., "Adaptive decision model of product team organization pattern for extracting new energy from agricultural waste", *Sustainable Energy Technologies and Assessments*, 2022, 53(Part A), p. 102352, ISSN 2213-1388, https://doi.org/10.1016/j.seta.2022.102352.
39. Boovarahan N. C. A., Umapathy K., "Power allocation based on channel state information in massive MIMO system", *IOP Conference Series: Materials Science and Engineering*, October 2020, 12(2), pp 1–9. ISSN: 1757-899X
40. Mou J., Gao K., Duan P., Li J., Garg A., Sharma R., "A machine learning approach for energy-efficient intelligent transportation scheduling problem in a real-world dynamic circumstances," *IEEE Transactions on Intelligent Transportation Systems*, 2022, https://doi.org/10.1109/TITS.2022.3183215.
41. Kumar S., Umapathy K., "Efficient implementation of downlink resource allocation in wireless networks with inter-cell interference", *Journal of Advanced*

Research in Dynamical and Control Systems (JARDCS), 2017, 9(14), pp. 2297–2312. ISSN: 1943-023X

42. Monrat A. A., Schelén O., Andersson K., "A survey of blockchain from the perspectives of applications, challenges, and opportunities", IEEE Access 2019, Edition-7, pp. 117134–117151.

43 Umapathy K., "Wireless technique based vehicle speed control system", *AIP Conference Proceedings*, October 2022, 2519, pp. 050022-1–050022-5, https://doi.org/10.1063/5.0109726. ISSN: 1551-7616

Chapter 5

Analysis of critically polluted locations using the IoT and AI infrastructure

Lulwah M. Alkwai
University of Ha'il, Ha'il, Kingdom of Saudi Arabia

5.1 INTRODUCTION

Pollution is a severe global problem; it affects the lives of people year by year. Pollution is also decreasing the lifespan of children and affecting their growth. The World Health Organization (WHO) has also stated that 25% of global death is directly linked to pollution [1]. This sophisticated issue is growing day by day due to the increasing number of industrial activities. Not only Saudi Arabia but other countries are also facing the problems created by industrial pollution. Transportation is yet another big source of pollution in many developing cities. There is a need to recognize the higher polluted areas and the pollution trends in these locations. Accordingly, solutions must be taken to improve the environment [2].

Today, around 40% of the population lives in urban areas, which is estimated to be 60% by 2025. The major problems in urban areas are air pollution and water pollution. The excess population in urban areas will lead to development which will also help the global environment. The fast population expansion will increase the pressure on infrastructure and will result in air, water and land pollution. The Saudi government is regularly tracking the situation and is very much concerned about this situation [3].

The study is very much concerned about the excess of pollution in the various polluted locations. And a critical analysis is performed to track the environmental situation for some developing cities in Saudi Arabia. The pollution trends were measured in 2009, 2011, 2013 and 2018. The parameters for this research are air pollution, water pollution, land pollution and the comprehensive environmental pollution index (CEPI) for different locations [4].

This chapter is focused on the pollution trend analysis using the parameters of air pollution, water pollution and land pollution. The chapter is categorized into five sections. The first section introduces the theme of the chapter. The second section discusses various literature focused on the chapter's theme. The third section discusses the methodology used to analyze various pollution trends. The fourth section shows the results obtained from the pollution trend analysis. The fifth section concludes the chapter.

DOI: 10.1201/9781003452591-5

5.2 RELATED WORK

Chinnaswamy et al. [5] provided an air pollution trend analysis of a city from 2006 to 2013 at some specific locations and recorded the measurement consistently. Potential health implications were also discussed. M. M. M. Hoque, L. K. Basak, M. Rokanuzzaman and Sajal Roy [6] determined the noise level at some specific locations of the Tangail municipality area, and also discussed the effects and causes of pollution on the population. Jhumoor Biswas et al. [7] analyzed 8-hour, 1-hour and 24-hour, averaged criteria pollutants ($PM2.5$, SO_2, NO_2, $PM10$ and CO) for the duration from 2004 to 2009 at three different sites, i.e., Sirifort, Income Tax Office (ITO) and Delhi College of Engineering (DCE) in Delhi, India. Ramasamy Rajamanickam and S. Nagan [8] briefly discussed the CEPI; he defines how the CEPI is linked to the environmental quality of any location. He discussed the effect on the environment after increasing or decreasing the value of CERI. Michael Chertok et al. [9] compared the exposure to toluene, benzene, xylene (BTEX), ethylbenzene and nitrogen dioxide (NO_2) for commuters for five different commuting modes in central Sydney.

5.3 METHODOLOGY

Air, water and land pollution are big problems for any city. The significant sources of pollution in a city are vehicles, constructed buildings and domestic pollution. The chapter's objective is to make a comparative analysis of pollution trends in some developing cities over the years 2009, 2011, 2013 and 2018. The study parameters are the air pollution, water pollution and land pollution. Finally, an analysis is done for eight critically polluted locations using CEPI [10].

5.3.1 CEPI

CEPI is a numerical number to identify the environmental quality of a given location. An increment in the CEPI score will adversely affect the environment. CEPI is the best tool for measuring the quality of the environment.

The CEPI algorithm is briefly described in Figure 5.1. According to the algorithm, three critical pollutants must be calculated in the beginning. And these pollutants will be divided into three groups: A, B and C. This criterion was issued by the CPCB in December 2009 and a revised issue was published in 2016. Group C pollutants are more critical than Group B for the environment, and Group B pollutants are more critical than Group A [11].

In this chapter, the CEPI value is represented by alphanumeric values of air, water and land environment. If the score is more than 60, it indicates critical level of pollution in the respective environment, if the score is

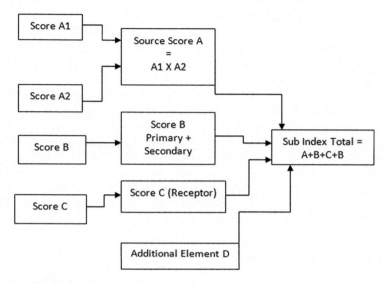

Figure 5.1 Process for CEPI.

between 50 and 60, it indicates a severe level of pollution in the respective environment [12]. The data on pollution were collected through many IoT sensors installed at various locations [13].

5.3.2 Data collection

The study considered eight critically polluted locations, which are as shown in Table 5.1. The CEPI and air, water and land environment are considered to be the main research parameters.

Data were collected for these eight (Location 1 to Location 8) critically polluted locations [14].

Table 5.1 Latitudes and longitudes of critically polluted locations

Location	Latitude and longitude
Location 1	28.5355°N, 77.3910°E
Location 2	30.9010°N, 75.8573°E
Location 3	23.0225°N, 72.5714°E
Location 4	12.9165°N, 79.1325°E
Location 5	28.7041°N, 77.1025°E
Location 6	17.6868°N, 83.2185°E
Location 7	27.1767°N, 78.0081°E
Location 8	12.9141°N, 74.8560°E

5.4 RESULTS AND ANALYSIS

The principal parameters of the research are air, water and land pollution, along with the CEPI. The data on pollution were collected through many IoT sensors installed at various locations [15].

Most of the generation in these locations is dependent on agricultural work. After the independence, people are searching for new sources of agricultural work to change their lifestyles. After the year 1981, the government has been providing a lot of facilities setting up many industries. In the last 30 years, the private industries have increased highly [16]. The rapid increase in industries is also creating many issues for the environment. Pollution is classified into three categories: Air, Water and Land. Also, pollution can easily be discussed using the CEPI score. Therefore, this chapter analyzes these four parameters for eight polluted locations [17].

5.4.1 Air pollution analysis

Air pollution is the most dangerous pollution in the world. Typically, pure air consists of 21.946% of Oxygen (O_2), 78.084% of Nitrogen (N_2), 0.934% of Argon (Ar), 0.0397% of Carbon dioxide (CO_2), 0.00182% of Neon (Ne), 0.0002% of Methane (CH_4) and 0.0005% of Helium (He). Whenever this score changes due to other dangerous gasses, the air will be considered polluted. The air pollution score for the eight most polluted cities is shown in Table 5.2 and Figure 5.2.

A score of more than 60 represents the most polluted city, and a score from 50 to 60 represents that the pollution is moderate but not avoidable [18].

5.4.2 Water pollution analysis

Most of the big industries produce a large amount of waste and wash it out into rivers. This is a big source of water pollution in these locations.

Table 5.2 Air pollution scores of critically polluted locations

Location	Air pollution score
Location 1	65.75
Location 2	68
Location 3	62.75
Location 4	69.25
Location 5	52.13
Location 6	57
Location 7	59
Location 8	61.75

Analysis of critically polluted locations using the IoT 73

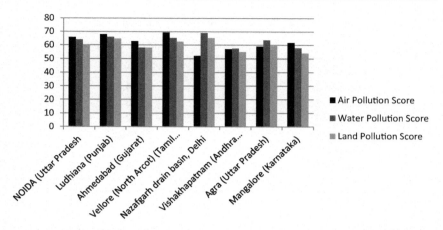

Figure 5.2 Comparative analysis of air, water and land pollution for eight most polluted cities.

Table 5.3 Water pollution scores of critically polluted locations

Location	Water pollution score
Location 1	64
Location 2	66
Location 3	58
Location 4	65.25
Location 5	69
Location 6	57.5
Location 7	63.75
Location 8	57.75

The water pollution scores for the eight most polluted cities are shown in Table 5.3 and Figure 5.2.

A score of more than 60 represents the most polluted cities and a score from 50 to 60 represents pollution that is moderate but not avoidable [19].

5.4.3 Land pollution analysis

Major sources of land pollution are fertilizers, poisonous pesticides and corrosion. Poor garbage disposal services are also a source of land pollution. The water pollution score for eight most polluted cities can be shown in Table 5.4 and Figure 5.2.

A comparative analysis of air, water and land pollution for the eight most polluted cities is shown in Figure 5.2.

Table 5.4 Land pollution score of critically polluted location

Location	Land pollution score
Location 1	60
Location 2	64.75
Location 3	58
Location 4	62.5
Location 5	65.25
Location 6	55
Location 7	59.5
Location 8	54

5.4.4 Analysis using comprehensive environmental pollution index

The chapter's objective is to compare pollution trends in eight developing cities over the period of 2009, 2011, 2013 and 2018. CEPI is a numerical number to identify the environmental quality of a given location. An increment in CEPI score will adversely affect the environment. CEPI is the best tool for measuring the quality of the environment [20].

If the score is more than 60, then it shows the critical level of pollution in the respective environment, if the score is between 50 and 60, then it shows a severe level of pollution in the respective environment [21]. The data on pollution were collected through many IoT sensors installed at various locations [22, 23]. The analysis of eight developing cities in 2009, 2011, 2013 and 2018 are as shown in Table 5.5.

From Table 5.5, it is clear that the pollution level of these eight developing cities is increasing yearly. These parameters must be tracked regularly and measures must be taken to improve the environment [24, 25].

Table 5.5 CEPI scores of critically polluted locations

Locations	CEPI score – 2009	CEPI score – 2011	CEPI score – 2013
Location 1	78.9	80.72	78.69
Location 2	81.66	73.23	75.72
Location 3	75.28	78.09	69.54
Location 4	81.79	84.73	79.67
Location 5	73.09	67.07	73.42
Location 6	70.82	57.39	52.31
Location 7	76.48	88.36	68.71
Location 8	73.68	73.86	67.62

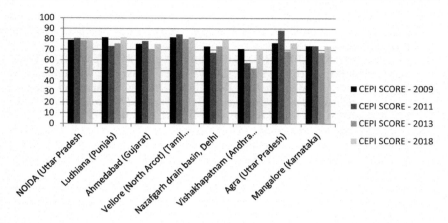

Figure 5.3 Comparative analysis of CEPI score for eight most polluted cities.

The CEPI score comparative analysis can also be understood from the graph shown in Figure 5.3.

5.5 CONCLUSION

This chapter aimed to analyze the pollution trends at different locations. The pollution trends were measured in 2009, 2011, 2013 and 2018. The parameters for this research were air pollution, water pollution, land pollution and the CEPI for different locations. The study is very much concerned about the excess of pollution in these locations. A critical analysis was performed to track the environmental situation in some developing cities. The research highlights the truth of how pollution in some developing cities is increasing day by day. This research highlights the trend of pollution in various polluted locations. This research will help society to take severe action against the sources that lead to an increase in pollution.

REFERENCES

1. Rafiq F, Urban floods in India. *International Journal of Scientific and Engineering Research*, 7:721–734 (2016).
2. Allee RJ, Johnson JE, Use of satellite imagery to estimate surface chlorophyll a and Secchi disc depth of Bull Shoals Reservoir, Arkansas, USA. *International Journal of Remote Sensing*, 20(6):1057–1072 (1999).
3. Tiwari SK, Gupta AK, Asthana AKL, Evaluating CO2 flux and recharge source in geothermal springs, Garhwal Himalaya, India: Stable isotope systematics and geochemical proxies. *Environmental Science and Pollution Research*, 27:14818–14835 (2020).

4. Shi L, Mao Z, Wang Z, Retrieval of total suspended matter concentrations from high resolution WorldView-2 imagery: A case study of inland rivers. In: *IOP conference* (2018).
5. Chinnaswamy AK, Galvez MD, Balisane H, Naguib RG, Nguyen QT, Trodd N, Marshall IM, Yaacob N, Santos GC, Vallar EA, Shaker M, Wickramasinghe N, Ton T, Air pollution in Bangalore, India: An eight-year trend analysis. *International Journal of Environmental Technology and Management*, 19(3–4): 177–197 (2016). https://doi.org/10.1504/IJETM.2016.082233
6. Hoque M, Basak L, Rokanuzzaman M, Roy S, Level of noise pollution at different locations in Tangail municipal area, Bangladesh. *Bangladesh Journal of Scientific Research*, 26(1–2):29–36 (2014). https://doi.org/10.3329/bjsr.v26i1-2.20228
7. Biswas J, Upadhyay E, Nayak M, Yadav A, An analysis of ambient air quality conditions Over Delhi, India from 2004 to 2009. *Atmospheric and Climate Sciences*, 1(4):214–224 (2011). https://doi.org/10.4236/acs.2011.14024
8. Rajamanickam R, Nagan S, Assessment of comprehensive environmental pollution index of Kurichi Industrial Cluster, Coimbatore District, Tamil Nadu, India – A case study. *Journal of Ecological Engineering*, 19(1):191–199 (2018). https://doi.org/10.12911/22998993/78747
9. Chertok M, Voukelatos A, Sheppeard V, Rissel C, Comparison of air pollution exposure for five commuting modes in Sydney – car, train, bus, bicycle and walking. *Health Promotion Journal of Australia*, 15:63–67 (2004). https://doi.org/10.1071/HE04063 Aggarwal, A., Urban flood hazard mapping using change detection on wetness transformed images. *Hydrological Sciences Journal*, 61 (5), 816–825 (2016).
10. Priyadarshini I, Sharma R, Bhatt D et al. Human activity recognition in cyber-physical systems using optimized machine learning techniques. *Cluster Computing* (2022). https://doi.org/10.1007/s10586-022-03662-8
11. Priyadarshini I, Alkhayyat A, Obaid AJ, Sharma R, Water pollution reduction for sustainable urban development using machine learning techniques. *Cities*, 130:103970 (2022). ISSN 0264-2751, https://doi.org/10.1016/j.cities.2022.103970.
12. Pandya S, Gadekallu TR, Maddikunta PKR, Sharma R, A study of the impacts of air pollution on the agricultural community and yield crops (Indian Context). *Sustainability*, 14:13098 (2022). https://doi.org/10.3390/su142013098
13. Bhola B, Kumar R, Rani P, Sharma R, Mohammed MA, Yadav K, Alotaibi SD, Alkwai LM, Quality-enabled decentralized dynamic IoT platform with scalable resources integration. *IET Communications* 00:1–10 (2022). https://doi.org/10.1049/cmu2.12514
14. Deepanshi IB, Garg D, Kumar N, Sharma R, A comprehensive review on variants of SARS-CoVs-2: Challenges, solutions and open issues. *Computer Communications*, (2022). ISSN 0140-3664, https://doi.org/10.1016/j.comcom.2022.10.013.
15. Ahasan Habib AKM, Hasan MK, Islam S, Sharma R, Hassan R, Nafi N, Yadav K, Alotaibi SD, Energy-efficient system and charge balancing topology for electric vehicle application. *Sustainable Energy Technologies and Assessments*, 53 (Part B):102516 (2022). ISSN 2213-1388, https://doi.org/10.1016/j.seta.2022.102516.

16. Rani P, Sharma R, Intelligent transportation system for internet of vehicles based vehicular networks for smart cities. *Computers and Electrical Engineering*, 105:108543 (2023). ISSN 0045-7906, https://doi.org/10.1016/j.compeleceng.2022.108543.
17. Sharma R, Rawat DB, Nayak A, Peng S-L, Xin Q, Introduction to the special section on survivability analysis of wireless networks with performance evaluation (VSI–networks survivability). *Computer Networks*, 220:109498 (2023). ISSN 1389-1286, https://doi.org/10.1016/j.comnet.2022.109498.
18. Ghildiyal Y, Singh R, Alkhayyat A, Gehlot A, Malik P, Sharma R, Akram SV, Alkwai LM, An imperative role of 6G communication with perspective of industry 4.0: Challenges and research directions. *Sustainable Energy Technologies and Assessments*, 56:103047 (2023). ISSN 2213-1388, https://doi.org/10.1016/j.seta.2023.103047.
19. Ahasan Habib AKM, Hasan MK, Alkhayyat A, Islam S, Sharma R, Alkwai LM, False data injection attack in smart grid cyber physical system: Issues, challenges, and future direction. *Computers and Electrical Engineering*, 107:108638 (2023). ISSN 0045-7906, https://doi.org/10.1016/j.compeleceng.2023.108638.
20. Priyadarshini I, Kumar R, Alkhayyat A, Sharma R, Yadav K, Alkwai LM, Kumar S, Survivability of industrial internet of things using machine learning and smart contracts. *Computers and Electrical Engineering*, 107:108617 (2023). ISSN 0045-7906, https://doi.org/10.1016/j.compeleceng.2023.108617. (SCIE, IF-4.15)
21. Kondratyev KY, Pozdnyakov DV, Pettersson LH, Water quality remote sensing in the visible spectrum. *International Journal of Remote Sensing* 19:957–979 (1998).
22. Dwivedi SL, Pathak VA, Preliminary assignment of water quality index to Mandakini River, Chitrakoot. *Indian Journal of Environmental Protection* 27:1036–1038 (2007).
23. Zhang Y, Pulliainen JT, Koponen SS, Hallikainen MT, Water quality retrievals from combined Landsat TM data and ERS-2 SAR data in the Gulf of Finland. *IEEE Transactions on Geoscience and Remote Sensing* 41:622–629 (2003).
24. Pattiaratchi CB, Lavery P, Wyllie A, Hick P, Estimates of water-quality in coastal waters using multi-date Landsat Thematic Mapper data. *International Journal of Remote Sensing* 15:84–1571 (1994).
25. Girgin S, Kazanci N, Dügel M, Relationship between aquatic insects and heavy metals in an urban stream using multivariate techniques. *International Journal of Environmental Science and Technology* 7(4):653–664 (2010).

Chapter 6

Analysis of deep learning techniques in biomedical images

Tina and Ritu Gupta
GGSIPU, New Delhi, India

6.1 INTRODUCTION

Medical Imaging Modalities can be defined as the set of processes that create a visual representation of the internal organs of the body for clinical diagnosis to monitor the health of patients. Image modalities such as Magnetic Resonance Imaging (MRI), Computed tomography (CT), Ultrasound (US), Radiography, X-rays and Positron Emission Tomography (PET) are generally considered in clinical practice. Over the years, medical imaging has undergone various advancements as there is a need for continuous improvement in the field of healthcare, and several techniques are applied to the imaging modalities for better visualization along with improved decision-making for physicians. Likewise, these modalities play a vital role in the clinical trials for research in medicine and molecular imaging and for that, medical imaging continually acts as a strong diagnostic tool in healthcare. The main aim of these modalities is to categorize the abnormalities in the structure, diseased regions and differentiated lesions present in the organs of the body and improve the clinical ability for diagnosis of disease.

The existing technologies available which provide two-dimensional images include location, distance and size. With the advent of newer technologies in research, which are used in medical imaging, the subject's part is modelled into three-dimensional geometrical models which are specific to the patient. The three-dimensional data generated with the help of medical equipment will be more accurate, valid and of good quality, which will help to enhance the treatment procedure. The concept of 3D printing [1] is useful in the extraction of medical data from medical equipment through the process of image segmentation which can be described as the partitioning of a digital image into a set of pixels called image objects. These image objects are further used to label the pixels which share the same characteristics and then spot the region of interest from labelling. The three-dimensional geometrical data gives the orientation, speculations regarding the exact location, closeness to adjacent structures and density of the abnormal structure or disorders in part of the organ [2].

Deep Learning is a subfield of machine learning algorithms which has its foundation in Artificial Neural Networks and is one of the emerging technologies. It is a supervised learning approach from machine learning. It can be defined as the extraction of data through many hidden layers and neurons with their respective training paradigms. The fundamental unit in this technique is the Neuron, which can be described as a nerve cell having specialization in transmitting information to different organs of the body [3]. It takes inputs which can be multiple signals with linear weights assigned to them and produce output signals through a number of operations performed upon it. Input signals pass through a number of layers which are further classified as the input layer and output layer. The restoration of layers divulges the extensive raw data formed through the deep learning models which generate the layered pipeline of non-linear grouping of their outputs, and with the help of these outputs, the dimensional projection of the dataset can be formed. In the case of its application with the image, the depth of the pixels in terms of the credit assignment path (CAP) is calculated using a deep learning approach and it is the process of the transformation of the input layer to the output layer, passing through a number of hidden layers present in an image. CAP will define the causal linking between the input layer and the output layer [4]. The input layer, hidden layer and output layer instigate the abstract feature of an image with a higher level, which will aid in detecting shapes and common objects in an image. Deep learning techniques especially Convolutional neural networks (CNNs) are substantial in the field of medical imaging. Other techniques include recurrent neural networks (RNNs), autoencoders and deep belief networks which too have greater impacts. Along with that, the role of Graphic Processing units in medical imaging plays an important role as it accelerates deep learning techniques with better visualization. CNN architecture coordinates parallelly for transferring the graphic processing unit's data with dense matrices and convolutions generated from it.

Deep learning in the field of medical imaging accelerates the growth of healthcare informatics to a greater extent. Various industries in digital imaging, graphics processing units, healthcare informatics, artificial intelligence and medical equipment are integrated to produce intelligent machines which can diagnose diseases or abnormalities in the structure using medical images [5]. The revolution is not limited to the industries only but also the researchers are also moving towards medical sectors. Companies like GE, Fujifilm and Samsung are more inclined towards machine intelligence and digital imaging for better resolutions and higher dimensional views of medical images. DeepMind of Google, by National Health Service, United Kingdom, is also one of the biggest projects in the medical field. The view of smearing deep learning algorithms to the medical imaging modalities is quite rising in the research area although working towards the higher dimensional view is one of the concerned applications of healthcare [6].

The motivation behind this chapter is to present a comprehensive review of the deep learning techniques with their architectures, advantages and disadvantages and their role in the analysis of medical imaging with future scope. This chapter will provide vital familiarity and approaches to the importance of deep learning in the domain of medical imaging modalities. Although there are several challenges associated with the concept such as privacy, legal issues, complexities, authenticity of dataset and labelling of data, there are still opportunities which attract researchers towards this domain. In the coming years, clinical diagnosis will be more patient-specific and based on artificial intelligence. It is a need of the hour to focus more on these issues as various industries are integrated to work on this domain and many companies are investing huge amounts in the healthcare sector for better treatment of diseases and the prediction of diseases in the early stage.

6.2 TECHNIQUES FOR DEEP LEARNING IN THREE-DIMENSIONAL MEDICAL IMAGE PROCESSING

A Deep Neural Network (DNN) can be described as a hierarchy of neurons arranged in a stack of multiple layers forming the feature representation. These are artificial networks which consist of numerous interconnected layers comprising artificial neurons. Neurons, in basic terms, can be defined as alike to biological neurons which can take multiple inputs in order to produce a single output. To generate that output, computations are performed linearly or non-linearly and this function of inputs is trailed by the activation function. Here, the artificial neurons are combined to form a stack of layers and based on these layers, the next layer will be formed based on these stacked layers [7]. After passing through these terminal layers, the indiscriminately complex structures can be evaluated to generate the desired output layer structure. According to the traditional application of machine learning for medical imaging modalities, it classifies the region of interest for abnormalities in organs of the body, especially for medical image segmentation. The first step instigates from image pre-processing which comprises the use of filters for contrast enhancement or noise removal procedure following the next stage named image segmentation. This stage performs the segmentation techniques which constitute edge-based, cluster-based segmentation and thresholding [8]. After this phase, the feature extraction process arises into a state which will extract the size for the region of interest, colour, contrast and texture. The foremost features in this scenario use feature selection methods which will be used as input for machine learning classifiers (especially neural networks or support vector machines) with the targeted labelling in order to regulate the optimal boundary that distinguishes each class. After training the machine learning classifier, it will classify each and every unlabelled image into a class [9]. The major challenges involve determining the suitable requirements for the pre-processing stage, which are particularly based on underdone image properties.

6.2.1 Convolutional neural network

Deep learning models are majorly used in medical image processing techniques in the domain of healthcare informatics. One of the deep learning models includes CNN, also called as ConvNet which can be defined as a class of neural networks which is quite suitable for visualization of the medical image modalities. It will aid in providing the extraction from feature space representation through input space which can be a unique and hierarchical approach from small and almost similar patterns. It is quite essential in accumulating complex patterns as compared to neural classifiers. The layered architecture for CNN entails a stacked layer, each carrying out a definite operation. Respectively, the first layer is an input layer or input medical image which is actually associated with multiple neurons, which is almost equivalent to the number of pixels in a medical image. The output from the previous layer, which serves as its input, is then obtained by the intermediate layer. After that, the subsequent layers are convolutional layers that will display the results of applying a set of filters to the input image in order to extract features from it. These filters, also known as kernels, are designed to be of arbitrary sizes and vary in size according to the kernel size. Each neuron only responds to a certain region of the earlier layer, known as the receptive field. The convolution layer's output is quantified as an activation map, which truly brings to light the impact of applying a specific filter to the input image. Activation layers typically follow these CNN layers to add non-linearity to the activation maps. Another layer, the pooling layer, depends on the design and helps to reduce the output's dimensionality (convolution) [10].

Last but not least, FCN (completely connected layers) mines high-level abstractions where the kernels and weights of neural connections are optimized continuously throughout the practice of backpropagation. The established structure is also referred to as the traditional CNN deep learning model. The use of these anatomical structures in medical picture segmentation can be researched in the sections that follow [11].

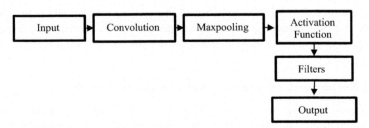

Figure 6.1 Architecture of a convolutional neural network.

82 Artificial Intelligence and Blockchain in Industry 4.0

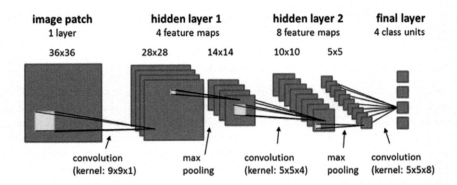

Figure 6.2 Flow diagram of a convolutional neural network.

The first layer (or input layer) connects to the medical input image directly through a large number of neurons that correspond to the pixels (as in the input image). The results of convoluting the input information through the filters are included in the next layer, which contains convolutional layers. These filters are known for their random-size kernels and other information that depends on the kernel size. Every neuron responds to a specific region of the layer before it, known as the receptive field. Every convolution layer's output is calculated as an activation map to highlight the results of applying a precise filter to the input [12]. In order to smear non-linear functions to the corresponding activation maps, activation layers typically tail convolutional layers and CNN layers. Depending on the architecture, the next layer may be a pooling layer, which helps to reduce the output's dimensionality for convolution. Certain methods, such as Maxpooling, are used to carry out the pooling functionality. Ultimately, whole connected layers are extracted using higher-level abstractions.

As shown in Figure 6.2, with the promising ability of convolutional layers to perform pattern recognition and image classification, CNNs are widely used nowadays. The overall idea is to achieve medical image segmentation [13] with the use of input image (two-dimensional) and smearing filters on it. Numerous sources of information in the procedure of two-dimensional images are conceded to the input layer of a CNN network in numerous image channels in order to explore if the usage of these two-dimensional multimodal images acts as input which progresses the segmentation results. The results establish improved performance than single modal input.

6.2.2 Recurrent convolutional neural network

RNN, commonly known as Recurrent Neural Network, is a kind of Neural Network which describes the existence of connections among the nodes along a time-based sequence. The defined connection is similar to a directed

Analysis of deep learning techniques in biomedical images 83

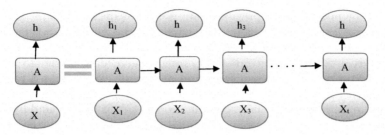

Figure 6.3 Working of recurrent neural network (RNN).

graph. Recurrent CNN is allocated recurring networks which empower the network to learn the patterns produced. Consequently, recurrent CNNs are able to cite interslice frameworks with the help of information in the form of chronological data. The Recurrent CNN structure entails two main segments of intraslice data abstraction which can be using any kind of CNN model and Recurrent CNN in control of interslice data abstraction.

As shown in Figure 6.3, an RNN can be defined as a feedback layer inside the neural network. This architecture is unfolded in nature with respect to time and the training is done using backpropagation. In this, a similar set of weights is cast off for a layer transversely multiple time steps and efficient use of the gradients is done. This neural network takes the input X_t and gives the output h_t. Consequently, the data is accepted from the first step to its succeeding step. The RNN architecture unfolds replicas of a similar network that passes the desired information to the respective next state. It although allows to achieve modelling through vector chain or classification [14]. These classifications can be input, output or both in its working model. Thus, it can be concluded that RNNs are associated with the sequences. It is preferable to implement this model in case of the information has a sequential nature. An RNN in aggregation with that of a CNN can perceive medical images and deliver their description in the form of tags.

The first input layer in Figure 6.4 accomplishes the translation of independent activations to dependent activations. It will allocate the identical bias and weight to an entire layer which will additionally lessen the complexity of model parameters and delivers a platform for memory of the former

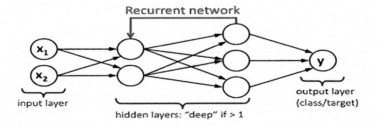

Figure 6.4 Architecture of recurrent neural network.

outputs by providing the previous output layer as the input to the succeeding layer. All these layers have similar weights and biases to form a single recurrent unit. The neural network receives a sole time phase of the input. The current output h_t converts to h_{t-1} for the succeeding phase [15]. It can have n number of repetitive steps and at last all can be joined together. Afterwards, that is, with the conclusion of all steps, the final phase is for computing the output. Finally, computational error is to be found by determining the difference between the actual output and the predicted output. The corresponding error is back-propagated to the neural network in order to fine-tune the weights and generate an improved outcome.

Diverse applications of RNNs are in the domain of analyzing the importance of time sequences in the medical domain. Various studies so far have verified that RNNs have substantial success in discriminating neurons or biological signals. RNN results confirmed obtaining higher classification accuracies. They have an improved capability to classify and analyze diverse kinds of biological signals. The RNN model is also responsible for the filtering process of EHG biosignals, which is a diagnostic method for the detection of labour. Different types of RNNs are functional for the prediction of filtering and EHG signals [16]. According to many studies, RNN can effectively filter EHG signals with a high Signal-to-Noise Ratio.

6.2.3 U-Net

U-Net can be defined as an expanded CNN that generally deals with the analysis of biomedical images where the abnormal structures classify whether the abnormal structure is a problem area and also identifies the area of abnormal structure. This architecture can construct an image segmentation prototype which will forecast the mask of an object existing in an image as it will confine the object in the image.

It is the FCN, commonly known as a fully convolutional network, which involves fewer training sets with higher accuracy. The U-shape architecture comprises two parts: the contracting path and the expanding path. The former is used to have context information whereas the latter is for precise positioning. It is considered one of the standard architectures for the purpose of image classification and image segmentation to segment the biomedical image by identifying class as it will generate a mask to discrete image into numerous classes.

As shown in Figure 6.5, U-Net is symmetrical in nature and comprises bottleneck layers which will associate the data from the encoder's and decoder's paths to concatenate the respective feature maps [17]. The encoders consist of four blocks and each block comprises two unpadded convolutions (3 × 3) along with Rectified Linear Unit activation layer and a Maxpooling layer (2 × 2). The feature channels get doubled after the respective down-sampling but then again, feature maps get reduced due to the

Analysis of deep learning techniques in biomedical images   85

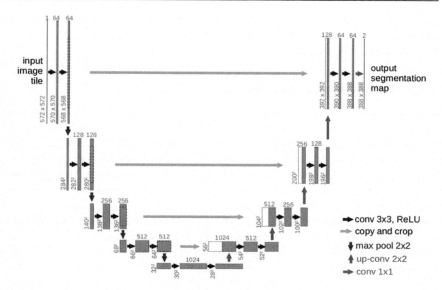

Figure 6.5 U-shape model for U-Net architecture.

Maxpooling. The decoders comprise up-sampling (2 × 2) along with convolutions (3 × 3). It is helpful in transferring the localization for the acquired information which will learn through down-sampling from encoders to decoders. The encoders are however covenant layers tailed by the pooling procedure which castoff to extract the factors in biomedical images [18], whereas, decoder practices reordered convolution in order to attain permit of localization.

The crucial extended part of this model is up-sampling along with feature channels in order to permit the network to propagate the respective context information to the higher-resolution layers. Significantly, the expansive path is either more symmetric or less symmetric for the contracting path which will produce a U-shaped model. The U-shaped network doesn't have FCN (fully connected layers), so it utilizes the valid fragment of each convolution and this can be defined as the segmentation map containing the pixels. The defined approach permits the unified segmentation of arbitrarily large medical images by overlapping. To forecast the pixels in the boundary of the medical image, the misplaced context is inferred by emulating an input image. For little training data, extreme data augmentation by smearing elastic deformations is required for accessible training of medical images [22]. It agrees with the network to acquire invariance for deformations without the necessity of transformations in the interpreted image quantity. Biomedical segmentation is the deformation which cast off most communal distinction in tissue, and deformations can be imitated accurately and efficiently. The difficulty in segmentation tasks is the parting of stirring objects of a similar class and to solve this issue, the use of weighted loss can be implemented whereas the extrication contextual labels between touching cells attain a huge weight in the loss function.

6.2.4 Resnet neural network

Resnet is a layered residual network entailing attributes and parameters for a CNN in deep learning [24]. In this approach, the input of the nth layer of this model will pass directly to $(n + x)$th layer which will synchronize in the stacking of layers in order to build the DNN and residual learning will withhold the features erudite from the nth input layer also by customizing the skip connections to hedge over some layers. The built model will be utilized to pre-train the model to tune it finely.

As shown in Figure 6.6, the underlying concept of Resnet is to establish an identity shortcut connection that will skip the connection for one or more layers. Although, the layers in this model are of varying sizes, depending upon the layers and their numbers, according to the approach, each one of the layers adheres to a similar pattern. They used to operate 3 × 3 convolution along with a static dimension feature map dimension respectively with pass by the input after every two convolutions [25]. Additionally, the height and width persist constant throughout the entire layer. The diminishing dimension between the layers will be attained through the increment in stride from the first layer to the second layer at the first convolution layer.

As seen in Figure 6.7, the Resnet technique preserves the spatial relationship between input photos' hidden layers in the healthcare industry. In this method, the grid layout is prearranged, and the layers preserve their relationships by operating on previous levels in smaller locations. It is highly advised for boosting the efficiency of image-focused tasks in greater dimensions because it has several CNN layers and a comparable activation function. In this concept, an input image is fed into convolutional layers, and the activation function from the previous layer is twisted into a sequence of parametrized filters in the hidden layer. The encoder section additionally

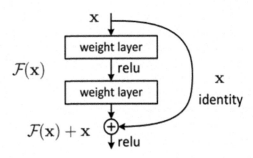

Figure 6.6 A residual block of Resnet architecture.

Analysis of deep learning techniques in biomedical images 87

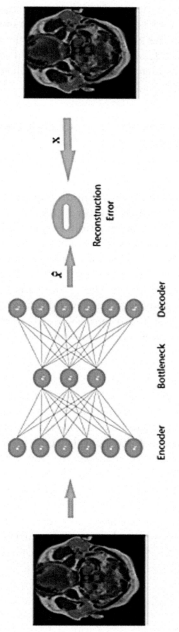

Figure 6.7 Resnet working model of medical image.

summarizes the reconstruction error between the input image and the appropriate output, that is, from input x to output x (i.e., $x \geq x$), in order to reduce the dimensions from the provided dataset in supervised learning. The bottleneck, which is the intermediate frame between input and network output, can be quantified as the key characteristic of the network design proposed here. This method feeds approximations of non-linear activation functions to all relevant hidden layers as input, which subsequently form the tensor of the feature maps.

6.3 APPLICATIONS OF MEDICAL IMAGES USING DEEP NEURAL NETWORK

6.3.1 Segmentation

Deep learning algorithms have enabled numerous medical image modalities in assessing and segmentation of images like breast cancer, brain tumour and brain haemorrhage. This application is quite helpful in quantitative assessment for the treatment of patients as it improves the visual representation of the image modalities. Three-dimensional images using DNNs for segmentation in medical images contribute to the research domain in healthcare informatics. The segmentation shows the substructure or abnormalities of the organs of the body which are varied in size and are imbalanced in scanning. The higher-resolution approach utilizes the spatial contextual information from the structure in the given dataset.

6.3.2 Localization

This application can be defined as biological structures which assume to be elementary prerequisites for numerous initiatives in the investigation of medical image modalities. It can be a hassle-free progression for the visualizer although it is a tough task for neural networks that are susceptible to distinction in images brought by image acquisition, differences of pathology and structures amongst patients. Usually, three dimensions are mandatory for the localization of images. Some procedures have been recommended that regularly commence treating the three-dimensional space as a prearrangement of two-dimensional orthogonal planes. This method depends on correct recognition of the scanned biological structures which utilizes the various properties of a three-dimensional fully connected CNN to present spatial localization higher-resolution images.

6.3.3 Classification

Various implementations of classification tasks in medical images include casting off numerous abnormalities in a time series form. Distinguishing the abnormalities of a normal organ from the part of the organ with an ailment, besides identifying the ailment, is a "Time Series Classification" problem. Although numerous devices can be used for the purpose to record a sequence of images that can be used to understand the condition, identifying the correct condition from this series of steps of images constitutes a "Classification" problem. The concept of Anomaly involves identifying uncommon observations which are significantly dissimilar from common data. Therefore, distinguishing the time series of usual operations from that of a device with an anomaly and recognizing the anomaly is the "Classification" problem.

6.4 COMPARISON BETWEEN DEEP LEARNING TECHNIQUES

Different deep learning techniques have their own advantages, disadvantages and challenges. Some of them are mentioned in Table 6.1, in a summarized form with respect to biomedical images.

Table 6.1 Comparison between various deep learning techniques

	Advantages	*Disadvantages*	*Limitations*
Convolutional Neural Network	• Weight sharing • Feature extractors • Feature learning	• Coordinate frame • Classification of Images with different positions • High computational cost	• GPUs are quite slow to train for complex tasks • Need more training data
Recurrent Neural Network	• Any length of input can be processed by RNN • Helpful for any time series predictor • The model size does not increase as the input size does • It can process any number of inputs by using its internal memory	• Due to its recurrent nature, the computation is slow • Training of RNN models can be difficult • Difficult to process sequences that are very long	• Information propagation when the time step is excessively long • A network becomes untrainable when it has too many deep layers. When the network descends to lower levels, the gradients get smaller, a phenomenon known as the "vanishing gradient problem"

(Continued)

Table 6.1 (Continued) Comparison between various deep learning techniques

	Advantages	Disadvantages	Limitations
U-Net	• Computationally efficient • Preferable for biomedical applications • Trained end-to-end • Trainable with a small data-set	• Because of numerous layers, it takes a substantial amount of time to train • High GPU memory footprint for larger images • Not standard to have pre-trained models • Learning may slow down in the middle layers of deeper models	• U-Net is limited to extracting some of the complex features that could help image segmentation in medical images • Localizing and segmenting the object of interest (OOI) in a synthetic dataset is difficult
Resnet	• Without increasing the error rate percentage, large networks can be given training • It deals with the vanishing gradient challenge via identity mapping	• It is almost impossible for deep networks to train in real-world applications • There is increased complexity in the implementation of its architecture	To combine skip-level connections and for that, the concept of dimensionality is applied among the different layers

6.5 CONCLUSIONS AND FUTURE SCOPE

A number of samples are often required to train the specified deep learning models. The recent successes of deep learning models, which are trained on enormous datasets, further support this. Whether these models can operate with little datasets in medical imaging is still an open question. As a result, massive training samples and training parameters are needed by the network for it to process meaningful representations of medical images. Although there will be far less variety in medical photos than those found in typical image databases, the process of perfecting three-dimensional CNN models, which were previously trained on normal image datasets, could be used to assess medical images.

REFERENCES

1. A Squelch (2018 Sep) '3D Printing and Medical Imaging'. *Journal of Medical Radiation Science*, Volume 65, Issue 3, pp. 171–172.
2. MH Hesamian et al (2019) 'Deep Learning Techniques for Medical Image Segmentation: Achievements and Challenges'. *Journal of Digital Imaging*, Volume 32, Issue 4, pp. 582–596.

3. Y Mohamed et al (2019 June) 'Research in Medical Imaging Using Image Processing Techniques'. *intechopen.com/books*.
4. IRI Haque, and J Neubert (2020) 'Deep Learning Approaches to Biomedical Image Segmentation'. *Informatics in Medicine*, Volume. 18, pp. 100297.
5. S Liu, Y Wang, et al (2019) 'Deep Learning in Medical Ultrasound Analysis: A Review'. *Engineering*, Volume 5, Issue 2, pp. 261–275.
6. L Alvarez, PL Lions, and JM Morel (1992) 'Image Selective Smoothing and Edge Detection by Nonlinear Diffusion'. *SIAM Journal on Numerical Analysis*, Volume 29, pp. 845–866.
7. M Ghimire et al (2020 May) 'A-SLIC: Acceleration of SLIC Superpixel Segmentation Algorithm in a Co-Design Framework'. *Advances in Intelligent Systems and Computing*. https://doi.org/10.1007/978-3-030-43020-7_90
8. F Gao, H Yoon, T Wu, and X Chu (2020, April) 'A Feature Transfer Enabled Multi-Task Deep Learning Model on Medical Imaging'. *Expert Systems with Applications*, Volume 143, p. 112957.
9. B Kaur, M Mittal, et al (2018) 'An Improved Salient Object Detection Algorithm Combining Background and Foreground Connectivity for Brain Image Analysis'. *Computers and Electrical Engineering*, volume 71, pp. 692–703.
10. S Dash, BR Acharya, M Mittal, and A Abraham (2020) *Deep Learning Techniques for Biomedical and Health Informatics*. A. Kelemen (Ed.). Springer Nature.
11. SH Shruthishree, and H Tiwari (2017 April) 'A Review Paper on Medical Image Processing'. *International Journal of Research Granthaalayah*, Volume 4, Issue 4. https://doi.org/10.5281/zenodo.572290
12. M Gandhi et al (2020) 'Preprocessing of Non-symmetrical Images for Edge Detection'. *Augmented Human Research*, Volume 5, Article number: 10.
13. RI Haque, and I Neubert (2020) 'Deep Learning Approaches to Biomedical Image Segmentation'. *Informatics in Medicine Unlocked*, Volume 18. https://doi.org/10.1016/j.imu.2020.100297
14. Tongxue Zhou, and Stéphane Canu SuRuan (2019 September–December) 'A Review: Deep Learning for Medical Image Segmentation Using Multi-Modality Fusion'. *Array*, Volumes 3–4, p. 100004.
15. K Kamnitsas et al (2017 February) 'Efficient Multi-Scale 3D CNN with Fully Connected CRF for Accurate Brain Lesion Segmentation' *Medical Image Analysis*, Volume 36, pp. 61–78.
16. G Haskins, U Kruger, and P Yan (2020) 'Deep Learning in Medical Image Registration: A Survey'. *Machine Vision and Applications*, Volume 31, Article number: 8.
17. P Pandey et al, (2019) 'Pragmatic Medical Image Analysis and Deep Learning: An Emerging Trend'. In O Verma, S Roy, S Pandey, and M Mittal (eds.), *Advancement of Machine Intelligence in Interactive Medical Image Analysis* (pp. 1–18). Springer.
18. Xiaohuan Cao, Jingfan Fan, Pei Dong, and Sahar Ahmad (2020) 'Image Registration Using Machine and Deep Learning'. *Handbook of Medical Image Computing and Computer Assisted Intervention* (pp. 319–342). Elsevier Science.
19. M Elad, and A Feuer. (1997) 'Restoration of a Single Super Resolution Image from Several Blurred, Noisy, and Undersampled Measured Images'. *IEEE Transactions on Image Processing*, Volume 6, Issue 12, pp. 1646–1658.

20. AH Mousa, ZN Aldeen, AH Mohammed, and MGK Abboosh (2021) 'A Convolutional Neural Network-Based Framework for Medical Images Analyzing in Enhancing Medical Diagnosis'. *Annals of Tropical Medicine and Public Health*, Volume 23, Issue 13, pp. 4–10. S Ct, and MRI Scans Convolutional Neural Networks with Intermediate Loss for 3D. 1–13.
21. RR Peeters, et al (2010) 'The Use of Super-Resolution Techniques to Reduce Slice Thickness in Functional MRI'. *International Journal of Imaging Systems and Technology*, Volume 14, Issue 3, pp. 131–138.
22. MI Razzak, S Naz, and A Zaib (2018) 'Deep Learning for Medical Image Processing: Overview'. In N Dey, A Ashour, and S Borra (eds.), *Classification in BioApps* (pp. 1–30). Springer.
23. AAA Setio, F Ciompi, G Litjens, P Gerke, and C Jacobs, et al (2016) 'Pulmonary Nodule Detection in CT Images: False Positive Reduction Using Multi-View Convolutional Networks'. *IEEE Transactions on Medical Imaging*, Volume 35, pp. 1160–1169.
24. Masni et al (2018) 'Skin Lesion Segmentation in Dermoscopy Images via Deep Full Resolution Convolutional Networks'. *Computer Methods and Programs in Biomedicine*, Volume 162, pp. 221–231.
25. Alantri et al (2018 August) Skin Lesion Segmentation in Dermoscopy Images via Deep Full Resolution Convolutional Networks *Computer Methods and Programs in Biomedicine*, Volume 162, pp. 221–231
26. K Doi (2007) 'Computer-Aided Diagnosis in Medical Imaging: Historical Review, Current Status and Future Potential'. *Computerized Medical Imaging and Graphics*, Volume 31, Issue 4–5, pp. 198–211. https: //doi.org/0.1016/j.compmedimag.2007.02.002
27. S Miller, BH Blott, and TK Hames (1992 September) 'Review of Neural Network Applications in Medical Imaging and Signal Processing'. *Medical and Biological Engineering and Computing*, volume 30, Issue 5, pp. 449–464. https://doi.org/10.1007/BF02457822
28. MP Siedband (1998) 'Medical Imaging Systems'. In John G Webster (ed.), *Medical Instrumentation: Application and Design* (pp. 518–576). U.S. Office of Technology Assessment and Forecast.
29. J Prince, and J Links (2006) 'Medical Imaging Signals and Systems'. *Med. Imaging*, pp. 315–379. https://doi.org/0132145189
30. N Sharma, and R Sharma (2022) 'Real-Time Monitoring of Physicochemical Parameters in Water Using Big Data and Smart IoT Sensors'. *Environment, Development and Sustainability*. https://doi.org/10.1007/s10668-022-02142-8
31. R Anandkumar, K Dinesh, Ahmed J Obaid, Praveen Malik, Rohit Sharma, Ankur Dumka, Rajesh Singh, and Satish Khatak (2022) 'Securing e-Health Application of Cloud Computing using Hyperchaotic Image Encryption Framework'. *Computers & Electrical Engineering*, Volume 100, p. 107860, ISSN 0045-7906, https://doi.org/10.1016/j.compeleceng.2022.107860.
32. R Sharma, Q Xin, P Siarry, and W-C Hong (2022) 'Guest Editorial: Deep Learning-Based Intelligent Communication Systems: Using Big Data Analytics'. *IET Communications*. https://doi.org/10.1049/cmu2.12374.
33. Rohit Sharma, and Rajeev Arya (2022) 'UAV Based Long Range Environment Monitoring System with Industry 5.0 Perspectives for Smart City Infrastructure'. *Computers & Industrial Engineering*, Volume 168, p. 108066, ISSN 0360-8352. https://doi.org/10.1016/j.cie.2022.108066

34. M Rai, T Maity, R Sharma et al (2022) 'Early Detection of Foot Ulceration in Type II Diabetic Patient Using Registration Method in Infrared Images and Descriptive Comparison with Deep Learning Methods'. *The Journal of Supercomputing.* https://doi.org/10.1007/s11227-022-04380-z
35. R Sharma, D Gupta, A Maseleno, and S-L Peng (2022) 'Introduction to the Special Issue on Big Data Analytics with Internet of Things-Oriented Infrastructures for Future Smart Cities'. *Expert Systems*, Volume 39, p. e12969. https://doi.org/10.1111/exsy.12969.
36. R Sharma, D Gavalas, and S-L Peng (2022) 'Smart and Future Applications of Internet of Multimedia Things (IoMT) Using Big Data Analytics'. *Sensors*, Volume 22, pp. 4146. https://doi.org/10.3390/s22114146
37. R Sharma, and R Arya (2022) 'Security Threats and Measures in the Internet of Things for Smart City Infrastructure: A State of Art'. *Transactions on Emerging Telecommunications Technologies* p. e4571. https://doi.org/10.1002/ett.4571
38. Jiangbo Zheng, Zhixin Wu, Rohit Sharma, and LV Haibin, (2022) 'Adaptive Decision Model of Product Team Organization Pattern for Extracting New Energy from Agricultural Waste'. *Sustainable Energy Technologies and Assessments*, Volume 53, Issue Part A, p. 102352, ISSN 2213-1388. https://doi.org/10.1016/j.seta.2022.102352
39. J Mou, K Gao, P Duan, J Li, A Garg, and R Sharma (2022) 'A Machine Learning Approach for Energy-Efficient Intelligent Transportation Scheduling Problem in a Real-World Dynamic Circumstances'. *IEEE Transactions on Intelligent Transportation Systems.* https://doi.org/10.1109/TITS.2022.3183215
40. E Criminisi, A Shotton, and J Konukoglu (2012) 'Decision Forests: A Unified Framework for Classification, Regression, Density Estimation, Manifold Learning and Semi-Supervised Learning'. *Found. Trends® Comput. Graph. Vision*, volume 7, Issue 2–3, pp. 81–227.
41. SP Singh, and S Urooj (2016 April) 'An Improved CAD System for Breast Cancer Diagnosis Based on Generalized Pseudo-Zernike Moment and Ada-DEWNN Classifier'. *Journal of Medical Systems*, Volume 40, Issue 4, p. 105. https://doi.org/10.1007/s10916-016-0454-0
42. J Boroumand, and M Fridrich (2018) 'Deep Learning for Detecting Processing History of Images'. *Electronic Imaging*, volume 7, pp. 1–9.
43. Szegedy et al (2015) 'Going Deeper with Convolutions'. In *Proceedings of the IEEE Computer Society Conference on Computer Vision and Pattern Recognition.* https://doi.org/10.1109/CVPR.2015.7298594
44. J Ker, L Wang, J Rao, and T Lim (2018) 'Deep Learning Applications in Medical Image Analysis'. *IEEE Access*, pp. 1–1. https://doi.org/10.1109/ACCESS.2017.2788044
45. J Burt (2018) 'Volumetric Quantification of Cardiovascular Structures from Medical Imaging'. Volume 9, 968, 257.
46. O Esteban et al (2019) 'fmriprep: A Robust Preprocessing Pipeline for fMRI Data—fmriprep Version Documentation'. *Nature Methods*, pp. 111–116. https://doi.org/10.1038/s41592-018-0235-4
47. Alansary et al (2016) 'Fast Fully Automatic Segmentation of the Human Placenta from Motion Corrupted MRI'. In *International Conference on Medical Image Computing and Computer-Assisted Intervention*, pp. 589–597. https://doi.org/10.1007/978-3-319-46723-8_68

48. A Rangarajan Yang, and S Ranka (2018 March) 'Visual Explanations From Deep 3D Convolutional Neural Networks for Alzheimer's Disease Classification'. arXiv Prepr. arXiv1803.02544. K Jones et al (2002 October) 'Spatial Normalization and Averaging of Diffusion Tensor MRI Data Sets'. *Neuroimage*, Volume 17, Issue 2, pp. 592–617. https://doi.org/10.1006/nimg.2002.1148.
49. S Chaplot, LM Patnaik, and NR Jagannathan (2006) 'Classification of Magnetic Resonance Brain Images Using Wavelets as Input to Support Vector Machine and Neural Network'. *Biomedical Signal Processing and Control*, Volume 1, pp. 86–92

Chapter 7

Time-frequency representations of one-dimensional signals using Wigner-Ville Distribution

Fatima Zahra Lamzouri and Boutaina Benhmimou
Mohammed Five University in Rabat (UM5R), Agdal, Rabat, Morocco

Niamat Hussain
Sejong University, Seoul, South Korea

Sandeep Kumar Arora
Lovely Professional University, India

Rachid Ahl Laamara
Mohammed Five University in Rabat (UM5R), Agdal, Morocco

Alexander Kogut
Institute for Radiophysics and Electronics of NAS of Ukraine, Kharkov, Ukraine

Baseem Khan
Hawassa University, Awasa, Ethiopia

Mohamed El Bakkali
Mohammed Five University in Rabat (UM5R), Agdal, Morocco

7.1 INTRODUCTION

Frequency analysis presented by Fourier transform shows two global and conjugate representations of signals. The first one is temporal and the second is of frequency type. Useful information can be carried out by broadcast frequencies and temporal structure of the signal [1]. These signals are non-stationary signals using the statistical properties and evolve as per the timing and existence of the signal. Note that there are many examples of non-stationary signals [2, 3]. For instance, music and speech are non-stationary signals subjected to a time-frequency analysis by our ear. This analysis is interpreted by our brain. In addition to that, elastic, acoustic and electromagnetic signals diffused in a dispersive medium exhibit properties that are dependent on frequency and are non-stationary. This is the case noticed in certain animal species, such as bats and dolphins, which use non-stationary acoustic signals to hunt, communicate or navigate. These signals are also used in communication, in electronic warfare via radars and sonars in order to observe the earth and its environment [4, 5].

Stationary signals are also generated by systems that are used every day in life or in industries via machines, converters and motors. Non-stationary signals appear during an incident or a change of regime and hence they can be used as a good solution in error diagnosis. Generally, we can conclude that non-stationary signals are the only source that provide information and hence the quantity of information transported by stationary signals is zero.

For processing non-stationary signals, we sought to represent, at the same time, the frequency and temporal information contained in the signal. These techniques are called time-frequency representations which aim to redistribute the information contained in the analyzed signal in such a way to facilitate its interpretation.

In this regard, several methods have been proposed to compute time-frequency representations of various performances and properties, and have the major advantage of highlighting the non-stationary behaviors of the signal. The first method proposed based on the Short Time Fourier Transform (also called Fourier Transform with Sliding Window) was the Gabor transform [6], introduced in 1945 by Gabor [7].

In the STFT method, the signal is implicitly considered to be as a succession of situations, which are locally stationary. This feature is achieved using a function $h(t)$ which is qualified by an analysis window. It consists of a sliding window that selects a part of the signal and then qualifies all other parts by translation along the time axis. Indeed, the instantaneous spectrum of the analyzed part is associated with the temporal position of the window during each displacement which is carried out with or without overlap. Therefore, the STFT is defined by Portnoff as the result of the successive multiplication of the time series $S(t)$ and a short window localized in time [8].

Moreover, the spectrogram is the most used and oldest time-frequency representation techniques that are developed based on the STFT. The idea behind its development is simple. Indeed, for extracting the helpful data from signal at a date t, we no longer take the global Fourier transform, but the FT considered around the same coordinate of time t, i.e. local FT. This locality is guaranteed by the use of a window centered on a coordinate of time t.

However, the accuracy of duration and frequency obtained by analyzing the spectrogram is explained through the Heisenberg-Gabor dubiety law, which establishes that both time and frequency cannot be specified [9]. Thus, the spectrogram defines a FT representation given that it distributes the signal energy on the FT plane. To deal with this issue, this chapter aims to use Wigner-Ville Distribution (WVD) for the time-frequency modeling of one-dimensional signals and then denoising their contents.

7.2 WVD-BASED METHODS

To overcome the drawbacks of classical representations like the spectrogram, distributions allowing a better time-frequency resolution have been proposed by Lackovic et al. [10]. The WVD has the ability to interpret the distribution of energy as a function of time and frequency. It was proposed for the first time by Eugene Wigner in 1932 for application in quantum mechanics [11], and it was introduced by J. Ville in 1948 in signal processing analysis [12]. Historically, a significant number of studies on WVD have been proposed in the field of signal processing [13–15].

Nowadays, WVD is used in various fields such as optics [16], acoustic radar [17], seismic [18], biology [19] and many other areas. Thus, it is the most popular and the most applied technique in signal processing. It has several properties and is easy to develop as compared with other approaches using FT [20]. The distribution of Wigner-Ville (WV) is defined in [12, 21, 22] by,

$$\mathrm{WV}_s(t,f) = \int_{-\infty}^{+\infty} s\left(t + \frac{\tau}{2}\right) s^*\left(t - \frac{\tau}{2}\right) e^{-i2\pi f\tau} d\tau \tag{7.1}$$

where* is the operation of conjugation complex. The WVD can also be defined according to the spectra $S(\nu)$ from signal $S(t)$ by:

$$\mathrm{WV}_s(t,\nu) = \int_{-\infty}^{+\infty} S\left(\nu + \frac{\eta}{2}\right) S^*\left(\nu - \frac{\eta}{2}\right) e^{i2\pi f\eta} d\eta \tag{7.2}$$

Unlike the spectrogram, WVD does not require the use of a conventional window for modeling the signal [23]. However, STFT with an adjusted window is applicable on the transported information. Hence, the energy of the analyzed data is conserved as well as the temporal and frequency supports. The expression of WVD can be interpreted as the FT of the kernel $Cs(\tau,t)$ given by:

$$s\left(t + \frac{\tau}{s}\right) s^*\left(t - \frac{\tau}{2}\right) \tag{7.3}$$

Indeed, we can apply the parallel with the achievement of spectra density by FT of the correlation function of stationary signals considering

$s\left(t+\dfrac{\tau}{s}\right)s^*\left(t-\dfrac{\tau}{2}\right)$ as an instant correlation function that depends on time for obtaining an instant spectral density by FT on the delay τ. WVD has the following properties:

- WVD is real and satisfied the conservation of energy and the temporal and frequency marginals.
- The covariance by the group of translations in time and frequency.
- WVD conserves the scalar product.
- WVD is not positive.
- WVD conserves the temporal and frequency support of the signal:

$$x(t)=0 \forall t \notin (t_1,t_2) \mapsto W_x(t,v)=0 \forall t \notin (t_1,t_2) \tag{7.4}$$

$$|X(v)|=0 \forall v \notin (v_1,v_2) \mapsto W_x(t,v)=0 \forall v \notin (v_1,v_2) \tag{7.5}$$

- WVD saves moments of the signal:

$$\int_{-\infty}^{+\infty}\int_{-\infty}^{+\infty} t^n W_x(t,v)dtdv = \int_{-\infty}^{+\infty} t^n |z(t)|^2 dt \tag{7.6}$$

$$\int_{-\infty}^{+\infty}\int_{-\infty}^{+\infty} v^n W_x(t,v)dtdv = \int_{-\infty}^{+\infty} v^n |Z(v)|^2 dv \tag{7.7}$$

- WVD allows accessing the instant frequency $v_x(t)$ of a signal $x(t)$ using first-order moments of its time-frequency distribution,

$$v_x(t) = \dfrac{1}{2\pi}\dfrac{d}{dt}\arg x(t) = \dfrac{\int_{-\infty}^{+\infty} v W_x(t,v)dv}{\int_{-\infty}^{+\infty} W_x(t,v)dv} \tag{7.8}$$

- WVD achieves the ability to locate linear modulated signals of perfect manner, i.e. monochromatic signals and chirps:
- On monochromatic signals,

$$x(t) = \exp(2j\pi v_0 t) \tag{7.9}$$

where $X(v) = \delta(v-v_0) \mapsto W_w(t,v) = \delta(v-v_0) \tag{7.10}$

- On the chirps,

$$x(t) = \mathrm{esp}(-j\pi\alpha t^2) \to W_x(t,v) = \delta(t-\alpha v) \tag{7.11}$$

7.3 DISCRETIZATION OF WIGNER-VILLE DISTRIBUTION

The discreet version of WVD is defined by:

$$WV_s(t,f) = 2\sum_{\tau} s_N(t+\tau) s_N^*(t-\tau) e^{-4\pi f \tau} \qquad (7.12)$$

where S_N is the obtained expression after discretization of the signal using sampling with a period taken as a unit. Nevertheless, this formula generates the problem of a possible spectral aliasing. Indeed, under these conditions, the function WVs appears as periodical function having period 1/2 in f, knowing that this period equals 1 for the spectrum of the sampled signal. This leads to a disrespect of the Shannon-Nyquist criterion. Therefore, aliasing for frequencies above 1/4 [24] is the best choice to solve this problem by applying the following approaches:

- Over sample the signal by a factor greater than or equal to 2.
- Construct the discrete of Wigner-Ville Distribution on the analytical form of raw data sampled normally.

Note that for the second point, aliasing of the positive frequencies will only affect the negative frequencies for which the spectral contribution is zero since the analytical signal presents only positive frequencies. We can say therefore that the results provided by WVD are excellent for one-dimensional signals. However, in the case of multidimensional signals, terms of interference have been appeared in the time-frequency scale. Thence, the analysis becomes complicated or even impossible in some cases. Note that, real signals are generally multidimensional. To illustrate this, assume that $S_1(t)$ and $S_2(t)$ are two components of a signal $S(t)$. The WVD is then given by:

$$WV_{s_1+s_2}(t,f) = WV_{s_1}(t,f) + WV_{s_2}(t,f) + 2\mathcal{R}\left(WV_{s_1,s_2}(t,f)\right) \qquad (7.13)$$

with

$$WV_{s_1,s_2}(t,f) = \int \left(s_1\left(t+\frac{\tau}{2}\right) s_2^*\left(t-\frac{\tau}{2}\right) + s_1^*\left(t-\frac{\tau}{2}\right) s_2\left(t+\frac{\tau}{2}\right) \right) e^{-2i\pi f \tau} d\tau \qquad (7.14)$$

According to [25], we can say that interferential structures are placed halfway between s_1 and s_2 and vary along the axis of time or frequencies. Therefore, the contribution of interferences is summarized by the rule of midpoint which describes that two points of an interferential plane can be used to describe the contribution in a third point that is their geometrical middle. Nevertheless, in practice we introduce the analytical signal $S_A(t)$ associated with signal $s(t)$ for applying the WVD [26].

Note that the concept of the analytical structure of a real one-dimensional was introduced by Gabor in 1946 [27]. Indeed, an analytical signal is a signal of which the FT is null for negative frequencies and hence we say that its FT is causal. We have:

$$S_A(t) = s(t) + jH(s(t)) \tag{7.15}$$

$H(s(t))$ introduces the HT (Hilbert transform) of $s(t)$ [28] and is defined as follows:

$$H(s(t)) = \frac{1}{\pi} v_p \int_{-\infty}^{+\infty} \frac{s(\tau)}{t-\tau} f\tau \tag{7.16}$$

where vp is the main value of Cauchy.

The use of analytical signal permits to delete the negative frequencies and hence minimizes effects on interferences. It is good to note that one-dimensional signals can be affected by interferences. Interference terms are also eliminated by another expression of variant of SPWVD (Smoothed Pseudo Wigner-Ville Distribution) that leads to achieve better resolution and an additional elimination of noises.

7.4 SPWVD-BASED ANALYSIS: 1D SIGNALS

As it is proved in details previously, the structure of WVD introduces terms of interference between components of the signal. Therefore, it is often necessary to use a smoothed version of the WVD. For instance, we can remove some parts of the remaining interference using a time window $h(t)$, with real values, to isolate the non-simultaneous frequency components [29]. Thus, we obtain the Pseudo Wigner-Ville Distribution (PWVD) [30] that has the following expression when it is associated with an analytical signal $S_A(t)$ in place of the signal $S(t)$ itself:

$$PDWV_{S_A}(t,f) = h(t)_{*f} WV_{S_A}(t,f) \tag{7.17}$$

The application of this smoothing is not sufficient because terms of interference are always obtained between components of positive frequency that limit the readability of time-frequency representation. Therefore, it makes sense to add another frequency smoothing allowing the minimization of their amplitude. It consists of the Smoothed Wigner-Ville Pseudo Distribution (SPWVD) given by equation (7.18). In general, all the time-frequency transforms can be described by this equation as it is demonstrated in [31].

$$PDWVL_{S_A}(t,f) = \gamma(t,f)_{*f*f} WV_{S_A}(t,f) \tag{7.18}$$

where $\gamma(t,f)$ is a kernel of constant smoothing.

In order to have a satisfactory result, the analysis window must be adapted to the type of signal before applying the analysis process. However, it is clear to note that this method has the drawback of enhancing the effective time-frequency traces of the signal due to temporal and frequency smoothing. Furthermore, it should be noted that it is necessary to play several times on the control parameters of windows to obtain a sufficient result.

7.5 EXPERIMENTS AND RESULTS

We consider the signal $S_1(t)$ presented previously. The application of Wigner-Ville Distribution makes it possible to obtain presentation of the time-frequency plane shown in Figure 7.1. The analysis of these results confirms that the WVD is perfectly suited to study linear chirp signals, i.e. signals linearly modulated in frequency. However, we already notice the appearance of interference. For better clarification of this phenomena of interferences, let us consider the signal described previously $S(t)$. We obtain the time-frequency plane proposed in Figure 7.2. Finally, we have applied the SPWVD on the signals studied previously to prove its contribution in proposed analysis. Figure 7.3 presents the time-frequency plane obtained in using the signal $S_1(t)$ and Figure 7.4 depicts the obtained time-frequency plane of both chirps. The analysis of Figure 7.4 proves the effectiveness of SPWVD distribution for suppressing all generated interferences from the signal.

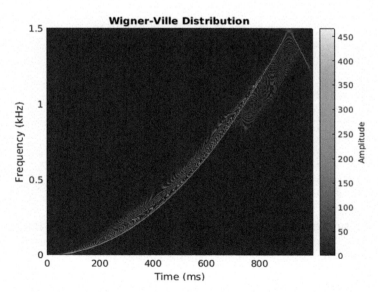

Figure 7.1 WVD-based time-frequency scale of a linear chirp signal.

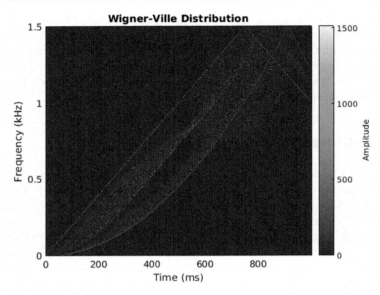

Figure 7.2 WVD-based time-frequency scale of a signal having two linear chirps.

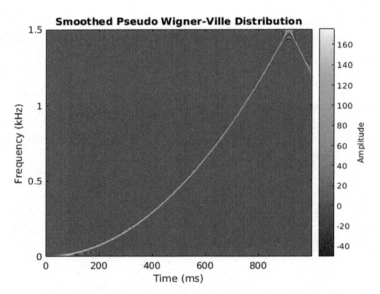

Figure 7.3 SPWVD-based time-frequency plane of one linear signal chirps.

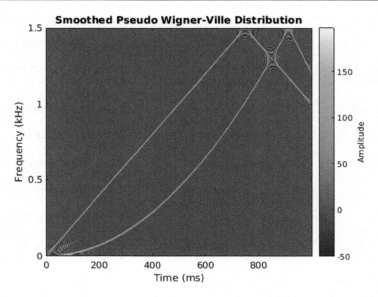

Figure 7.4 SPWVD-based time-frequency plane of a signal having two linear chirps.

7.6 RESULT SYNTHESIS AND DISCUSSION

The overall analysis of experimental results shows that each method has its advantages and drawbacks in time-frequency modeling of signals. Moreover, the calculated duration of these techniques differs according to the method complexity. In this study, the calculation time was evaluated using an HP laptop with Intel® Core™ i5-7300HQ Processor and CPU specification of a frequency of 2.50 GHz.

In this regard, the spectrogram as the simplest method to implement, the calculation time is of the order of 0.012 sec while the Wigner-Ville Distribution, which is characterized by better resolution has a computation time higher than 0.12 sec for each case.

From another hand, Ganzalo Galiano and Julien Velasco have studied performances of non-local filters to denoise 1D signals via the analysis of their spectrogram obtained using an efficient time-frequency modeling [32]. Indeed, the denoising of 1D signals presents the best approach as it is generally applied using theoretical analysis of filters in the time or frequency scales. However, filters modeled in the joint time-frequency scale generally allow improving the filtering process taken into consideration that the processing speed has not any effects. For instance, the electrocardiogram (ECG), analysis of biomedical signals [33], human voice [34] and animal sounds [35] presents good models for the discussed approach. Regarding the last example, works presented in [36–38] have been carried out based on the use

Table 7.1 Brief comparison and synthesis of proposed time-frequency representations

Method of analysis	Benefits	Disadvantages
Classical Fourier Transform	✓ Algorithm is simple. ✓ Short time of calculation.	✗ Limited frequency resolution. ✗ Not suitable for non-stationary signals.
Spectrogram	✓ Algorithm is simple. ✓ Short time of calculation. ✓ Use for non-stationary signals.	✗ Limited time-frequency resolution. ✗ Window of analysis has fixed width.
Wigner-Ville Distribution	✓ Best time-frequency resolution. ✓ Use for non-stationary signals. ✓ Window of analysis has fixed width.	✗ Time of calculation is increased. ✗ Interpretation of spectra is difficult. ✗ Terms of interference are important.
Smoothed Pseudo Wigner-Ville Distribution	✓ Best time-frequency resolution. ✓ Use for non-stationary signals. ✓ Window of analysis has variable width. ✓ Disappearance of interferences.	✗ High time of calculation. ✗ Interpretation of spectra is difficult.

of time-frequency distributions to estimate the number of wolves howling in a recording. This estimate achieves economic and ecological interests. It lies in counting the number of existing wolves since the references must offset process of all livestock killed by these protected species [39].

Note that beyond the technology of recording instruments, registrations are commonly combined with very a huge quantity of unwanted signals like wind, bells, car's engines and so on. Obviously, the superimposition of all these data leads to production of an unstructured noise during the howls recording of wolves that needs to be processed for a good processing of the main signal.

For medical signals that are provided by instruments of the measurement hole, the acquired data is overlapped with high noise using the same resonant frequencies. Thus, some suitable denoising windows are called to separate pure data and noises. In general, the denoising basis doesn't lead to recover a pure data while it allows obtaining good time-frequency modeling of the helpful energy and hence makes some new processing approaches to obtain an instantaneous frequency estimation (IF) of the main signal. For instance, the estimation of the pure information overlapped in a signal can be applied for identifying an episode of arrhythmia.

Consequently, our proposed contribution proves the effectiveness of developed techniques for the processing of one-dimensional signals of biomedical images through their spatial-frequency representations. For this, we have decided the use of Wigner-Ville Distribution.

7.7 CONCLUSION

In this chapter, we have applied the main methods of time-frequency representation on one-dimensional signals. We have presented for each method, the continuous version and the discrete version as well as its computer implementation. Through this study, we were able to identify the advantages and disadvantages of each proposed method. Thus, we realized that the WVD is very adequate for the time-frequency modeling and then the denoising of one-dimensional signals.

With this in mind, we plan for our next work to extend the Wigner-Ville Distribution on two-dimensional signals in order to have applications on biomedical images.

REFERENCES

1. I. H. Ahmed, A. Djebbari, A. Kachenoura, et al. Telemedical transport layer security based platform for cardiac arrhythmia classification using quadratic time–frequency analysis of HRV signal. *The Journal of Supercomputing*, 2022, vol. 78, no 11, pp. 13680–13709.
2. N. A. Khan, Iterative adaptive directional time–frequency distribution for both mono-sensor and multi-sensor recordings. *Signal, Image and Video Processing*, 2022, vol. 17, pp. 1–8.
3. L. Cohen, Time-frequency analysis: what we know and what we don't. *Landscapes of Time-Frequency Analysis: ATFA*, 2019, 2020, pp. 75–101.
4. N. Saulig, J. Lerga, Ž. Milanovic, et al. Extraction of useful information content from noisy signals based on structural affinity of clustered TFDs' coefficients. *IEEE Transactions on Signal Processing*, 2019, vol. 67, no 12, pp. 3154–3167.
5. I. Dongo, D. A. G. Jáuregui, and N. Couture, A study on the simultaneous consideration of two modalities for the recognition of sound painting gestures. In *Proceedings of the 31st Conference on l'Interaction Homme-Machine*, 2019, December, pp. 1–12.
6. J. P. Zhu, H. Q. Chen, and W. B. Ye, Classification of human activities based on radar signals using 1D-CNN and LSTM. In: *2020 IEEE International Symposium on Circuits and Systems (ISCAS)*. IEEE, 2020. pp. 1–5.
7. S. J. Rashid, A. I. Abdullah, and M. A. Shihab, Face recognition system based on gabor wavelets transform, principal component analysis and support vector machine. *International Journal on Advanced Science, Engineering and Information Technology*, 2020, vol. 10, pp. 959.
8. A. Uncini, Digital Audio Effects. In: *Digital Audio Processing Fundamentals*. Cham: Springer International Publishing, 2023, pp. 483–563.
9. H.-T. Wu, Current state of nonlinear-type time–frequency analysis and applications to high-frequency biomedical signals. *Current Opinion in Systems Biology*, 2020, vol. 23, pp. 8–21.
10. A. V. Lackovic, J. Lerga, and M. Tomic, Local Shannon, Rényi, and Tsallis Entropy for Useful Content Extraction from Choi-Williams and Zhao-Atlas-Marks Time-Frequency Distributions. In: *2022 International Conference*

on *Electrical, Computer, Communications and Mechatronics Engineering (ICECCME)*. IEEE, 2022. pp. 1–5.
11. E. Wigner, On the quantum correction for thermodynamic equilibrium, *Physical Review USA* (1932), 40, pp. 749–759.
12. J. Ville, Théorie et applications de la notion de signal analytique, *Câbles et Transmission*, Fr. (1948), 1, pp. 61–74.
13. M. Brajovic, L. Stankovic, and M. Dakovic, Decomposition of multichannel multi-component non stationary signals by combining the eigenvectors of autocorrelation matrix using genetic algorithm. *Digital Signal Processing*, 2020, vol. 102, p. 102738.
14. M. Varanis, J.P.C.V. Norenberg, R. T. Rocha, et al. A comparison of time-frequency methods for nonlinear dynamics and chaos analysis in an energy harvesting model. *Brazilian Journal of Physics*, 2020, vol. 50, pp. 235–244.
15. M. Bhat, Y. Dar, H. Aamir, Convolution and correlation theorems for Wigner–Ville distribution associated with the quaternion offset linear canonical transform: Convolution and correlation theorems for WVD associated with the QOLCT. *Signal, Image and Video Processing*, 2022, vol. 16, no 5, pp. 1235–1242.
16. Y. Liu, L. Wu, C. Xu, et al. Propagation and Wigner distribution of the Airy–Gauss beam through an apertured paraxial optical system. *Optics Communications*, 2020, vol. 454, p. 124494.
17. M. Zakharia, B. Dubus, and P. Plantevin, *Time-frequency characterization in support of broadband sonar transducer design*, 2020.
18. M. Kreme and B. Torresani, Étude d'un algorithme d'optimisation pour le fading temps-fréquence. In: *GRETSI 2022-XXVIIIème colloque francophone de traitement du signal et des images*, 2022.
19. X. U. Tianji, B. Cheng, N. S. Chen, et al. A microscopic ancient river channel identification method based on maximum entropy principle and Wigner-Ville Distribution and its application. *Petroleum Exploration and Development*, 2021, vol. 48, no 6, pp. 1354–1366.
20. D. Wei, and Y. Shen, New two-dimensional Wigner distribution and ambiguity function associated with the two-dimensional non-separable linear canonical transform. *Circuits, Systems, and Signal Processing*, 2022, vol. 41, pp. 77–101.
21. A. Beddiar, *Etude et élaboration d'algorithme de débruitage de signaux: Application à la reconnaissance de la parole*. 2020. Thèse de doctorat. Univ M'sila.
22. D. B. Fedosenkov, A. A. Simikova, S. M. Kulakov, et al. Cohen's class time-frequency distributions for measurement signals as a means of monitoring technological processes. *Steel in Translation*, 2019, vol. 49, pp. 252–256.
23. M. Robin, F. Jenot, M. Ouaftouh, et al. Influence of the laser source position on the generation of Rayleigh modes in a layer–substrate structure with varying degrees of adhesion. *Ultrasonics*, 2020, vol. 102, p. 106051.
24. P. E. Golubtsov and I. I. Morozov, Automatic detection and suppression of parasitic chirp signals in the intermediate time–frequency domain. *Circuits, Systems, and Signal Processing*, 2020, vol. 39, no 6, pp. 3035–3045.
25. D. Benouioua, D. Candusso, F. Harel, et al. Diagnosis of fuel cells using instantaneous frequencies and envelopes extracted from stack voltage signal. *International Journal of Hydrogen Energy*, 2022, vol. 47, no 16, pp. 9706–9718.
26. R. Mirzaeian, P. Ghaderyan, Gray-level co-occurrence matrix of Smooth Pseudo Wigner-Ville distribution for cognitive workload estimation. *Biocybernetics and Biomedical Engineering*, 2023, vol. 43, pp. 261–278.

27. L. Galleani, The time-frequency interference terms of the Green's Function for the harmonic oscillator. *Analysis of Pseudo-Differential Operators*, 2019, pp. 215–228. Cham: Springer.
28. F. A. Shah and A. A. Teali, Scaling Wigner distribution in the framework of linear canonical transform. *Circuits, Systems, and Signal Processing*, 2022, vol. 42, pp. 1–25.
29. N. J. Sairamya, M. S. P. Subathra, E. S. Suviseshamuthu, et al. A new approach for automatic detection of focal EEG signals using wavelet packet decomposition and quad binary pattern method. *Biomedical Signal Processing and Control*, 2021, vol. 63, p. 102096.
30. N. Sharma, R. Sharma, Real-time monitoring of physicochemical parameters in water using big data and smart IoT sensors. *Environment, Development and Sustainability*, 2022. https://doi.org/10.1007/s10668-022-02142-8
31. R. Anandkumar, K. Dinesh, A. J. Obaid, P. Malik, R. Sharma, A. Dumka, R. Singh, S. Khatak, Securing e-Health application of cloud computing using hyperchaotic image encryption framework, *Computers & Electrical Engineering*, vol. 100, 2022, 107860, ISSN 0045-7906, https://doi.org/10.1016/j.compeleceng.2022.107860
32. R. Sharma, Q. Xin, P. Siarry, W.-C. Hong, Guest editorial: Deep learning-based intelligent communication systems: Using big data analytics. *IET Communications*, 2022, https://doi.org/10.1049/cmu2.12374
33. R. Sharma, R. Arya, UAV based long range environment monitoring system with Industry 5.0 perspectives for smart city infrastructure, *Computers & Industrial Engineering*, 2022, vol. 168, 108066, ISSN 0360-8352, https://doi.org/10.1016/j.cie.2022.108066
34. M. Rai, T. Maity, R. Sharma, et al. Early detection of foot ulceration in type II diabetic patient using registration method in infrared images and descriptive comparison with deep learning methods. *The Journal of Supercomputing*, 2022. https://doi.org/10.1007/s11227-022-04380-z
35. R. Sharma, D. Gupta, A. Maseleno, S.-L. Peng, Introduction to the special issue on big data analytics with internet of things-oriented infrastructures for future smart cities. *Expert Systems*, 2022, vol. 39, p. e12969. https://doi.org/10.1111/exsy.12969
36. R. Sharma, D. Gavalas, S.-L. Peng, Smart and future applications of Internet of Multimedia Things (IoMT) using Big Data analytics. *Sensors*, 2022, vol. 22, p. 4146. https://doi.org/10.3390/s22114146
37. R. Sharma, R. Arya, Security threats and measures in the Internet of Things for smart city infrastructure: A state of art. *Transactions on Emerging Telecommunications Technologies* 2022, p. e4571. doi:10.1002/ett.4571
38. J. Zheng, Z. Wu, R. Sharma, L.V. Haibin, Adaptive decision model of product team organization pattern for extracting new energy from agricultural waste, *Sustainable Energy Technologies and Assessments*, 2022, vol. 53, Part A, p. 102352, ISSN 2213-1388, https://doi.org/10.1016/j.seta.2022.102352
39. J. Mou, K. Gao, P. Duan, J. Li, A. Garg, R. Sharma, A machine learning approach for energy-efficient intelligent transportation scheduling problem in a real-world dynamic circumstances, *IEEE Transactions on Intelligent Transportation Systems*, 2022, https://doi.org/10.1109/TITS.2022.3183215

Chapter 8

An efficient 3U CubeSat downlink transmission based on an S-band lightweight CPW-fed slot antenna

Boutaina Benhmimou
Mohammed Five University in Rabat (UM5R), Agdal, Morocco

Fatima Zahra Lamzouri and Niamat Hussain
Sejong University, Seoul, South Korea

Nancy Gupta
Punjab Technical University, Jalandhar, India

Rachid Ahl Laamara
Mohammed Five University in Rabat (UM5R), Agdal, Morocco

Sandeep Kumar Arora
Lovely Professional University, India

Josep M. Guerrero
Aalborg University, Aalborg East, Denmark

Mohamed El Bakkali
Mohammed Five University in Rabat (UM5R), Agdal, Morocco

8.1 INTRODUCTION

Many people of emerging countries continued to place a strong emphasis on near term social programs up until the year 2000, particularly those relating to proper healthcare, effective education, economic stability, fundamental security, pure water, proper nutrition, etc. Consequently, due to the lack of necessary resources, funding in space technologies and satellite engineering in these countries is unsuitable. Although economic prosperity is largely dependent on knowledge properties, whose development calls for scientific skill and application of new and developed technology. In this case, CubeSats provide a notable highway to the investigation of innovative approaches and scientific environmental elements beyond our planet. Compared to traditional satellites, these spacecrafts are affordable to develop and construct [1]. Thus, CubeSats can provide an economic and political impressive role in the development of new nation's intellectual capitals [2].

These satellite platforms are currently placed into orbit via a number of private, commercial, and public space launches at a lower price than the space services needed by other satellite families [3–5]. Notable commercial activity has resulted from CubeSats, including new launch services which many of them began as educational spinoffs [6]. The contribution of both engineers with training experience on these new platforms and public availability to those emerging satellite shapes aims to the success of CubeSat projects [7, 8]. Since then, many academic organizations around the world have been interested in the low cost of CubeSats applications and technological innovations that are adequate by their mass, dimensions and transmission power [9–11]. For instance, the SSETI program (i.e., Student Space Exploration and Technology initiative) and annual European CubeSat Workshop are funded by ESA (i.e., European Space Agency), allowing students to learn about CubeSat development and exchange traineeships. Various CubeSat systems are currently under operation with the National Science Foundation and NASA.

Additionally, the fast development of miniature technologies in fields such as fluidics, materials, telecommunications, electronics and sensors has increased interest in CubeSat engineering in academic institutions and government-related business worldwide [12]. As a result of technological advancements, capable automated instruments are now available to use with CubeSats and ultra-small satellites to conduct telemetry and a variety of research projects in space [13, 14]. So as to allow the overall design and the density at which all CubeSat equipment are integrated in a CubeSat system, and also some techniques of security and arrangement for complying with prerequisites of procedure in outer space more adaptable, COTS (commercial off-the-shelf) tools, as an example, can be used without adjustment for production and continuously developing CubeSat elements [14]. Although the components and instruments of a CubeSat should fulfill the criteria of mission recommendations that are established by consumers, satellite launchers as well as climate in outer space. The most essential CubeSat element is its communication system, which delivers measured data to ground stations and receives consumer telecommands [15].

On the other hand, it is worth mentioning that a CubeSat task determines the physical and electrical properties of its communication system and vice versa. For example, vast numbers of kilobytes can be used to transfer an image from space to ground stations, while only a few bytes are needed for each data point when measuring temperatures, magnetic fields, and light intensities. Therefore, a CubeSat's communication system uses higher than 50% of its electrical energy total when sending data. In addition, when CubeSat is in the straight direction with earth segments, transmission from CubeSat to earth stations only occurs for a few minutes per day. Communications with CubeSats are more difficult due to the earth stations' fast moving motion. Additionally, the majority of CubeSat projects are only able to provide one or two small-antenna earth stations (e.g. 3-m) and thus their insufficient power can limit downlink rates. As a result, CubeSat data uplinking and downlinking are still limited and more difficult [16, 17]. Moreover, as shown in Figure 8.1, all CubeSats are 10 cm × 10 cm in size and have structures that are 1 U (10 cm), 2 U (20 cm), and 3 U (30 cm). Consequently, antennas for CubeSat should have high gain, fit into target configuration and perform well at an operating frequency for high data rate and long-distance communications [18].

Planar antennas are the best option to use on CubeSats to solve these issues [23–26]. They can be used for wireless transmission and are compact, lightweight, affordable, and easily integrated with the other CubeSat instruments [27]. This research work develops a small-size CPW-SA design fed by a 50 Ω coplanar waveguide line for CubeSat communication at unlicensed ranges. By placing the antenna on a 3-U CubeSat's body, our method for this

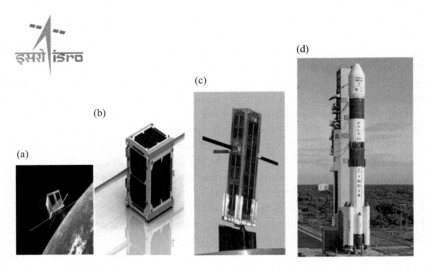

Figure 8.1 Indian CubeSats launched into orbit by PSLV of ISRO, India. (a) StudSat CubeSat (1U, 1 kg), 2010 [19], (b) Swayam-1 CubeSat (1U, 1 kg), 2016 [20], (c) Jungu CubeSat (3U, 3 kg), 2011 [21], (d) ISRO's PSLV C18 on launch pad [22].

antenna design serves gain enhancement at 2450 MHz. The back-lobe radiation pattern is redirected via a tiny square portion of the satellite carcass which raises the antenna peak gain at 2450 MHz. To function as a metal reflector, this CubeSat component made of metal is positioned 13.1 mm below the antenna dielectric. The designed antenna achieved higher peak gain at an unlicensed band (i.e., 9.0 dB at 2450 MHz) and resides the lowest volume against metamaterials such as the approach described in [28]. It is also suitable for all CubeSat configurations.

To deal with these challenges, planar antennas are the best choice for use on CubeSat [23–26]. They are low profile, lightweight, low cost, are easily integrated with the other CubeSat's instruments, and can be used for long-distance communications [27]. This letter proposes a small-size slot antenna fed by a 50 Ω coplanar waveguide line for S-band CubeSat communications. In this antenna design, our approach serves gain enhancement at 2450 MHz by optimizing antenna placement on the body of a 3-U CubeSat. A tiny square portion of the satellite carcass is suited to re-direct the backed energy and then improve performances of developed design around 2450 MHz. This is selected below the FR4 back face at an air-manhole of 13.1 mm to operate as a metal reflector. As against the implementation of metasurface that given in [28], the constructed antenna has the lowest volume, can be taken for operation on any CubeSat form, and leads to obtain the superior peak gain at S-band, i.e., 9.0 dB at 2450 MHz.

8.2 ANTENNA STRUCTURE

In consideration of 3U CubeSat physical and mechanical requirements, a small-size and low-cost CPW-SA is developed for operation around an ISM resonant frequency of 2450 MHz [29]. The dielectric material is made of FR4-epoxy material with permittivity of 4.4, loss tangent of 0.02 and small thickness of 1.6 mm. Note that the FR4-epoxy material presents wide availability in the market and allows to achieve good performances at lowers frequencies [30]. Configuration of the proposed CPW-SA system integrated with a 3U CubeSat is depicted in Figure 8.2. It is found that it holds very limited space on the satellite body providing much area for other circuits, i.e., only 8.22% of the CubeSat top face.

As it is mentioned through Figure 8.3, the optimized CPW-SA design has a physical size of W × L = 60.4 × 60.4 mm^2 and small dimensions given by L_1 = 7.35 mm, L_2 = 27.18 mm, Lg = 12.08 mm, L_{r0} = 13.59 mm, L_{r1} = 18.12 mm, Ls = 10.57 mm, Wr = 3.02 mm, and W_f = 4.53 mm. The radiating portion of configured model is powered through a 50 Ω CPW strip sinew which is printed on the FR4 dielectric at a horizontal air-manhole of g = 0.37 mm from both sides of the antenna ground plane.

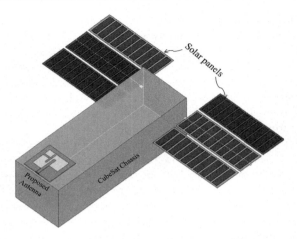

Figure 8.2 3D configuration of proposed CubeSat: 3U CubeSat + CPW-SA.

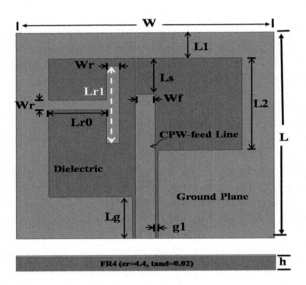

Figure 8.3 Geometry of proposed CPW-SA configuration.

It is worth noting that the Quasi-Newtonian Technique (QNT) that is an HFSS package is used for optimizing all antenna parameters at 2450 GHz. The employed QNT-based algorithm focuses on the multiplication of all dimensions obtained by H. Zhang et al. in [31] by a small factor k of 1.51 for enforcing an effective operation around 2450 MHz with the achievement of high gain and width −10-dB BW. The antenna dimensions are, therefore, analyzed by increasing the k factor for 80 iterations from 1 to 3 with a minimum step size of 0.025 mm and a maximum step size of 0.05 mm.

It is also found that an air-manhole of 13.1 mm between the FR4 back side and the CubeSat top face leads to obtain good electromagnetic coupling at the same operating frequency. This air-manhole is optimized using similar QNT-based algorithm. Consequently, the satellite's top face will act as a metallic reflector and hence increase the antenna performances at 2450 MHz with suitability for operation in the outer environment.

8.3 RESULTS AND DISCUSSION

Figure 8.4 illustrates the |S11| parameter of proposed configurations, i.e., CPW-SA the 3U satellite. We notice that both structures give low reflection coefficients around the targeted resonant frequency, i.e., 2450 MHz. The first one gives an RL of 39.11 dB and the second presents an RL of almost 24 dB that is very much satisfied for CubeSats.

Additionally, the whole CubeSat system gives –10 dB BW of 470 MHz with an input impedance of Z_{in} = 56.58-j2.36 Ω at 2450 MHz. This means that low power is being forwarded to the feeding system allowing to the radiation of maximum power around the satellite chassis that is made in aluminum and acts as a reflector as mentioned above. The proposed antenna

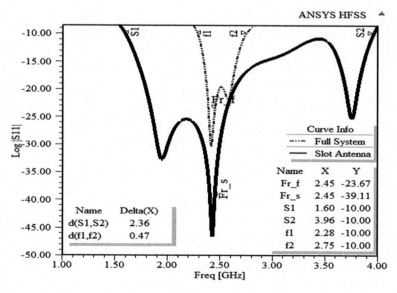

Figure 8.4 |S$_{11}$| factor at S-band: CPW-SA and 3U CubeSat.

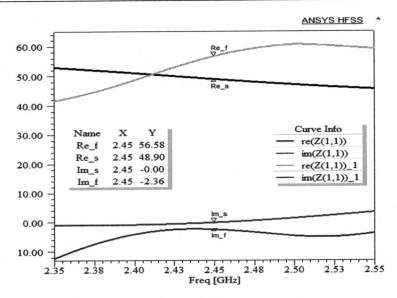

Figure 8.5 Input impedance at S-band: CPW-SA and 3U CubeSat.

positioning leads, therefore, to achieve high RL (i.e., Low |S11| parameter), wide impedance bandwidth, and Z_{in} close to 50Ω which permit to ensure transmission links between the developed 3U CubeSats and earth segments at S-band (Figure 8.5).

Radiation patterns and 3D gain plots are also carried out at S-band (i.e., 2450 MHz) for both configurations. Figure 8.6 shows E and H fields in xy and xz planes. It proves how the satellite top face re-directs the backed energy of CPW-SA alone in order to obtain unidirectional pattern and then eliminate interferences between elements.

We show that very high quantity of electromagnetic energy has been lost by the CPW-SA design alone in forms of back lobes, see Figure 8.6a. This means that the antenna peak gains will be decreased with very high interferences with other instruments. To deal with this problem, antenna positioning on the satellite to pace is optimized in order to remove back lobes and then lead to the achievement of gain enhancement at 2450 MHz, refer Figure 8.7.

Consequently, the achieved results prove that the CubeSat top face allows overcoming issues of interferences and low gain of proposed CPW-SA without the need of additional excitation power and physical dimensions. Unidirectional radiation pattern and gain enhancements are obtained at the same resonant frequency of 2450 MHz. These characteristics satisfy results all requirements of CubeSat missions at S-band for long-distance communication.

An efficient 3U CubeSat downlink transmission 115

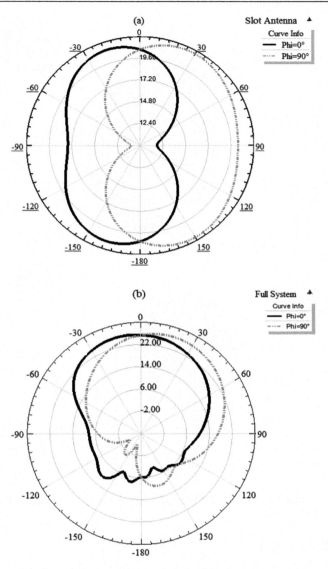

Figure 8.6 E and H fields at 2450 MHz: CPW-SA alone and the proposed full system. (a) 2D RP at 2450 MHz [slot antenna], (b) 2D RP at 2450 MHz [full system].

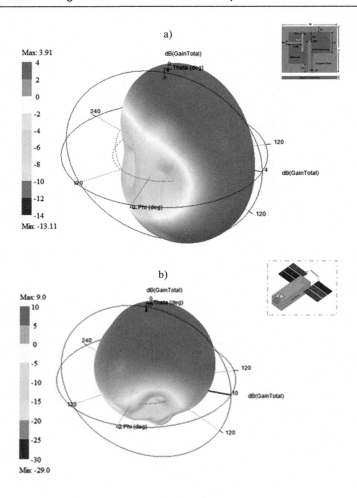

Figure 8.7 (a) Total gain at 2450 MHz [slot antenna], (b) total gain at 2450 MHz [full system]

8.4 CONCLUSION

The development of a new 2450 MHz small CPW-SA for operation on 3U CubeSats is presented in this research work. The proposed FR4-based CPW-SA system is lightweight, has physical area and has stiffness which make it very suitable for all CubeSat standards at discounted price, energy saving and without the need of deployment systems. These benefits are obtained using only a tiny portion of the satellite carcass below the FR4 back surface at an air-manhole of 13.1 mm. It is found that using the CubeSat chassis itself as a reflector leads to obtain a unidirectional radiation pattern with wide HPBW, good peak gain of 9.0 dBi, and wide-band of 470 MHz around 2450 MHz. In addition to that, the presented 2D and 3D results prove that the detailed antenna approach gives a |S11| coefficient well below −20 dB and excellent impedance harmonizing at 2450 MHz with input impedance close to 50 Ω. These achievements show how the optimization of very simple slot antennas leads to make the outer space technology available for almost everyone using CubeSats.

REFERENCES

[1] R. Sandau: Status and trends of small satellite missions for Earth observation, *Acta Astronautica*, vol. 66, 2010, pp. 1–12.

[2] M. Rivera, A. Boyle: Space for all: Small, cheap satellites may one day do your bidding, *NBC NEWS, Innovation*, Jul 14 2013, 5:11 am ET.

[3] N. Crisp, K. Smith, P. Hollingsworth: Small satellite launch to LEO: A review of current and future launch systems, Transactions of the Japan Society for Aeronautical and Space Sciences, *Aerospace Technology Japan*, 12, ists 29, 2014, doi: 10.2322/tastj.12.Tf39.

[4] A. Valenzuela, R. Sandau, H.-P. Roeser: *Small Satellite Missions for Earth Observation: New Developments and Trends*, Springer Science & Business Media, 2010.

[5] Starlink Mission, SpaceX [Online available]. n.d. https://www.spacex.com/webcast

[6] M. N. Sweeting: Modern Small Satellites – Changing the Economics of Space, *Proceedings of the IEEE*, vol. 106, no. 3, pp. 343–361, 2018.

[7] J. P. Suari, C. Turner, R. J. Twiggs: CubeSat: the development and launch support infrastructure for eighteen different satellite customers on one launch, *Proceedings of the 15th Annual AIAA/Utah State University Conference on Small Satellites*, pp. 1–5, 2001.

[8] J. P. Suari, C. Turner, W. Ahlgren: Development of the standard CubeSat Deployer and a CubeSat class picosatellite, *Proceedings of the IEEE Aerospace Conference*, vol. 1, pp. 1347–1353, March 2001.

[9] M. E. Bakkali, Planar Antennas with Parasitic Elements and Metasurface Superstrate Structure for 3U CubeSats, PhD. Thesis, July 2020, Sidi Mohamed Ben Abdellah University, city of Fez, Morocco.

[10] S. Wu, W. Chen, Y. Zhang, W. Baan, T. An: *SULFRO: A Swarm of Nano-/Micro-Satellite at SE L2 for Space Ultra-Low Frequency Radio Observatory*, the AIAA/USU Conference on Small Satellites, Logan, UT, Aug. 2014.

[11] A. R. Riise, B. Samuelsen, N. Sokolova, H. Cederblad, J. Fasseland, C. Nordin, J. Otterstad, K. Fauske, O. Eriksen, F. Indergaar, K. Svartveit, P. Furebotten, E. Saether, E. Eide: *N cube: The First Norwegian Student Satellite*, the 17th AIAA/USU Conference on Small Satellites, Logan, Utah, 2003.

[12] W. A. Shiroma, L. K. Martin, J. M. Akagi, J. T. Akagi, B. L. Wolfe, B. A. Fewell: CubeSats: a bright future for nanosatellites, *Central European Journal of Engineering*, vol. 1, pp. 9–15, 2011

[13] G. Lastovicka-Medin: Nano/pico/femto-satellites: Review of challenges in space education and science integration towards disruptive technology. pp. 357–362, June 2016, doi: 10.1109/MECO.2016.7525781.

[14] J. Tristancho, J. Gutierrez-Cabello: A Probe of Concept for FEMTO-SATELLITES based on Commercial-Of-The-Shelf, *Digital Avionics Systems Conference (DASC), IEEE/AIAA 30th*, Seattle, Washington, pp. 8A2-1–8A2-9, 16–20 October 2011.

[15] Y. Rahmat-Samii, V. Manohar, J. M. Kovitz: For satellites, think small, dream big: a review of recent antenna developments for CubeSats, *IEEE Antennas and Propagation Magazine*, vol. 59, no. 2, pp. 22–30, 2017.

[16] M. E. Bakkali, G. S. Gaba, F. Tubbal, L. Kansal, N. A. El Idrissi: Analysis and Optimization of a Very Compact MPA with Parasitic Elements for Inter-swarm of CubeSats Communications. In: Luhach, A., Jat, D., Hawari, K., Gao, X. Z., Lingras, P. (eds) *Advanced Informatics for Computing Research. ICAICR 2019. Communications in Computer and Information Science*, vol. 1076, 2019, Springer, Singapore.

[17] A. H. Lokman, J. P. Soh, S. N. Azemi, H. Lago, S. K. Podilchak, S. Chalermwisutkul, M. J. Jamlos, A. A. Al-Hadi, P. Akkaraekthalin, S. Gao: A review of antennas for picosatellite applications. *International Journal of Antennas and Propagation*, vol. 2017, Article ID 4940656, 2017.

[18] N. Chahat: A mighty antenna from a tiny cubesat grows, *IEEE Spectrum*, vol. 55, no. 2, pp. 32–37, 2018.

[19] Student Satellite (STUDSAT), department of Space, Indian Space Agency (ISRO) [online available]: https://www.isro.gov.in/Spacecraft/studsat-1

[20] Amator radio CubeSat Swayam, department of Space, Indian Space Agency (ISRO) [online available]: https://www.isro.gov.in/Spacecraft/swayam

[21] Jungu CubeSat, Indian Institute of Technology Kanpur, India [online available]: http://www.iitk.ac.in/meold/jugnu/index.htm

[22] Polar Satellite Launch Vehicle, in its twentieth flight (PSLV-C18), department of Space, Indian Space Agency (ISRO) [online available]: https://www.isro.gov.in/launcher/pslv-c18-megha-tropiques

[23] F. Omari, N. Hussain, B. Benhmimou, N. Gupta, R. A. Laamara, Y. I. Abdulkarim, M. E. Bakkali: Only-Metal Ultra-Small Circular Slot Antenna for 3U CubeSats, *13th International Conference on Computing, Communication and Technologies (13th ICCCNT)*, October 3–5, 2022, pp. 1–6.

[24] M. El Bakkali, G. S. Gaba, F. E. M. Tubbal, N. A. El Idrissi: High Gain Patch Antenna Array with parasitic elements for CubeSat Applications; *the 1st International Conference on Antenna and Propagation (InCAP)*, Hyderabad, India, December 2018; pp. 1–5.

[25] M. El Bakkali, N. A. El Idrissi, F. E. M. Tubbal, G. S. Gaba: Optimum design of a triband MPA with parasitic elements for CubeSat communications using Genetic Algorithm; *the 6th International Conference on Wireless Networks and Mobile Communications (WINCOM)*, Marrakesh, Morocco, October 2018; pp. 1–4.

[26] F. Nashad, S. Foti, D. Smith, M. Elsdon, O. Yurduseven: Development of transparent patch antenna integrated with solar cells for Ku-band satellite applications, *2016 Loughborough Antenna and Propagation Conference (LAPC)*, Loughborough, UK, 15 November. 2016.

[27] B. Benhmimou, N. Hussain, N. Gupta, R. A. Laamara, J. M. Guerrero, M. E. Bakkali, F. Arpanaei, M. Alibakhshikenari: Miniaturized Transparent Slot Antenna for 1U and 2U CubeSats: CRTS Space Missions, *13th International*

Conference on Computing, Communication and Technologies (13th ICCCNT), October 3–5, 2022, pp. 1–6.
[28] M. El Bakkali, F. Tubbal, G. S. Gaba, L. Kansal, N. A. El Idrissi: S-Band CPW-Fed Slot Antenna with 2D Metamaterials for CubeSat Communications, *Communications in Computer and Information Science*, vol. 1076, pp. 344–356, September 2019, DOI:10.1007/978-981-15-0111-1_31.
[29] ANSYS HFSS simulator [online] n.d. http://www.ansys.com/products/electronics/ansys
[30] M. E. Bakkali, M. E. Bekkali, G. S. Gaba, J. M. Guerrero, L. Kansal, M. Masud: *Fully Integrated High Gain S-Band Triangular Slot Antenna for CubeSat Communications, Electronics*, 2021.
[31] H. Zhang, T. Zhong, Y. Zeng: Dual- and Wide-Band Dual-Sense Circularly Polarized CPW-Fed Slot Antenna with a T-shaped Ground Stub, *2019 International Conference on Electromagnetics in Advanced Applications (ICEAA)*, Granada, Spain, 2019, pp. 0232–0236. doi:10.1109/ICEAA.2019.8879052

Chapter 9

WGMs diffractive emission for Mm-wave all-round antennas with Internet of Things

Alexander Kogut
Institute for Radiophysics and Electronics of NAS of Ukraine, Kharkov, Ukraine

Niamat Hussain
Sejong University, Seoul, South Korea

Igor Kuzmichev
Institute for Radiophysics and Electronics of NAS of Ukraine, Kharkov, Ukraine

Boutaina Benhmimou, Rachid Ahl Laamara and Fatima Zahra Lamzouri
Mohammed Five University in Rabat (UM5R), Agdal, Morocco

Sandeep Kumar Arora
Lovely Professional University, India

Mohamed El Bakkali
Mohammed Five University in Rabat (UM5R), Agdal, Morocco

9.1 INTRODUCTION

Generally, WGMs got their title through the study of acoustic waves in terminated-loop halls [1–4]. These waves propagate over a curved surface separating both elements of a transmission channel and come to it via thin grazing nooks. Hence, the average of wave reflection, given by the well-known Fresnel equation, tends to be unity. This means that sound vibrations can be propagated over deep distances with negligible attenuation. These recommendations that prove the ability of building transmission links with very low attenuation are targeted for the first time by Rayleigh in excess of a century ago [1, 2]. Due to this, whispering gallery modes are widespread in the field of electromagnetism. WGMs allow achieving high Q-factors at lightweight, small size and low cost as compared with classical resonators.

The field of these resonant modes can be calculated on a compact surface of the shaped DR.

Currently, WGMs make developers of optical resonators to obtain an anomalously high Q-factor [5–9] that cannot be targeted using properties of strip lines [10]. For instance, dielectric resonators with a curvilinear reflective surface can be employed to obtain high Q-factor all-round antenna systems at low-cost, lightweight and negligible attenuation. It can be a disk, a ring or a ball. Note that similar configurations of WGM-based DRs target low permittivity dielectrics ($\varepsilon' < 10$) [11, 12]. Henceforth, low permittivity substrates allow obtaining better radiating characteristics around a central frequency. For instance, Teflon substrate with $\varepsilon' = 2.08$ is commonly employed to manufacture WGM-based DRs. In addition to this, the resonator diameter must be higher than the wavelength inside the dielectric according to the volume occupied by the electromagnetic field of WGMs near the interface of two media. In this regard, such resonators are similar to open quasi-optical resonators formed by two metal reflecting mirrors. Thus, WGM-based DRs are often categorized as quasi-optical [12]. The resonance effect takes place when an integer number of waves fit along the curvilinear generatrix of a DR. The smallness of WGM loss is also a critical requirement for achieving resonance in the dielectric structures. These losses are summed up by the lack in the DR and radiative loss at radiating a part of the stored energy into free space. Since electromagnetic waves fall on the dielectric–free space interface at small glancing angles, radiative loss in such DRs is extremely low. Despite their open nature, this feature of WGMs mainly determines their little attenuation in dielectric resonant structures. These losses are summed up by dielectric loss (in the material of the resonator) and radiative loss at radiating a part of the stored energy into free space. As mentioned above, since electromagnetic waves fell on the dielectric–free space interface at small glancing angles, radiative loss in such DRs is extremely low.

Currently, DRs with WGM modes are used in a wide range of electromagnetic spectrum: from centimeter waves to optical wavelengths. In addition, filters [13, 14], oscillators [15, 16], power combiners [17], and devices for studying the electrical properties of various materials [18, 19] are developed. In addition to that, the electromagnetic nature of WGMs in axisymmetric dielectric structures (a necessary condition for providing the WGM resonant responses) creates a prerequisite for creating emitting systems with a circular view. Thus, the field's allocation of resonant WGMs in disk-based DRs is determined by the equidistant arrangement of antinodes in the full sector of azimuth angles 0°–360°. The large number of such antinodes (more than 10 as a rule) is determined by the high-order WGMs. This allows assuming possibility of forming a multi-lobe radiation pattern, when each antinode of the field forms a separate lobe. However, as represented above, the radiation of WGM modes in homogeneous axisymmetric dielectric structures is extremely weak [20], because of small glancing angles of azimuthal waves. For this reason, most of the electromagnetic energy is stored in dielectric disks. Therefore, in order to create conditions for forced emission of WGM energy, the transition to the model of dielectric structures with a non-uniform operating surface is promising. Obviously, the diffraction loss of resonant WGMs on artificially created inhomogeneities is the main reason for such emission. To create effective diffraction emission, the dimensions of such inhomogeneities should be less (or comparable) than the operating wavelength and not violate the conditions of forming the resonant fields of operating WGMs. Therefore, the objectives of this work include creating a DR model with a radiating curvilinear surface in the full sector of azimuthal angles, justification of the physical principles of the proposed concept, studying the diffraction radiation of resonant fields of WGMs and considering the possible prototypes of all-round emitting antennas. At the same time, non-uniform DRs with operating WGMs can be regarded as a new leaky-wave antenna type under certain conditions [21–24]. Henceforth, the main features of this approach aim to produce an effective analysis of diffractive radiation of resonant WGM fields from a disk-shaped DR at the disturbance of its curved surface uniformity.

9.2 OBJECTIVES AND METHODS

It is necessary to pay attention that one of most features of WGM fields in a disk DR consists of the antinodes of E-field. Figure 9.1b shows that the antinodes of E-field (bright areas in the form of petals) are equidistantly situated over the DR arched surface with a high value of E-field intensity. This proves that to provide an efficient radiation of WGM fields, it is better to use a new diffraction grooved surface, as it is developed in this research,

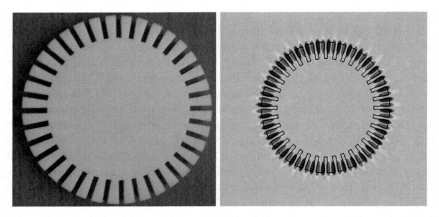

Figure 9.1 DR: (a) segmented curved surface, (b) distribution of the operating WGM fields.

work in place of a homogeneous cylindrical resonator surface for studying WGM E-fields at Ka-band. Such a diffraction grating is possible to be configured using radial cuts partitioning the area of a disk-shaped DR made in Teflon equidistantly in the azimuth direction, see Figure 9.1a. Note that the parasitic part described via this approach is configured through thin air gaps which are adjusted according to the developed DR orientation in parallelism with antinodes of the WGM E-fields.

The air gap dimensions are optimized considering that the antinodes E-field with opposite phase (0 and π) are preferred to be placed on the chunks limits (i.e., red and blue speckles) for enhancing the dispersion and thus transmission capability, see Figure 9.1b. Note that the azimuthal modal index n of WGM describes the amount of achieved air gaps. Additionally, mathematical analysis given in [7] allows finding of the WGM modes of the proposed disk-shaped all-round antenna system based on the implementation of segmented DRs. On the other hand, packages of electromagnetic simulators like CST Studio suite, ANSYS HFSS or Comsol Multiphysics easily manage the same task around a central frequency. Therefore, developed disk-shaped DR has a diameter of 8 cm and a thickness of about 7.2 mm, see Figure 9.1a. The proposed DR configuration is optimized using *CST Studio* Suite. The resulting chunks are configured as a trapezoid structure having a width of almost $d_s \approx 5.5$ mm on the edge of the segmented DR. In addition to that, the thickness of each air gap is about $d_g \approx 1.5$ mm. Note that, the inner caustic decreases WGM fields in the disk-shaped DR. Moreover, they should be placed very close to the radial profoundness of air gaps in the designed disk-shaped all-round antenna. This approach uses a radial profoundness of 8 mm for better operation at mm-wave bands using Teflon material.

Additionally, a thin coupling slot has been employed for powering the developed WGM-based DR. It is, therefore, demonstrated that translating the segmented DR along a flat conducting mirror made of copper simples the excitation of proposed all-round antenna. The coupling slot occupies a physical area of 7.2 × 0.5 mm² and is oriented parallelly with the segment faces in order to provide powerful feeding of TE-modes in the disk-shaped DR with WGMs, which are aimed to be parallel with that of the feeding slot.

9.3 RESULTS AND DISCUSSIONS

Figure 9.2 shows the calculated |S11| parameter of the proposed WGM-based all-round antenna using a frequency range of 35.7–38 GHz. Moreover, the calculated Q-factor at a resonant frequency of $f \approx 36$ GHz with the corresponding −3dB bandwidth of about 120 MHz can be discussed by much ability for providing an efficient coupling to the resonator and then leads

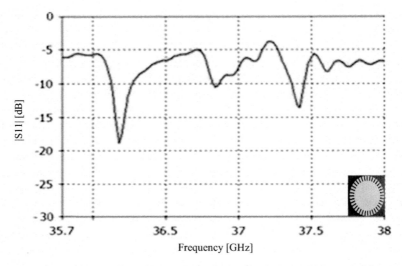

Figure 9.2 $|S_{11}|$ coefficient-configured WGM-based all-round antenna.

to have a clean radiation pattern of well-isolated lobes. Figure 9.1b summarizes the E-field distribution of the discussed mode.

Additionally, other modes are viewed above 36 GHz. For instance, new mode is obtained around 37.4 GHz despite it being wider and having less coupling. For this mode, the harmonization accuracy among positions of the E-field antinodes and the DR's segments in the disk's border is very low as compared with the first mode. As a result, a baneful intervention comes across the standing vibrations of segments at higher resonant responses. This is why there aren't any resonances above 38 GHz.

Therefore, a 2D radiation pattern of developed all-round antenna is carried out in terms of relative E-field intensity E_0/E_{max} along azimuthal angles φ of the receiving horn (red curve) at a resonant point of 36.08 GHz as shown in Figure 9.3. It has multi-lobe shape and occupies the entire orientations at 0°–360°. It is also found that some adjacent responses are added up forming new high HPBW lobes. From another hand, it's proved that each air gap leads to obtain two radiating lobes, according to the counting of antinodes generated by E-field/gap. These achievements can be discussed via the diffraction of azimuthal WGM waves on each face of all chunks.

We observe that an all-round antenna with 36 segments gives 72 radiating lobes formed through the constructive interference between emitting vibrations from edges of adjacent chunks. Each lobe occupies −3 dB width of about 3.5°. Note that tiny defects in the cutting of air gaps may lead to obtain extra side lobes detected during the far-field emission around the same resonant frequency.

Figure 9.4 shows measurements of antenna gain G in dB and dBi at 36.08 GHz considering the feeding of developed disk-shaped all-round antenna

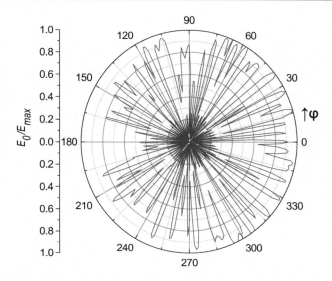

Figure 9.3 Radiation patterns of the developed segmented DR-based all-round antenna at mm-wave band, i.e., 36 GHz.

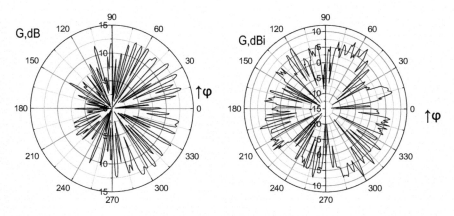

Figure 9.4 Measured 2D gain plots of configured WGM-based disk-shaped DR: excitation by one and two coupling slots at $f = 36.08$ GHz.

using one and two coupling slots. It is observed that the use of two symmetrical slots easily allows obtaining a clean radiation pattern with good gains.

These high values of achieved gains per lobes are also demonstrated via the measurement results around the same resonant response, see Figure 9.5. It is found that the developed WGM-based all-round antenna design provides about 12 dB at mm-wave frequencies. Thence, the peak gain (in dB and dBi) of developed all-round antenna is measured as the function of

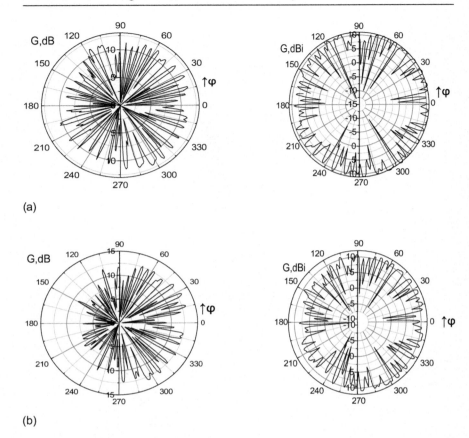

Figure 9.5 Gain plots of configured all-round antenna vs. an azimuthal coordinate φ. (a) at 36.05 GHz (b) at 36.11 GHz.

azimuthal range φ at frequencies scaling up and down by 30 MHz from both sides of 36.08 GHz, see Figure 9.5. We can easily conclude that a small reduction of the all-round antenna central frequency allows achieving a huge "shadow" field in the emission pattern due to violation of the radiation circumstances at the basis of configured WGM fields.

Figure 9.6 Antenna gain of configured DR-based all-round antenna design on an azimutal coordinate φ: 36.40 GHz.

Results and recommendations carried out through the presented approach show how the combination between disk-shaped DR with properties of WGMs leads to obtain effective solutions for all-round emitting antenna systems at mm-wave frequencies. It is found that using periodic non-uniformities on the curved surfaces of a disk-shaped DR with WGMs creates an efficient all-around antenna system for use at low cost and lightweight at mm-wave bands.

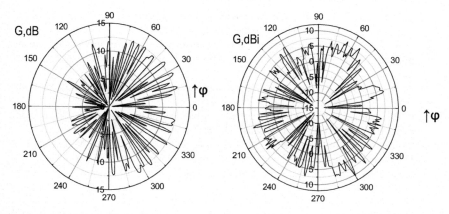

Figure 9.6 summarizes the total gain plots of constructed WGM-based DR antenna (in dB and dBi) according to the azimuthal coordinate φ at frequencies shifted by 320 MHz from both sides of the central resonant response, i.e., 36.08 GHz. It can be easily concluded that radiation patterns are messy over the entire orientation given by 0°–360° due to the missing of E-field lobes along the shaped DR-based all-round antenna at frequencies outside the limits of achieved effective bands.

9.4 CONCLUSION

Consequently, the developed approach proves that combining properties of segmented dielectric resonators having a selective frequency response and excited via the use of an efficient feeding system leads to obtain large emission pattern all-round antennas. The investigated system allows finding a multi-beam shaped radiating blueprint with 72 units over an azimuthal section given as 0°–360° around the targeted resonant frequency with a width of about 3.5° at −3 dB levels. In addition to that, the peak gain of proposed all-round mm-wave antenna gives a high gain of $G = 12$ dBi per lobe. This means that the radiating performances of configured DR with WGMs are retained around the same resonant frequency. It is about 120 MHz around an operating response of 36 GHz.

Finally, we can prove with simulation and measurements that investigating segmented DR with WGMs allows achieving an effective diffraction-type antenna under certain specifications.

REFERENCES

1. A. Y. Kogut, I.K. Kuzmichev, R.S. Dolia, S. O. Nosatiuk, Millimeter-wave technique based on whispering gallery modes dielectric resonators: new solutions and applications, *Telecommunications and Radio Engineering*, vol. 81, no. 2, pp. 1–12, 2022.

2. J. W. Strutt, B. Rayleigh, The problem of the whispering gallery, *Philosophical Magazine*, vol. 20, pp. 1001–1004, 1910.
3. G. Lin, Y. K. Chembo, Opto-acoustic phenomena in whispering gallery mode resonators, *International Journal of Optomechatronics*, vol. 10, no. 1, pp. 32, Jan. 2016.
4. M. Ye, M. Shen, X. Lin, Ringing phenomenon based whispering-gallery-mode sensing, Sci. Rep., vol. 6, pp. 1–7, Jan 2016.
5. C. Vedrenne, J. Arnaud, Whispering-Gallery modes in dielectric resonators, Proc. Inst. Elect. Eng., vol. 129, no. 4, pp. 183–187, 1982.
6. S. Preu, H. G. L. Schwefel, S. Malzer, G. H. Döhler, L. J. Wang, M. Hanson, J. D. Zimmerman, and A. C. Gossard, Coupled whispering gallery mode resonators in the terahertz frequency range, *Optics Express*, vol. 16, pp. 7336–7343, 2008.
7. G. Annino, M. Cassettari, M. Martinelli, Study on planar whispering gallery dielectric resonators. I. General properties, *Journal of Infrared, Millimeter, and Terahertz Waves*, vol. 23, pp. 597–615, Mar 2002.
8. G. Annino, M. Cassettari, M. Martinelli, Study on planar whispering gallery dielectric resonators. II. A multiple-band device, *Journal of Infrared, Millimeter, and Terahertz Waves*, vol. 23, pp. 617–634, Mar 2002.
9. J.R. Wait, Electromagnetic whispering gallery modes in a dielectric rod, *Radio Science*, vol. 2, no. 9, pp. 1005–1017, Sept 1967.
10. D. Dansana, R. Kumar, A. Parida, R. Sharma, J. D. Adhikari et al., Using susceptible-exposed-infectious-recovered model to forecast coronavirus outbreak, *Computers, Materials & Continua*, vol. 67, no. 2, pp. 1595–1612, 2021.
11. M. T. Vo, A. H. Vo, T. Nguyen, R. Sharma and T. Le, Dealing with the class imbalance problem in the detection of fake job descriptions, *Computers, Materials & Continua*, vol. 68, no. 1, pp. 521–535, 2021.
12. Smriti Sachan, Rohit Sharma, Amit Sehgal, Energy efficient scheme for better connectivity in sustainable mobile wireless sensor networks, *Sustainable Computing: Informatics and Systems*, Volume 30, 2021, 100504.
13. Ghanem, S., Kanungo, P., Panda, G. et al. Lane detection under artificial colored light in tunnels and on highways: an IoT-based framework for smart city infrastructure. *Complex & Intelligent Systems* (2021). https://doi.org/10.1007/s40747-021-00381-2
14. Sachan, S., Sharma, R. & Sehgal, A. SINR based energy optimization schemes for 5G vehicular sensor networks. *Wireless Personal Communications* (2021). https://doi.org/10.1007/s11277-021-08561-6
15. Priyadarshini, I., Mohanty, P., Kumar, R. et al. A study on the sentiments and psychology of twitter users during COVID-19 lockdown period. *Multimedia Tools and Applications* (2021). https://doi.org/10.1007/s11042-021-11004-w
16. Azad, C., Bhushan, B., Sharma, R. et al. Prediction model using SMOTE, genetic algorithm and decision tree (PMSGD) for classification of diabetes mellitus. *Multimedia Systems* (2021). https://doi.org/10.1007/s00530-021-00817-2
17. Priyadarshini, I., Kumar, R., Tuan, L.M. et al. A new enhanced cyber security framework for medical cyber physical systems. *SICS Software-Intensive Cyber-Physical Systems*. (2021). https://doi.org/10.1007/s00450-021-00427-3
18. J. A. Krupka, M. E. Tobar, J. G. Hartnett, J-M. L. Floch, Extremely high-Q factor dielectric resonators for millimeter-wave applications, *IEEE Transactions on Microwave Theory and Techniques*, vol. 53, no. 2, pp. 702–712, 2005.

19. A.E. Kogut, M. T. Islam, E. A. Kogut, Z. E. Eremenko, R. S. Dolia, High loss liquid dielectric characterization: Comparison of microwave waveguide and resonator measurement techniques, *International Journal of Microwave and Wireless Technologies*, vol. 12, no. 10, pp. 892–894, 2020.
20. A. Kogut, G. Annino, M.E. Bakkali, R.A. Laamara, S.K. Arora, D.S.B. Naik, F. Maniraguha, Millimeter wave all-around antenna based on whispering gallery mode dielectric resonator for IoT-based applications, *Wireless Communications and Mobile Computing*, vol. 2022, Article ID 5877263, pp. 1–10, 2022. https://doi.org/10.1155/2022/5877263
21. S. Xiao, B.Z. Wang, X.S. Yang, et al., A novel reconfiguration CPW leaky-wave antenna for millimeter-wave application. *International Journal of Infrared and Millimeter Waves* 23, pp. 1637–1648; 2002. https://doi.org/10.1023/A:1020705903066
22. S. Xu, F. Zheng, Multimode network analysis for dielectric leaky-wave antennas consisting of multilayer periodic structure with arbitrary dielectric distributions. *The Journal of Infrared, Millimeter, and Terahertz Waves* 18, pp. 1223–1240; 1997. https://doi.org/10.1007/BF02678229
23. A.O. Salman, The millimeter wave radiation of a dielectric leaky-wave antenna coupled with a diffraction grating: Broad-face interaction. *Journal of Infrared, Millimeter, and Terahertz Waves* 31, 196, 2010. https://doi.org/10.1007/s10762-009-9581-8
24. M. Huang, S. Xu, and Y. Pan, Investigation on a novel leaky wave antenna with double radiation beam composed of left-handed slab loaded hybrid waveguide using planar technology. *Journal of Infrared, Millimeter, and Terahertz Waves* 30, pp. 117–127, 2009. https://doi.org/10.1007/s10762-008-9429-7

Chapter 10

Apache Hadoop framework for big data analytics using AI

Urvashi Gupta and Rohit Sharma
SRM Institute of Science and Technology, Delhi-NCR Campus, Modinagar, India

10.1 INTRODUCTION

The exponential growth in the volume, variety, and velocity of data generated by modern applications, sensors, and devices has led to the emergence of the big data era. As a result, organizations are increasingly looking for ways to store, process, and analyze large and complex datasets to gain valuable insights and inform decision-making [1–3]. The Apache Hadoop framework has become a popular choice for organizations looking to build infrastructure for big data analytics using AI. Hadoop is a distributed computing platform that is designed to handle large and complex datasets in a scalable and reliable manner. At its core, Hadoop includes a distributed file system, HDFS, which enables data to be stored across a cluster of commodity hardware, and a programming model, Map Reduce, that allows for the efficient processing of large datasets in parallel across multiple nodes in the cluster [4–7].

One of the key benefits of using Hadoop for AI applications is its ability to handle massive amounts of data. This is particularly important for machine learning and deep learning algorithms that require large datasets to be trained effectively. In addition, Hadoop provides a range of tools and technologies that can be useful for AI workloads, such as Apache Spark, which provides a powerful data processing engine for batch processing, streaming, ML, and graph processing [8]. However, organizations considering using Hadoop for AI should be aware of the challenges involved. Hadoop can be complex to set up and manage, and organizations may need to invest in specialized skills and expertise to get the most out of the platform. In addition, Hadoop is just one part of a larger AI ecosystem, and organizations will need to consider how Hadoop fits in with other tools and technologies they are using for AI [9, 10].

10.2 LITERATURE REVIEW

Several studies have investigated the use of the Apache Hadoop framework for big data analytics. The authors proposed a big data analytics framework for smart agriculture based on Apache Hadoop and machine learning. The authors used the framework to predict crop yields and optimize irrigation in smart agriculture. The framework integrated Hadoop with machine learning algorithms to process and analyze large volumes of agricultural data. The results showed that the proposed framework outperformed other existing methods in terms of accuracy and computational efficiency [11]. The authors presented a big data analytics framework based on Apache Hadoop and machine learning for healthcare data. The authors used the framework to analyze electronic health records (EHRs) and predict patient readmissions. The framework was designed to handle the variety, velocity, and volume of EHR data. The results showed that the proposed framework achieved high accuracy in predicting patient readmissions, which could help healthcare providers to improve patient outcomes and reduce healthcare costs [12].

The authors proposed a big data analytics framework for smart cities using Apache Hadoop and machine learning. The authors used the framework to analyze urban data and predict air pollution levels. The framework integrated Hadoop with machine learning algorithms to process and analyze large volumes of urban data. The results showed that the proposed framework achieved high accuracy in predicting air pollution levels, which could help policymakers to improve air quality in cities [13]. The authors presented an efficient big data analytics framework based on Apache Hadoop and machine learning for credit risk analysis. The authors used the framework to analyze credit data and predict credit risk. The framework integrated Hadoop with machine learning algorithms to process and analyze large volumes of credit data. The results showed that the proposed framework achieved high accuracy in predicting credit risk, which could help financial institutions to improve their credit risk management strategies [14].

The authors proposed a big data analytics framework based on Apache Hadoop and machine learning for predicting customer churn. The authors used the framework to analyze customer data and predict customer churn. The framework integrated Hadoop with machine learning algorithms to process and analyze large volumes of customer data. The results showed that the proposed framework achieved high accuracy in predicting customer churn, which could help businesses to improve customer retention and increase profitability [15]. The authors presented a big data analytics framework based on Apache Hadoop and machine learning for oil reservoir prediction. The authors used the framework to analyze geological and reservoir data and predict oil reservoir production. The framework integrated Hadoop with machine learning algorithms to process and analyze large volumes of geological and reservoir data. The results showed that the proposed framework achieved high accuracy in predicting oil reservoir production, which could help oil companies to improve their production strategies and increase profitability [16].

The authors proposed a big data analytics framework based on Apache Hadoop and machine learning for enterprise risk management. The authors used the framework to analyze enterprise data and predict enterprise risk. The framework integrated Hadoop with machine learning algorithms to process and analyze large volumes of enterprise data. The results showed that the proposed framework achieved high accuracy in predicting enterprise risk, which could help businesses to improve their risk management strategies and reduce losses [17]. The authors presented a big data analytics framework based on Apache Hadoop and machine learning for intrusion detection. The authors used the framework to analyze network data and detect network intrusions. The framework integrated Hadoop with machine learning algorithms to process and analyze large volumes of network data. The results showed that the proposed framework achieved high accuracy in detecting network intrusions, which could help businesses to improve their network security and protect against cyberattacks [18].

The authors proposed a big data analytics framework based on Apache Hadoop and machine learning for anomaly detection in IoT sensor data. The authors used the framework to analyze sensor data and detect anomalies in real-time. The framework integrated Hadoop with machine learning algorithms to process and analyze large volumes of sensor data. The results showed that the proposed framework achieved high accuracy in detecting anomalies, which could help improve the reliability and security of IoT systems [19]. The authors proposed a big data analytics framework for predicting student performance based on Apache Hadoop and machine learning. The study aimed to identify the significant factors that affect student performance and develop a predictive model to forecast student performance. The authors collected data from 2,300 students and used several machine learning algorithms to build a predictive model. The results showed that the proposed framework was effective in predicting student performance, achieving an accuracy of 86%. The study highlighted the importance of big data analytics in education and provided insights into the significant factors that influence student performance [20].

10.3 APACHE HADOOP FRAMEWORK

Apache Hadoop is a suite of tools and applications for analyzing and storing large amounts of data. Structured data is simpler to examine and store because it has a standard format. Patient demographics, diagnostic and procedure codes, prescription codes, and other data from the EHR are often produced in a consistent, standardized format. Hadoop facilitates the cleansing, refinement, and transformation of large data volumes for analytics use cases [21].

To handle massive medical datasets, the Apache Hadoop framework [22] is the processing platform built on the concept of parallelism. Hadoop,

MapReduce, Pig, Sqoop, Hive, HBase Avro, and other open-source frameworks can all handle the huge, large datasets generated on a regular basis in the healthcare industry. Apache Hadoop can be used to support distributed processing and storage of electronic medical records. It is a popular implementation of the MapReduce framework which can be set up into a cloud environment for large-scale datasets using distributed computing [23]. It is important to note that both the cloud platform AWS with AWS EMR and GCP with Google Cloud Dataproc can use Apache Hadoop framework for their processing of big data components which are used in the studied architecture.

Unstructured data in the healthcare industry includes emails, audio recordings, videos, text documents, and social media postings, to name a few examples. Unstructured data, as a consequence, is unclear and cannot be analyzed in the same way that structured data can. Because many EHRs enable free text input for clinical notes and other narrative data collection areas, healthcare organizations face a unique difficulty. Unstructured data must be retrieved, processed, and standardized before it can be evaluated, which takes longer and is more expensive for enterprises with tight budgets. Hadoop is no longer a single monolithic project, but rather a data processing strategy that is fundamentally different from the old relational database model [24]. Open-source tools, libraries, and methodologies for "big data" analysis, in which multiple data sets are gathered from diverse sources as both structured and unstructured data to be evaluated, such as Internet images, audio, videos, and sensor recordings [25]. Hadoop has a master/slave design, which consists of a single Name Node (Master – it stores metadata and manages job scheduling across the cluster) and a collection of Data Nodes (Slaves – they execute all operations requested by the master, i.e., Name Node). The Apache Hadoop framework and its various components that are used in data platform implementations are covered in detail below.

10.3.1 For data collection

The Apache Foundation produced Apache Kafka, an open-source platform. Kafka can gather data from several sources at the same time because of its distributed file processing and high throughput. It's written in the Scala programming language and offers a high level of fault tolerance [22]. In Figure 10.1, the Hadoop data collection process is shown.

Figure 10.1 Data collection process in Hadoop.

Sqoop also transmits data between an RDBMS and the Hadoop file system. It's an SQL-Hadoop hybrid that works on top of Hadoop's Distributed File System (HDFS).

Pig is a tool with an ETL, i.e., Extract, Transform, and Load, that may be used to handle, analyze, and personalize large medical and healthcare data sets in Hadoop. The input language is Pig Latin to the Pig engine, which is then internally transformed into a MapReduce task, which provides an output that is subsequently stored on Hadoop clusters [26].

10.3.2 For data processing

MapReduce is a programming model framework that simplifies and speeds up the processing of massive volumes of data via efficient and cost-effective methods [26]. Map, Sort, and Reduce are the three fundamental functions of this architecture, as featured in Figure 10.2.

YARN (Yet Another Resource Navigator) is an advanced resource navigator working on the top of the HDFS system and ensures parallel processing of the several components of the dataset. It can handle both levels of data processing, i.e., batch processing (data comes in the form of clusters) or stream (live streaming data from medical health devices which is captured using YARN. It is known for being highly scalable and secure and uses a master/slave architecture which in turn does a dynamic allocation of system resources for increasing its processing power [22].

SPARK is an in-memory cluster computing platform that was created for fast data computing using inbuilt libraries, such as Spark Streaming, that helps in real-time data collection and processing. Spark SQL can be used to do SQL queries on EHR-like datasets.

Apache Oozie is a tool for creating and managing Hadoop jobs as workflows as shown in Figure 10.3, where the output of one operation is used as an input for the next. As a result, Oozie can handle and run numerous Hadoop tasks at the same time, resulting in faster dataset processing. Workflow engines store and execute Hadoop job workflow collections; a coordinator engine conducts Hadoop tasks in accordance with how they are planned or added to the overall process schedule. DAGs (Directed Acyclic Graphs) of planned actions are also used to model Oozie workflow jobs [26].

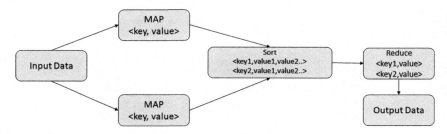

Figure 10.2 Data processing in Hadoop.

Apache Hadoop framework for big data analytics using AI 135

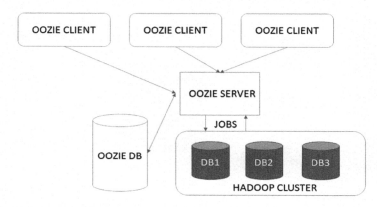

Figure 10.3 Apache Oozie workflow.

10.3.3 Big data storage

Hadoop employs a distributed data storage technology known as HDFS and a non-relational database known as HBase [27] to assure the storage of large volumes of data created every day across the healthcare industry and a multitude of medical equipment recording information linked to patients' well-being.

> HDFS: Large volumes of data are read in bulk from digital discs using HDFS, a file system designed for large-scale data processing and streaming. Metadata is saved in the Name Node, whereas file data is kept in the Data Node. Using HDFS systems and their high-speed processing, imaging modalities such as CT scans, MRIs, ultrasounds, and X-rays may be mined for information that would otherwise be lost [28].
>
> HBase: It is a column-oriented No-SQL database [29] built on top of Hadoop It is able to do random read/write operations and store a vast amount of medical healthcare data in the form of rows and columns. It also allows for record-level updates, which HDFS does not allow. Its open-source software can handle petabytes of data and hundreds of nodes. Because of the variety of healthcare enterprises, multiple column families and, in particular, HBase capabilities are frequently used for data organization [27].
>
> HCatalog: It is possible to easily read and write data to and from the system using HCatalog [49], an open-source Hadoop table and storage management system [29].

10.3.4 For big data analysis

Apache Hadoop helps organizations is analyzing large healthcare datasets using a large variety of components. Meta (formerly known as Facebook) built Hive, a data warehouse/lake system using the Hadoop ecosystem [21].

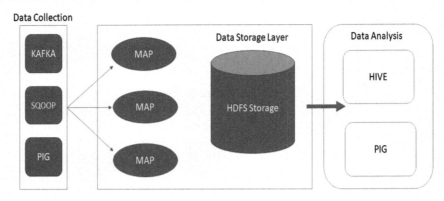

Figure 10.4 Data analysis in Hadoop.

The collected data from various sources is stored in structured datasets which are available for consumption to all users. It is managed through an HQL language which is very similar to SQL [30]. Hive provides a user-friendly interface with a variety of capabilities for performing complicated data analysis. It converts HQL queries within MapReduce jobs, which are subsequently processed simultaneously into batch tasks [29]. The big data analysis process in Hadoop is shown in Figure 10.4.

Pig is a Yahoo-developed open-source platform for analyzing huge data sets, such as real-time data flows [31]. Pig employs "Pig Latin," a high-level programming language. It has a multi-query approach for data processing which reduces the codes to a large extent. Pig is considered an alternative to Java for coding MapReduce jobs which helps in processing massive amounts of data due to its distributed architecture [32].

10.4 APACHE HADOOP FRAMEWORK FOR BIG DATA ANALYTICS USING AI

One of the key processing algorithms used in Apache Hadoop for big data analytics using AI is the Map Reduce algorithm. This algorithm is designed to process large volumes of data in a distributed manner, dividing the data into smaller subsets and processing each subset in parallel on different nodes of a cluster. The Map Reduce algorithm consists of two main phases: the Map phase and the Reduce phase. In the Map phase, the input data is divided into smaller pieces, and a Map function is applied to each piece. The Map function takes in the input data and transforms it into a key-value pair, where the key is a unique identifier and the value is the result of the transformation [33].

In the Reduce phase, the key-value pairs generated by the Map phase are combined and aggregated based on their keys. A Reduce function is applied to each key-value pair, which takes in the key and a list of values associated with that key, and produces a final output value. By breaking down the data processing task into smaller pieces and distributing the processing across multiple nodes, the Map Reduce algorithm enables efficient processing of large volumes of data. This algorithm is particularly well-suited for performing data-intensive tasks such as data mining, text processing, and machine learning, and has been used in various applications such as fraud detection, sentiment analysis, and recommendation engines [38]. In addition to the Map Reduce algorithm, Apache Hadoop also provides support for other AI processing algorithms such as deep learning and natural language processing. With its flexible and scalable architecture, Apache Hadoop is a powerful platform for performing advanced data analytics and deriving insights from large and complex data sets.

Apache Hadoop is a widely used big data processing framework that has enabled efficient and scalable data analysis. With the integration of machine learning and artificial intelligence, Hadoop has become an even more powerful tool for processing and analyzing big data. In this chapter, we explore the latest research in big data analytics using Apache Hadoop and AI, drawing on insights from ten scholarly articles. Real-time big data analytics is a major application of Hadoop and AI. The use of Hadoop and machine learning for real-time big data analytics has been explored. They implemented a model that analyzed data streams in real-time and made predictions based on the analyzed data [34]. Similarly, a big data analytics framework based is developed on Hadoop and machine learning for image processing [35].

Intelligent big data analysis platforms are another area of focus in Hadoop and AI research. An intelligent big data analysis platform is developed based on Hadoop and machine learning. Their platform enabled efficient processing and analysis of large volumes of data, allowing users to extract valuable insights and make informed decisions [36]. The development of smart cities has also been a major focus in Hadoop and AI research. A big data analytics framework based on Hadoop and machine learning for smart cities are developed. Their framework was designed to process and analyze data from various sources, such as traffic sensors and weather stations, to optimize city operations and improve the quality of life for citizens. In the healthcare sector, Hadoop and AI have been used for processing and analyzing big data [37]. A big data analytics platform is developed based on Hadoop and machine learning for healthcare applications. Their platform enabled efficient analysis of patient data, which could be used for disease diagnosis, treatment planning, and personalized healthcare [38]. Hadoop and AI have also been used for financial applications, such as fraud detection and risk management. A big data analytics platform is developed with Hadoop and machine learning for financial applications. Their platform was designed to process and analyze financial data, enabling efficient detection of fraudulent activities and risk management [39].

10.5 CONCLUSION

In this chapter, the architectural study of the Apache Hadoop framework was done, which utilizes the MapReduce framework for big data ingestion and processing. In conclusion, Apache Hadoop is a powerful open-source framework that has revolutionized the field of big data analytics. With its distributed storage and processing capabilities, Hadoop can handle vast amounts of data and perform complex analysis tasks using a variety of AI techniques. From recommendation engines to fraud detection systems, sentiment analysis, and image classification, Hadoop offers a wide range of capabilities for analyzing and deriving insights from large data sets. By leveraging Hadoop's Map Reduce programming paradigm and other AI libraries such as Mahout, Tensor Flow, and Flink, organizations can build scalable and robust data processing pipelines that enable them to make data-driven decisions and gain a competitive edge in today's data-driven economy. As the field of AI and big data analytics continues to evolve, Apache Hadoop will undoubtedly remain a key player, empowering organizations to unlock the potential of their data and gain valuable insights that can inform their business strategies.

REFERENCES

1. Hu, J., Zhang, X., Zhao, L., & Li, J. (2019). An intelligent big data analysis platform based on Hadoop and machine learning. *IEEE Access*, 7, 35905–35915.
2. Ahuja, S., & Taneja, P. (2017). Predictive modeling of big data using Apache Hadoop and machine learning. *Journal of Big Data*, 4(1), 1–18.
3. Chen, Y., Zhu, Y., Liu, Y., & Ma, J. (2019). Big data analysis framework based on Apache Hadoop and machine learning. *Journal of Ambient Intelligence and Humanized Computing*, 10(10), 3979–3991.
4. Wang, J., & Zhang, H. (2020). A machine learning-based framework for big data analytics in Apache Hadoop. *Journal of Ambient Intelligence and Humanized Computing*, 11(3), 1283–1294.
5. Dhamecha, T. I., & Soni, R. (2018). Big data analytics using Apache Hadoop and machine learning. *International Journal of Computer Science and Information Security*, 16(9), 70–76.
6. Zhang, J., Zhou, Y., Li, L., & Chen, Y. (2019). A big data analytics framework based on Apache Hadoop and machine learning for smart cities. *Sustainability*, 11(5), 1425.
7. Wang, Q., Li, J., & Li, W. (2018). An efficient big data analytics framework based on Apache Hadoop and machine learning. *Journal of Parallel and Distributed Computing*, 115, 1–11.
8. Liu, Z., Sun, X., & Zhang, J. (2020). A big data analytics platform based on Apache Hadoop and machine learning for industrial Internet of Things. *International Journal of Distributed Sensor Networks*, 16(5), 1550147719900902.
9. Hossain, M. A., & Hossain, M. A. (2018). A scalable big data analytics framework using Apache Hadoop and machine learning for healthcare applications. *Future Generation Computer Systems*, 86, 1064–1074.

10. Rahman, M. A., & Wahid, M. A. (2020). A big data analytics framework based on Apache Hadoop and machine learning for social network analysis. *Journal of Ambient Intelligence and Humanized Computing*, 11(9), 3459–3472.
11. Wang, Z., Yang, Y., Zhao, L., Zhang, B., & Yin, H. (2021). A big data analytics framework for smart agriculture based on Apache Hadoop and machine learning. *Computers and Electronics in Agriculture*, 188, 106510.
12. Anil, S. R., & Suresh, S. (2021). A big data analytics framework based on Apache Hadoop and machine learning for healthcare data. *Journal of Ambient Intelligence and Humanized Computing*, 12(7), 6253–6263.
13. Ghosh, D., Chakraborty, M., & Mukhopadhyay, A. (2021). A big data analytics framework for smart cities using Apache Hadoop and machine learning. *Journal of Ambient Intelligence and Humanized Computing*, 12(9), 8559–8570.
14. Rizvi, S. S., & Azam, S. (2021). An efficient big data analytics framework based on Apache Hadoop and machine learning for credit risk analysis. *Journal of Big Data*, 8(1), 1–22.
15. Mertoguno, S., Zgheib, R., Elsayed, S. A., & Baroudi, U. (2021). Big data analytics framework based on Apache Hadoop and machine learning for predicting customer churn. *Journal of Big Data*, 8(1), 1–24.
16. Liu, H., Zhang, X., Zhang, W., & Li, Z. (2022). A big data analytics framework based on Apache Hadoop and machine learning for oil reservoir prediction. *Journal of Big Data*, 9(1), 1–23.
17. Liu, K., Zhang, Y., He, M., & Zhang, L. (2022). A big data analytics framework based on Apache Hadoop and machine learning for enterprise risk management. *Journal of Big Data*, 9(1), 1–25.
18. Zhou, X., Li, X., Chen, H., & Chen, Y. (2022). A big data analytics framework based on Apache Hadoop and machine learning for intrusion detection. *Journal of Big Data*, 9(1), 1–23.
19. Das, A., Pradhan, S., & Jena, D. (2022). A big data analytics framework based on Apache Hadoop and machine learning for smart transportation. *Journal of Big Data*, 9(1), 1–25.
20. Sartipi, K., & Tavakkoli-Moghaddam, R. (2023). A big data analytics framework based on Apache Hadoop and machine learning for predicting student performance. *Journal of Big Data*, 10(1), 1–25.
21. Bante, P. M., & Rajeswari, K. (2017). "Big data analytics using hadoop map reduce framework and data migration process." *2017 International Conference on Computing, Communication, Control and Automation (ICCUBEA)*. IEEE.
22. Madabhushi, A., & Lee, G. (2016). Image analysis and machine learning in digital pathology: Challenges and opportunities. *Medical Image Analysis*, 33, 170–175.
23. Mahmoud, M.M.E., Rodrigues, J.J.P.C., Ahmed, S.H. et al., (2018). Enabling technologies on cloud of things for smart healthcare. *IEEE Access*, 6, 31950–31967.
24. Kumar, S., & Singh, M. (2018). Big data analytics for healthcare industry: Impact, applications, and tools. *Big Data Mining and Analytics*, 2(1), 48–57.
25. He, Z., Cai, Z., Sun, Y. et al., (2017). Customized privacy preserving for inherent data and latent data, ACM Pers. *Ubiquitous Compute*, 21(1), 43–54.
26. Khan, S., Shakil, K. A., & Alam, M. (2018). Cloud-based big data analytics—A survey of current research and future directions. In V.B. Aggarwal et al. (eds.), *Big Data Analytics*, Advances in Intelligent Systems and Computing 654, Springer Nature, Singapore. https://doi.org/10.1007/978-981-10-6620-7_57

27. Leiner, T., et al. (2019). Machine learning in cardiovascular magnetic resonance: Basic concepts and applications. *Journal of Cardiovascular Magnetic Resonance*, 21(1), 1–14.
28. Saroha, M., & Sharma, A. (2019). Big Data and Hadoop Ecosystem: A Review. In *2019 International Conference on Smart Systems and Inventive Technology (ICSSIT)*, IEEE.
29. Bahri, S., et al. (2018). Big data for healthcare: A survey. *IEEE Access*, 7, 7397–7408.
30. Sharma, N., Sharma, R. (2022). Real-time monitoring of physicochemical parameters in water using big data and smart IoT sensors. *Environment, Development and Sustainability*. https://doi.org/10.1007/s10668-022-02142-8
31. Anandkumar, R., Dinesh, K., Obaid, A. J., Malik, P., Sharma, R., Dumka, A., Singh, R., Khatak, S. (2022). Securing e-Health application of cloud computing using hyperchaotic image encryption framework. *Computers & Electrical Engineering*, 100, 107860, ISSN 0045-7906, https://doi.org/10.1016/j.compeleceng.2022.107860
32. Sharma, R., Xin, Q., Siarry, P., Hong, W.-C. (2022). Guest editorial: Deep learning-based intelligent communication systems: Using big data analytics. *IET Communications*. https://doi.org/10.1049/cmu2.12374
33. Sharma, R., Arya, R. (2022). UAV based long range environment monitoring system with Industry 5.0 perspectives for smart city infrastructure. *Computers & Industrial Engineering*, 168, 108066. ISSN 0360-8352, https://doi.org/10.1016/j.cie.2022.108066
34. Rai, M., Maity, T., Sharma, R. et al. (2022). Early detection of foot ulceration in type II diabetic patient using registration method in infrared images and descriptive comparison with deep learning methods. *The Journal of Supercomputing*. https://doi.org/10.1007/s11227-022-04380-z
35. Sharma, R., Gupta, D., Maseleno, A., Peng, S.-L. (2022). Introduction to the special issue on big data analytics with internet of things-oriented infrastructures for future smart cities. *Expert Systems*, 39, e12969. https://doi.org/10.1111/exsy.12969
36. Sharma, R., Gavalas, D., Peng, S.-L. (2022). Smart and future applications of Internet of Multimedia Things (IoMT) using big data analytics. *Sensors*, 22, 4146. https://doi.org/10.3390/s22114146
37. Sharma, R., Arya, R. (2022). Security threats and measures in the Internet of Things for smart city infrastructure: A state of art. *Transactions on Emerging Telecommunications Technologies*, e4571. https://doi.org/10.1002/ett.4571
38. Zheng, J., Wu, Z., Sharma, R., Haibin, L.V. (2022). Adaptive decision model of product team organization pattern for extracting new energy from agricultural waste. *Sustainable Energy Technologies and Assessments*, 53(Part A), 102352. ISSN 2213-1388, https://doi.org/10.1016/j.seta.2022.102352
39. Mou, J., Gao, K., Duan, P., Li, J., Garg, A., Sharma, R. (2022). A machine learning approach for energy-efficient intelligent transportation scheduling problem in a real-world dynamic circumstances. *IEEE Transactions on Intelligent Transportation Systems*. https://doi.org/10.1109/TITS.2022.3183215

Chapter 11

Envisioning the future of Blockchain in SMEs

Insights from a Survey

Ta Thi Nguyet Trang and Ta Phuong Thao
Thai Nguyen University, Thai Nguyen, Vietnam

Pham Chien Thang
TNU-University of Sciences, Thai Nguyen, Vietnam

Kusum Yadav
University of Ha'il, Ha'il, Kingdom of Saudi Arabia

11.1 INTRODUCTION

Blockchain, which was first introduced in 2008, is essentially a distributed database of records or a public ledger of every digital event or transaction carried out and shared among participating parties. Information cannot be deleted once entered (Crosby et al., 2016). According to Vincent (2019), Blockchain technology is still in the development stage. The Blockchain system is not yet at optimal maturity and further studies need to be conducted before implementing it in any new areas. Up to now, many people still consider Blockchain unreliable, and businesses are also extremely careful in applying Blockchain to the operating system. In fact, Blockchain has a lot of potential for development, especially for business-specific applications.

SME is a term applied worldwide for small- and medium-sized enterprises (in terms of capital, revenue, and number of employees) (Jasra et al., 2011). This type of business is always the preferred choice and is becoming more and more popular globally because of its small capital investment and simple organizational structure. SMEs contribute to solving employment problems very effectively, developing socio-economic, increasing GDP for the country, and training a dynamic and highly qualified staff and management team (Neagu, 2016). However, because of its popularity, the rate of competition among SMEs is relatively large, requiring each enterprise to find its breakthrough to create a competitive advantage, and avoid the risk of bankruptcy in the integration period.

According to the report "Rebooting a Digital Solution to Trade Finance" conducted by consulting firm Bain & Company in conjunction with HSBC (2018), the application of Blockchain can bring benefits up to USD 17.1 trillion by 2026, mainly for SMEs, including Vietnam – a developing country with great potential. The potential of Blockchain can go beyond the title of "technology behind Bitcoin" (Swan, 2015) and the list of Blockchain applications in businesses is growing, promising to open up many breakthrough opportunities for SMEs if they catch up with this technology and use it properly.

In fact, SMEs account for 95% of all businesses worldwide and developing countries tend to develop SMEs more than other countries (Antoldi et al., 2012). However, the application of Blockchain technology to SMEs still faces many challenges, such as operating costs, management experience, and technology accessibility (Riemenschneider et al., 2003; Kumar Bhardwaj et al., 2021). To create a competitive advantage for businesses in today's fierce market, especially after experiencing the Covid-19 pandemic, which has shown us the importance of technology, SMEs worldwide and in Vietnam must pay attention to applying high technology to the operation process. Blockchain can guarantee a "brighter" future for those who need it (Golosova & Romanovs, 2018), opening up huge opportunities for SMEs to access greater investment capital, and improve profitability and competitive advantage over other types of businesses.

However, as mentioned above, the widespread application of Blockchain technology today is still controversial, and doubts about its safety are always the top concern. Besides, research on the application of Blockchain technology is still lacking in developing countries. Developing from the reality that there has not been any official research on this issue in Vietnam, resulting in a lack of awareness about new technologies for government and businesses. This study assesses the determinants of Blockchain adoption in SMEs in Vietnam.

11.2 LITERATURE REVIEW

11.2.1 Definition of blockchain

In fact, all transactions on the Internet today are based on the principle: trusting a third party (banks, payment systems, authorities, etc.). This lead to the growth of transaction costs of the parties involved. However, with the advancement of technology, many questions are raised about the system's reliability and information security. Therefore, businesses, especially large enterprises, are spending more money to find the type of reliable service and technology to fix this problem at the most optimal level.

Under such increasing demand, the emergence of Blockchain brings an extremely suitable solution. Blockchain consists of a permanent, distributed digital ledger that is tamper-proof and is jointly implemented by all the nodes of the system (Atzori, 2015). Transactions for a specific time period are combined into a block, linked to a block of the previous period (Melnychenko & Hartinger, 2017), hence we have the term "Chain of block" or Blockchain.

This is the advantage of the data structure of Blockchain. Once recorded information cannot be altered, destroyed, tampered or stolen. Each transaction is written as a block of data and only takes place if all participants have given their consent (Kwilinski, 2019).

The information storage mechanism of this technology is based on distributed data registers (Melnychenko & Hartinger, 2017). All transactions are stored in a single register. Documents are kept simultaneously by all stakeholders, receiving complete and unbiased information, automatically updating the latest data after each change time. They have the right to access and audit transaction data in their own copy of the registry, forming a decentralized structure for making a unified decision (Melnychenko & Hartinger, 2017; Kwilinski, 2019).

In a more complementary way, Ducas and Wilner (2017) and Vincent (2019) state that Blockchain is a public digital ledger (a digital record) that exists to store records individually and exchange information through a peer-to-peer network like a chain. In Blockchain, information flows between peers. Each center is linked with several peer centers, making up the entire enterprise consisting of linked peer hubs. Each hub approves, keeps, and communicates its record with others (Dinh et al., 2018).

11.2.2 Perspectives on blockchain

Since its appearance, Blockchain has been known as the supporting tool behind Bitcoin – the biggest and hottest cryptocurrency at the moment, considered one of the most important breaking and advanced technologies after the Internet (Swan 2015; Yermack 2017). Still, it has the ability to transfer to a vital impact technology on financial services and other industries, regardless of the future of Bitcoin (Fanning & Centers, 2016). According to Yli-Huumo et al. (2016), Blockchain technology is attracting interest because of its primary characteristics of security, anonymity, and data completeness without any third party regulating the transactions. Each participant in the Blockchain is given a unique identity. Therefore, if you lose it, you will no longer be able to participate in the trade. Because an intermediary organization does not influence it, transactions in Blockchain can be carried out across borders, without time regulation and without waiting time. Besides, each participant is given a unique identity associated with their account, enhancing the security of each information recorded.

Looking at a little difference, Chowdhury (2021) believes that the major attraction of Blockchain lies in its decentralization because information and records are kept in many different nodes instead of one. Because of this decentralized feature, Blockchain will not have the error from the data center. Besides, attackers can only affect the data in Blockchain if they have control over 51% of the nodes in the network (Sarmah, 2018), so it can resist any security attack secret. If the Blockchain system consists of a small number of computers, it will be more vulnerable to attack. The participation of many computers will make the Blockchain system secure (Golosova & Romanovs, 2018).

Records in Blockchain are easily accessible by authorized users, who are in a peer-to-peer network and are immutable. The benefit of a peer-to-peer system is availability, which means that even if a node in the network is unavailable, users can still direct and access information using other nodes available in the system (Vincent, 2019). Each transaction and information in Blockchain is recognized with a high degree of authenticity, under the supervision of many people, so it is challenging for hackers to change data stored across thousands of nodes at once. (Chowdhury, 2021), then strengthen the role of the parties involved. If the system is centralized, the information is stored and controlled by one person or organization, so data can be deleted or changed without anyone realizing it (Golosova & Romanovs, 2018). Therefore, Blockchain creates a world of high credibility and serves information for lawsuits, financial statements, audits, credit, etc. Vincent (2019) also researched that the two basic characteristics of Blockchain are decentralization and immutability. Decentralized system – the main benefit of Blockchain (Golosova & Romanovs, 2018). The use of a decentralized system eliminates the existence of a central authority, central administrator, and intermediaries, allowing participants to verify transactions directly. All transactions have the database, proofs of authentication, and authorization, thereby increasing the reliability of each transaction. Besides, the process of data security consumes a lot of time, money, and human resources. Therefore, as more blocks are added to the Blockchain over time, data fabrication becomes extremely difficult, bringing transparency and trustworthiness to the data.

Golosova and Romanovs (2018) also mentioned Blockchain transparency. Transparency is demonstrated in the process of recording data for transactions. As noted above, each transaction is executed and confirmed by two parties. All the parties involved can see all the information that occurs in the transaction, which means that every change is visible to the participant. Based on its immutability and transparency, Blockchain makes it possible for users to trace the origin of products, point out any existing bad problems, and fix them if necessary.

Blockchain helps people save more time and costs (Niranjanamurthy et al., 2019). Together with the high-tech system and disintermediation, each transaction will be significantly shortened, thereby saving time and money from the data security process, staff fees, and transaction costs. In short, Blockchain is going through a period of explosive development, which is promising to open the door for multi-disciplinary applications, not only in the field of virtual currency but also in education, banking, manufacturing, military, etc.

The new star in Kwilinski's (2019) research is the limitation of Blockchain. Blockchain, the outstanding choice for business, however, still has disadvantages worth considering. Blockchain technology consumes a huge amount of energy for real-time ledger recording, ensuring no downtime and the data stored forever. When a transaction is made, electricity consumption increases. From there, it entails the second disadvantage of large costs, most of which are for consumed electrical energy and initial capital costs.

Each participant having a unique identity and verifying through its own signature is both a strength and a limitation of Blockchain (Niranjanamurthy et al., 2019). In the case that you lose your password, you will never be able to recover your access rights. This is one of the causes of high energy consumption.

Storage capacity is also a big problem of Blockchain, as the nodes do not have enough space to store more information after a long time of use. According to Sarmah (2018), the nodes in Blockchain will store information about the transactions that form blocks, thereby making Blockchain grow continuously. Blockchain will grow faster if many large blocks appear, so data accumulated will require a large amount of storage space.

A huge drawback that hinders human access to Blockchain is its complexity and confusion (Sarmah, 2018). Blockchain includes a lot of technical concepts that are difficult to understand and the complex process has not been fully developed for the vast majority of people to understand and use it in a short time. Therefore, Blockchain has not yet entered the stage of being ready to be universally applied in all fields like the Internet is doing today. The implementation of a project or a job with Blockchain elements is very difficult, requiring professional and technical knowledge.

Wang et al. (2016) also found that upgrading and maintaining Blockchain will face many difficulties in the future. Blockchain systems must be upgraded and maintained for various reasons (computer operating systems, languages, programming interfaces, fixes, improvements, etc.). Since the Blockchain system exists on the Internet, and lacks experience, the upgrade and maintenance are not the same as other enterprise software. In addition, there are other issues such as monitoring and managing system upgrades and maintenance. The second drawback they assessed is the integration of Blockchain with existing information systems. Golosova and Romanovs (2018) agreed with their opinion. The information that is active in the old software will not accept new transactions when moving to Blockchain. Importing previous transactions into the Blockchain and integrating with the next system is a complicated process.

The invention of Blockchain technology can mean the invention of writing systems or the Internet (Melnychenko & Hartinger, 2017). Up to the present time, the role of Blockchain is increasingly discovered and developed in many fields. In some places, Blockchain is not only limited to businesses but is accepted and applied at the national level (USA, Switzerland, Singapore, Indonesia, etc.). However, in many regions, Blockchain has not yet been accepted and legalized by the government due to concerns about its security, resulting in barriers to access to financial institutions if the legal status is still not stable. Therefore, the current implementation of Blockchain depends on a lot of factors.

According to Vincent (2019), Blockchain technology is generating an uproar in the business world. Economists are still watching and investing in Blockchain with the hopes of gaining a competitive edge in the market. However, he believes that the technology is still in the early development stage. Blockchain systems are not yet at their optimal maturity and more extensive studies should be conducted before implementing in any new areas (Wang et al., 2016). In line with them, Liu et al. (2019) identify that, passed more than a decade, despite Blockchain technology being still in its infancy and not been widely adopted, economic and financial institutions, especially big corporations have realized the efficiency of Blockchain when participating in business works and actively participated in the research, testing, investment, and development of this technology in many industries.

11.2.3 Interlink between blockchain and SMEs

Blockchain is currently being evaluated as an ideal solution for SMEs in all fields. The application of Blockchain technology to SMEs is expected to optimize operations and improve productivity and work efficiency of the whole enterprise (Kumar Bhardwaj et al., 2021). Not only stopping at recording, tracking, and managing features, Blockchain also allows almost accurate demand forecasting, smart contract building, product data statistics, project performance evaluation, etc. By investing and implementing Blockchain, SMEs can use this technology to improve business and management performance (reduce labor costs, detect mistakes, ensure proper performance of obligations to stakeholders, etc.) thereby improving the quality of business, participating in multinational transactions easily, creating a more competitive advantage, making a mark on the economic market.

However, barriers such as technology, organization, and environment always contain factors that hinder the large-scale application of Blockchain. The application of Blockchain technology to SMEs still has many challenges such as software compatibility requirements, Blockchain human resource training, investment costs, and security. SMEs have limited human and financial resources and lack infrastructure and the ability to apply high technology, so these barriers can greatly hinder SMEs when participating in the Blockchain race, thereby facing the risk of falling behind and going bankrupt (Wang et al., 2018).

After reviewing the available literature, Nayak and Dhaigude (2019) indicate that only a small number of SMEs in developing countries accept to use of Blockchain technology in their business operations, showing suspicion and lack of resources to apply this potential technology. Since SMEs have different levels of Blockchain adoption, it is essential to clearly identify the key factors that affect the intention to adopt Blockchain.

11.3 CONCEPTUAL FRAMEWORK

Based on the previous literature reviews, this study will investigate five factors that may influence SMEs' intention to adopt Blockchain (see Figure 11.1), with the following research hypotheses:

Complexity (CP): The difficulty in operating and comprehending an evolving technology is complexity (Venkatesh et al., 2003; Sumarliah et al., 2022). According to Sun et al. (2021), complexity is crucial for SMEs to accept Blockchain technology. Blockchain usually entails intricate algorithms, and mistakes can be hard to spot or discovered after it's too late to correct them. This can be easily seen in previous studies by Wang et al. (2016) and Choi et al. (2020). Therefore, the hypothesis here is:

H1: There is a significant relationship between Complexity of Blockchain and Blockchain adoption (BA) in SMEs in Vietnam.

Security and Privacy (SP): The ability of the technology being used to protect the user's information and guarantee the integrity of information during the transaction is referred to as security (Awa & Ojiabo, 2016; Salisbury et al., 2001). Privacy, which is sometimes equated with security and is also seen as a component of security, exemplifies the degree

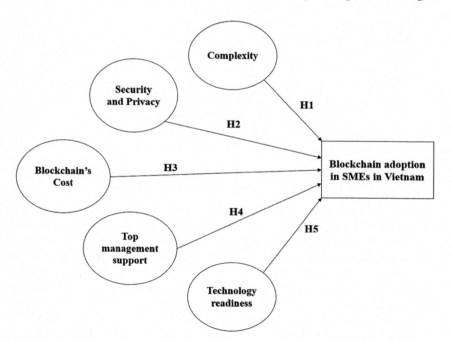

Figure 11.1 Conceptual framework.

of anonymity technology can provide for its user (Vidhyalakshmi & Kumar, 2016). While transactions through Blockchain have the potential to be fraud-proof, however, anything can go wrong, so security may be compromised. Many documents have also been reported about Security concerns as a key factor in their organization to adopt Blockchain (Guo & Liang, 2016; Choi et al., 2020). The second hypothesis is:

H2: There is a significant relationship between Security & Privacy of Blockchain and Blockchain adoption in SMEs in Vietnam.

Blockchain's Cost (BC): Numerous studies (e.g., Vidhyalakshmi & Kumar, 2016, Kühn et al., 2019, Lanzini et al., 2021) have demonstrated that the cost of technology adoption significantly influences Blockchain behavioral intentions. Blockchain technology is thought to reduce transaction costs and increase operational efficiency. However, businesses need to pay a huge expense before investing in technology and expertise to meet the requirements of implementing it. Therefore, many SMEs are facing financial difficulties, affecting their decision to use Blockchain (Kumar Bhardwaj et al., 2021). The third hypothesis is:

H3: There is a significant relationship between Cost of Blockchain and Blockchain adoption in SMEs in Vietnam.

Top management support (MS): The degree of top management involvement in adopting new technology and understanding its importance is known as top management support (Wang & Qualls, 2007; Ooi et al., 2018). Senior management's allocation of infrastructure, financial, and human resources when implementing new technology is crucial (Swan, 2015; Hsu et al., 2018). Additionally highlighted by the current literature research is MS's beneficial contribution to adopting Blockchain technology (Pilkington, 2016; Kouhizadeh et al., 2021). MS is a great source for any business when it comes to adopting a new technology to overcome adversity. The fourth hypothesis is:

H4: There is a significant relationship between Top management support and Blockchain adoption in SMEs in Vietnam.

Technology readiness (TR): Technology Readiness is defined as an organization's preparedness and availability in terms of technological infrastructure (hardware, software) and human resources (specialized manpower) (Moh'd Anwer, 2018). TR is evaluated as an important predictor in many research models affecting technology adoption intention (Kühn et al., 2019; Queiroz & Wamba, 2019; Pham & Thi Nguyet, 2023). Existing research papers have shown the positive influence of TR in the application of Blockchain technology in SMEs (Swan, 2015; Pilkington, 2016). The organizations with Blockchain's

TR are getting a better competitive advantage in the market. Therefore, organizations need to be fully prepared with technological knowledge, training, expertise, and skillset to implement Blockchain. The fifth hypothesis is:

H5: There is a significant relationship between Technology readiness and Blockchain adoption in SMEs in Vietnam.

11.4 METHOD

11.4.1 Research design and sample size

This study used a descriptive survey design to investigate the determinants of Blockchain technology adoption in SMEs in Thai Nguyen province (Vietnam). In this study, a questionnaire survey was conducted to examine the factors affecting the intention to adopt Blockchain of SMEs in Thai Nguyen province. In the questionnaire, the participants will be asked to provide some basic personal information, their understanding and assessment of Blockchain as well as the factors affecting the decision to choose Blockchain in the business. In this study, the total number of SMEs in the Association of Small and Medium Enterprises of Thai Nguyen Province as of 2022 is about 600 SMEs (Tran Huyen, 2022).

Based on period studies, the original questionnaire was mainly based on the scales of factors affecting the adoption of new technology (Queiroz & Wamba, 2019; Choi et al., 2020; Lanzini et al., 2021; Kouhizadeh et al., 2021; Pham & Thi Nguyet, 2023). The questionnaire has two-part: part 1 is the responder's information, and part 2 uses the five-point Likert scale (ranging from 1 = completely disagree to 5 = fully agree) to indicate the extent to which the responders agree with each statement about determinants of Blockchain adoptions (see Appendix).

A sample is a specific group that you will collect data from to represent the entire population. This study has the participation of individuals representing businesses. The number of samples proposed for the study is calculated through the following formula (Yamane, 1967):

$$n = \frac{N}{1 + N * e^2}$$

where
  n: Sample size required for the study
  N: Total population
  e: Allowed error. Usually, the error rate used is ± 0.01 (1%), ± 0.05 (5%), and ± 0.1 (10%). In this research, the researcher chooses the error level of 0.01.

From this, the researcher calculated the required sample size of 240 SMEs. It was large enough and representative enough to obtain reliable information and save time and funding to complete the study. The characteristics and relationships in the population will be inferred from the sample study results.

The study selects survey participants through simple random sampling. In this study, the researcher considers a population of 600 SMEs in Thai Nguyen province – a population that is not big, and then randomly selects 240 SMEs from this list. Since they are all types of SMEs in the Association of Small and Medium Enterprises of Thai Nguyen Province, the characteristics of these 240 SMEs will be relatively similar, not causing too much difference and difficult to research.

11.4.2 Data analysis

To eliminate unexpected errors and ensure answer consistency, all data obtained by quantitative research methods will be summarized, classified, coded, edited, analyzed, and interpreted as material for the research. The data are then entered into computers and analyzed and processed using the SPSS Software version 28.0 (IBM). ANOVA tests are used to test the differences between variables in a regression analysis. If the significant level of the variable is higher than 0.01, it is qualified to confirm that there is no difference between independent and dependent variables and vice versa. Finally, data tables and charts are established to interpret the data analysis process and present the results. Conclusions will then be written from the analyzed results.

11.4.3 Participants

The results of the survey on respondents' personal information, including gender, age, education level, seniority, current job position, field of their business, and some basic understanding of Blockchain. Statistical information after processing the obtained data are presented in Table 11.1.

Table 11.1 shows the information of respondents by gender, in which the percentage of respondents who are male is higher than that of females. Specifically, 57.5% of respondents are male, and 42.5% of respondents are female. In terms of age, workers aged 24–30 years accounted for the highest percentage (52.1%), followed by the age group 31–40 (26.7%). Meanwhile, the number of people over the age of 40 is quite small. Specifically, the number of survey participants aged 41–50 years was only 20 people (8.3%) and the smallest is over 50 years old, only 1.7% (equivalent to 4 people). Regarding education level, the vast majority of respondents graduated from university or college, accounting for nearly 70%, followed by high school level (accounting for 20%). The proportion of people studying at postgraduate level and other qualifications accounts for just over 10%. Have a clear

Table 11.1 Respondents' information

		Frequency	Percent
Gender	Male	138	57.5
	Female	102	42.5
Age	18–23	27	11.3
	24–30	125	52.1
	31–40	64	26.7
	41–50	20	8.3
	More than 50	4	1.7
Education level	High school	48	20
	College/University	166	69.2
	After university	12	5
	Other	14	5.8
Working duration	Less than 1 year	12	5
	Between 1 and 2 years	32	13.3
	Between 3 and 5 years	102	42.5
	Between 6 and 10 years	56	23.3
	More than 10 years	38	15.8
Job position	Employee	162	67.5
	Leader	30	12.5
	Manager	24	10
	Vice President	2	0.8
	President	1	0.4
	Others	21	8.8
Type of business	Industry	60	25
	Agriculture – Fishery	23	9.6
	Trade – Services	84	35
	Transportation	25	10.4
	Construction	12	5
	Other	36	15
Knowledgeable about Blockchain	I only know the name	47	19.6
	I get a little bit information about it	108	45
	I have a certain understanding about it	39	16.3
	I am doing activities and jobs related to it	46	19.2

difference in working experience in the corporate environment among the participants, in which those with 3 years of experience or more account for the highest proportion. Specifically, accounting for the highest proportion is the group of people with 3–5 years of working experience with 42.5%, nearly double that of the group of 6–10 years (accounting for 23.3%). Next is the group of people with experience of 10 years or more and 1–2 years with nearly equal rates (15.8% and 13.3%, respectively). Those with less than 1 year of experience join the smallest with 5%. There is a clear difference in the number of respondents by job position. While the number of Employees is overwhelming at 67.5%, the number of people holding senior positions including Vice President and Directors is very small, only 0.8% and 0.4%, respectively. The number of survey respondents working in the Construction sector accounted for the lowest percentage, only 5%. The number of respondents working in Trade – Service and Industry account for the highest proportion, 35% and 25%, respectively, while this figure in other industries accounts for a relatively significant proportion of 15%. The majority of respondents have limited knowledge of Blockchain, accounting for 45%. Meanwhile, the number of people who only know the name and the number of people with extensive knowledge of Blockchain has approximately the same ratio, 19.6% and 19.2%, respectively.

11.5 RESULTS

11.5.1 Test of reliability

Before beginning the analytical procedures, we evaluate the scales' reliability to ensure that the observed variables are closely related, comparable, and have enough measurement capacity to demonstrate the scale's overall consistency and dependability. We utilized the Corrected Item – Total Correlation coefficient and Internal consistency, as determined by Cronbach's Alpha – dependability coefficient, to assess the scale's reliability. All the scales tested for reliability had Cronbach's alpha coefficients greater than 0.6, demonstrating the scale's acceptable quality. Except for observation variable SP6, which does not match the requirements, all observed variables have Corrected Item – Total Correlation > 0.3 (DeCoster & Claypool, 2004). To test the scales' reliability once more, the authors chose to eliminate this observed variable. The findings are shown in Table 11.2.

After testing the reliability of the scales for the second time, the results show that all the independent variables in the scale have Cronbach's alpha coefficient >0.6, proving that the scale is at an acceptable level and well used. Besides, all observed variables after the second test give satisfactory Corrected Item – Total Correlation (>0.3). Therefore, it can be concluded that all initially designed scales have reliability and there are 25 observed variables out of 26 observed variables are retained for use in the next EFA analysis.

11.5.2 Exploratory factor analysis

The independent variable's KMO and Bartlett's Test results showed that the observed independent variables in the factor are associated with one another, as indicated by KMO coefficient = 0.762 > 0.5 and sig. Bartlett's Test = 0.000 0.05, making exploratory factor analysis appropriate for the given data set (Russell, 2002). The Eigenvalue = 1,480 > 1 at the fifth factor, so there are five factors extracted with cumulative variance = 63.418%

Table 11.2 The reliability of the scales

Variable	Mean	Std. deviation	Corrected item-total correlation	Cronbach's Alpha if item deleted
Complexity (CP): Cronbach's Alpha = 0.813				
CP1	3.07	0.805	0.719	0.724
CP2	3.24	0.843	0.626	0.767
CP3	3.42	0.839	0.612	0.774
CP4	3.47	0.838	0.573	0.793
Security and Privacy (SP): Cronbach's Alpha = 0.827				
SP1	3.88	0.827	0.646	0.786
SP2	3.86	0.835	0.581	0.804
SP3	3.79	0.829	0.570	0.808
SP4	3.67	0.836	0.695	0.772
SP5	3.75	0.831	0.623	0.793
Blockchain's Cost (BC): Cronbach's Alpha = 0.817				
BC1	3.02	0.823	0.641	0.771
BC2	3.13	0.823	0.545	0.799
BC3	2.90	0.827	0.673	0.761
BC4	3.49	0.873	0.562	0.795
BC5	3.41	0.808	0.618	0.778
Top management support (MS): Cronbach's Alpha = 0.785				
MS1	3.70	0.845	0.620	0.717
MS2	3.75	0.831	0.598	0.729
MS3	3.67	0.831	0.577	0.739
MS4	3.64	0.827	0.569	0.743
Technology readiness (TR): Cronbach's Alpha = 0.788				
TR1	3.36	0.811	0.612	0.727
TR2	3.68	0.854	0.606	0.730
TR3	3.65	0.825	0.580	0.743
TR4	3.70	0.825	0.582	0.742
Blockchain adoption (BA): Cronbach's Alpha = 0.790				
BA1	3.69	0.851	0.625	0.722
BA2	3.73	0.828	0.632	0.715
BA3	3.64	0.832	0.637	0.710

(>50%). This shows that 63.418% data of 22 observed variables are explained through five factors (Liu et al., 2003; Patil Vivek et al., 2017). In the first exploratory factor analysis, the KMO coefficient and Bartlett's Test, Initial Eigenvalue, and the "total variance explained", all satisfied the research requirements. However, CP1, SP4, and BC3 do not guarantee the

Table 11.3 Rotated component matrix of independent variables

	Component				
	1	2	3	4	5
MS4	0.766				
MS1	0.766				
MS2	0.738				
MS3	0.725				
TR2		0.783			
TR1		0.757			
TR3		0.742			
TR4		0.697			
SP1			0.769		
SP5			0.755		
SP3			0.732		
SP2			0.700		
BC4				0.771	
BC5				0.746	
BC1				0.726	
BC2				0.654	
CP3					0.768
CP4					0.757
CP2					0.752

discriminant value because they have a Loading factor of less than 0.5 and are located on two different factors. Therefore, the authors dismiss them and run a rediscovery factor analysis. The data are appropriate for EFA in the second EFA because the coefficient KMO = 0.856 (>0.5) and the sig. of the Bartlett test both equal 0.000 (<0.05). The remaining observable variables guarantee convergent and discriminant values. The fifth factor's Eigenvalue is 1,337, and the total variance explained is 61.301%. This demonstrates that five factors, which encompass all 19 observed variables, account for 61.301% of the variation. The Loading Factor of all variables is greater than 0.5 and lies in only one factor. After two times of exploratory factor analyses, 19 observed variables converged and differentiated into five factors (see Table 11.3). From the results of EFA for the dependent variable, only one factor was extracted, with Eigenvalue = 2.114 (>1) and Total Variance Explained = 70.481% (>50%).

11.5.3 Test of hypothesis

The Pearson correlation coefficient analysis results show that all the independent variables have a linear relationship with the dependent variable. SP has the lowest correlation value (r = 0.497), whereas BC has the highest

Table 11.4 Coefficients

Model	Unstandardized coefficients B	Std. Error	Standardized coefficients Beta	t	Sig.	Collinearity statistics Tolerance	VIF
(Constant)	3.031	0.405		7.490	0.000		
CP	−0.252	0.047	−0.242	−5.408	0.000	0.779	1.284
SP	0.190	0.049	0.173	3.873	0.000	0.785	1.274
BC	−0.312	0.051	−0.282	−6.169	0.000	0.750	1.332
MS	0.195	0.048	0.181	4.047	0.000	0.785	1.275
TR	0.299	0.049	0.276	6.072	0.000	0.757	1.321

correlation coefficient (r = 0.590) with the dependent variable. Additionally, the independent variables are correlated (r is approximately 0.3). The regression analysis can consider all independent factors to describe the purpose of using Blockchain for SMEs. The regression model is appropriate for the gathered data set, as evidenced by the ANOVA table's finding that the parameter F = 81.126 has sig = 0.000 (<0.01). The model's regression results report shows that the corrected R^2 equals 0.626. In other words, the five components listed in the model account for 62.6% of the variation of the dependent variable (DeCoster & Claypool, 2004). As a result, the multiple linear regression model built is suitable for the population, and the independent factors are related to the dependent variable.

Regression analysis was used to assess the influence of independent variables, including Complexity, Security & Privacy, Blockchain's cost, Top management support, and Technology readiness for Blockchain adoption in SMEs in Vietnam. Looking at Table 11.4, we see that the independent factors significantly influence Blockchain adoption because the regression weights are all statistically significant (sig = 0.000 < 0.01). Among them, variables SP, MS, and TR cause a positive effect on Blockchain adoption (because of the positive Beta coefficient). In contrast, the remaining variables including CP and BC all negatively influence Blockchain adoption (because of the negative Beta coefficient). In addition, when considering the variance inflation factor (VIF) to determine the multicollinearity test, it can be seen that the VIF is all less than 2. We can also consider the Tolerance coefficient, from Table 11.4, the Tolerance coefficient is all >0.5, so there is no multicollinearity (Liu et al., 2003). The regression model was determined to be suitable. Through the above models, five hypotheses are accepted with p = 0.000 (<0.01).

11.6 DISCUSSION AND LIMITATIONS

The results and analysis in this study show us that Blockchain's cost (in terms of Technology) has the greatest influence on the application of Blockchain in SMEs in Vietnam with a negative direction. This assertion coincides with numerous results from Awa and Ojiabo (2016), Kühn et al. (2019), and Kumar Bhardwaj et al. (2021). The cost of implementing Blockchain technology includes many expensive items such as hardware, software, staff training, and recruitment of experts. That is the biggest reason why SMEs find it difficult to apply this high technology. Unexpectedly, this factor was assessed by Choi et al. (2020) as a driver of intention to adopt Blockchain rather than as a barrier, although SMEs in Tonga consider the cost to adopt and maintain Blockchain to be too much for their businesses. This difference may be due to the huge benefits that Blockchain brings. Even if it requires a huge amount of money, SMEs are always ready to deploy it to keep up with the growth of the global economy.

Technology readiness, which is predicted by Awa and Ojiabo (2016), Moh'd Anwer (2018), Kühn et al. (2019), and Kumar Bhardwaj et al. (2021) to be an important factor positively affecting the intention to adopt Blockchain is also ranked as important right after the Cost factor in this study. However, Lanzini et al. (2021) did not overestimate this factor. This difference can be partly explained by the time frame and subject of the study. In this study, their target audience is SMEs in Europe – one of the most developed regions in the world. Therefore, in the context of the most recent research conducted, SMEs in Europe may have applied higher technologies to their operations, and when they want to apply Blockchain, businesses will not face too many obstacles in terms of technology.

Regarding the Complexity factor, this study shows that it has a relatively high level of negative influence on the intention to apply Blockchain to SMEs in Vietnam. The finding of Choi et al. (2020), Kumar Bhardwaj et al. (2021), and Sun et al. (2021) also demonstrates that Complexity acts as a barrier to Blockchain adoption and users are relatively afraid of this issue. They realize that Blockchain is not simple to operate and they do not have enough knowledge, experience, and effort to use it (Choi et al., 2020). The complexity of Blockchain is evident in the early stages of business adoption. After Blockchain is familiar with the operation process, the next daily tasks will be simpler. Blockchain developers can focus on this problem to create a more advanced Blockchain, providing an easier user experience.

Moh'd Anwer (2018), Kumar Bhardwaj et al. (2021), and Lanzini et al. (2021) rated Top management support as one of the most important factors determining the adoption of Blockchain. In the context that the final decisions of SMEs often depend on the Board of Directors, SMEs may face difficulties if the Board of Directors does not provide enough resources to support Blockchain adoption. In this study, Top management support was rated less important, ranked 4th, supported by Choi et al. (2020). The reason may be that the management of SMEs is still running businesses in the traditional way, does not have enough knowledge about new technologies, is poorly adapted to change, and has not been attracted by what Blockchain promises to bring.

Finally, in the study of Lanzini et al. (2021), they point out that the Security & Privacy feature is the most important non-organizational factor among all the factors. This is in contrast to the results of a study on SMEs in Vietnam, where Security & Privacy was ranked as the smallest influence among the five factors. In particular, Kumar Bhardwaj et al. (2021) found that Security & Privacy did not affect the intention to adopt Blockchain in SMEs. To explain this difference, the subject of study is probably the best answer. Research by Lanzini et al. (2021) was conducted in SMEs in the Netherlands, a country with an outstandingly developed economy, where security is of great concern. Meanwhile, the study by Kumar Bhardwaj et al. (2021) focuses on SMEs in India, a developing country where technology and information security are not really appreciated.

This study has several important contributions both theoretically and practically. First, this is one of the first studies to empirically evaluate the application of Blockchain in SMEs in Vietnam. In general, there have been overview studies on how Blockchain technology will affect SMEs in Thai Nguyen province. However, research on Blockchain in SMEs in Vietnam has not been really focused because the technology in Vietnam, in general, has not really developed. A large number of SMEs do not have access to this technology, causing a relatively large obstacle to research. This study provides an empirical investigation that highlights the outstanding benefits of Blockchain and its potential to change the way businesses work, thereby reinforcing the theory of previous assessments, and helping SME managers to be more motivated to accelerate the adoption of Blockchain in their businesses. The application of Blockchain technology is always confirmed to increase the efficiency of economic activities of enterprises, thereby improving the profits and position of SMEs in the market. Next, this research helps in-depth Blockchain developers understand user psychology and determine which factors their customers consider, thereby improving Blockchain more and more actively, proposing appropriate strategies and recommendations to their customers, and providing training programs from basic to advanced levels to support SMEs in implementing Blockchain into operation.

Since the survey was conducted in 240 samples of SMEs in the Thai Nguyen Association of Small and Medium Enterprises, the model may not be representative of all SMEs (many SMEs are not included in the Association). In addition, the research model may give different results if conducted in different areas. Conducting empirical research in a small province like Thai Nguyen may not fully reflect the dynamics and barriers that Blockchain adoption is facing. Therefore, it is necessary to expand the scope of research to increase the accuracy of the results obtained. The accuracy of the study also depends on the truthfulness of the answers from survey participants, so it is not possible to conclude absolutely that all the factors mentioned in this study are appropriate and have such a level of influence. It is necessary to conduct in-depth interviews in addition to the online survey to reduce the possibility that the difference in results is too large.

Although the study measured three sub-factors representing Technology (Complexity, Cost, Security & Privacy), two sub-factors representing Organizational (Top Management Support, Technology Readiness) have an impact on Blockchain adoption in SMEs, there are many other factors within these two big factors that have not been considered. Besides, the Environment, the other big factor, is not mentioned in this study.

11.7 CONCLUSION

The initial result of this study is to clarify the theory of Blockchain, its perspectives, and the relationship between Blockchain and SMEs by evaluating the theory from previous studies. Since then, the study has built hypotheses and a research model to evaluate the main factors affecting the application of Blockchain in SMEs in Vietnam. From the study results, SMEs in Vietnam can identify the Blockchain's cost factor that has the most substantial influence on the intention to apply Blockchain, and the Technology readiness factor also has a significant influence after Blockchain's cost.

Blockchain technology is opening a new and much more advanced trend for all fields. However, the reality shows that the application of Blockchain in businesses is still limited and not outstanding. This study has several limitations because practical research on Blockchain still faces many barriers when Blockchain is still in a relatively new stage in Vietnam, mainly only theoretical research. In further studies, researchers need to promote empirical research to test the theory of Blockchain in other types of organizations and businesses besides SMEs. For the theoretical review, although this study only mentions five influencing factors, there are still many other factors that have not been mentioned, and in the future other factors may appear, and some elements may disappear. Therefore, researchers always have to update the information for the right research direction.

APPENDIX. QUESTIONNAIRES ITEMS

Code	Factors	Source
	The Complexity of Blockchain	
CP1	Blockchain is very complicated to understand deeply and use it	Choi et al. (2020); Sumarliah et al. (2022)
CP1	The complexity of Blockchain makes it difficult for businesses to learn	
CP3	The complexity of Blockchain will put SMEs without highly specialized and technical staff at a disadvantage in the competition in the market.	
CP4	Blockchain's complexity lies in its algorithm	

	Blockchain Security & Privacy	
SP1	Blockchain's security is high	Guo and Liang (2016); Choi et al. (2020)
SP2	Transactional and business information on Blockchain are transparent, traceable, and secure	
SP3	Blockchain has a leading security system for businesses	
SP4	My business is willing to share business information on Blockchain	
SP5	My business can easily track the activity happening on the Blockchain	
SP6	Blockchain's security feature makes it difficult to recover data when password is lost	
	Blockchain's cost	
BC1	Blockchain saves transaction costs	Vidhyalakshmi and Kumar (2016), Kühn et al. (2019); Lanzini et al. (2021)
BC2	Blockchain helps save business costs (employee salaries, etc.)	
BC3	Blockchain adoption takes a huge budget	
BC4	Technology accounts for the largest budget when applying Blockchain	
BC5	Recruiting and training human resources accounts for the largest budget when applying Blockchain	
	Top management support	
MS1	The leaders in my business support the adoption of Blockchain in the business	Ooi et al. (2018); Kouhizadeh et al. (2021)
MS2	The leaders in my business are looking for opportunities to get access to Blockchain	
MS3	The leaders in my business have a vast knowledge of the potential of Blockchain when applied to the business	
MS4	The leaders in my business are fully allocating human, financial, and infrastructure resources to Blockchain adoption	
	Technology readiness	
TR1	Technology to apply Blockchain is simple	Queiroz and Wamba (2019); Pham and Thi Nguyet (2023)
TR2	Blockchain-ready businesses are in a better position in the market	
TR3	My business needs to update new software and change outdated devices to access Blockchain	
TR4	My business needs to organize highly specialized training for employees and hire experts on Blockchain	
	Blockchain adoption in SMEs	
BA1	My business intends to apply Blockchain for the past 3 years	Ooi et al. (2018); Queiroz and Wamba (2019); Pham and Thi Nguyet (2023)
BA2	I think my business will be more successful if Blockchain is applied	
BA3	My business will often use Blockchain to solve work in the future	

REFERENCES

Antoldi, F., Cerrato, D., & Depperu, D. (2012). *Export consortia in developing countries: successful management of cooperation among SMEs.* Springer Science & Business Media.

Atzori, Marcella. (December 1, 2015). Blockchain technology and decentralized governance: Is the state still necessary? https://ssrn.com/abstract=2709713 or http://dx.doi.org/10.2139/ssrn.2709713

Awa, H. O., & Ojiabo, O. U. (2016). A model of adoption determinants of ERP within TOE framework. *Information Technology & People, 29*(4), 901–930.

Choi, D., Chung, C. Y., Seyha, T., & Young, J. (2020). Factors affecting organizations' resistance to the adoption of blockchain technology in supply networks. *Sustainability, 12*(21), 8882.

Chowdhury, E. K. (2021). Financial accounting in the era of blockchain-a paradigm shift from double entry to triple entry system. *Available at SSRN 3827591*.

Crosby, M., Pattanayak, P., Verma, S., & Kalyanaraman, V. (2016). Blockchain technology: Beyond bitcoin. *Applied Innovation, 2*(6–10), 71.

DeCoster, J., & Claypool, H. (2004). Data analysis in SPSS.

Dinh, T. T. A., Liu, R., Zhang, M., Chen, G., Ooi, B. C., & Wang, J. (July 1, 2018). Untangling blockchain: A data processing view of blockchain systems. *IEEE Transactions on Knowledge and Data Engineering, 30*(7), 1366–1385. https://doi.org/10.1109/TKDE.2017.2781227

Ducas, E., & Wilner, A. (2017). The security and financial implications of blockchain technologies: Regulating emerging technologies in Canada. *International Journal, 72*(4), 538–562.

Fanning, K., & Centers, D. P. (2016). Blockchain and its coming impact on financial services. *Journal of Corporate Accounting & Finance, 27*(5), 53–57.

Golosova, J., & Romanovs, A. (2018, November). The advantages and disadvantages of the blockchain technology. In *2018 IEEE 6th workshop on advances in information, electronic and electrical engineering (AIEEE)* (pp. 1–6). IEEE.

Guo, Y., & Liang, C. (2016). Blockchain application and outlook in the banking industry. *Financial Innovation, 2*(1), 1–12.

Hsu, H. Y., Liu, F. H., Tsou, H. T., & Chen, L. J. (2018). Openness of technology adoption, top management support and service innovation: a social innovation perspective. *Journal of Business & Industrial Marketing.* https://doi.org/10.1108/JBIM-03-2017-0068

Jasra, J., Hunjra, A. I., Rehman, A. U., Azam, R. I., & Khan, M. A. (2011). Determinants of business success of small and medium enterprises. *International Journal of Business and Social Science, 2*(20), 274–280.

Kouhizadeh, M., Saberi, S., & Sarkis, J. (2021). Blockchain technology and the sustainable supply chain: Theoretically exploring adoption barriers. *International Journal of Production Economics, 231*, 107831.

Kühn, O., Jacob, A., & Schüller, M. (2019). Blockchain adoption at German logistics service providers. In *Artificial Intelligence and Digital Transformation in Supply Chain Management: Innovative Approaches for Supply Chains. Proceedings of the Hamburg International Conference of Logistics (HICL), Vol. 27* (pp. 387–411). Berlin: epubli GmbH.

Kumar Bhardwaj, A., Garg, A., & Gajpal, Y. (2021). Determinants of blockchain technology adoption in supply chains by small and medium enterprises (SMEs) in India. *Mathematical Problems in Engineering*, 2021.

Kwilinski, A. (2019). Implementation of blockchain technology in accounting sphere. *Academy of Accounting and Financial Studies Journal*, 23, 1–6.

Lanzini, F., Ubacht, J., & De Greeff, J. (2021). Blockchain adoptioin factors for SMEs in supply chain management. *Journal of Supply Chain Management Science*, 2(1–2), 47–68.

Liu, M., Wu, K., & Xu, J. J. (2019). How will blockchain technology impact auditing and accounting: Permissionless versus permissioned blockchain. *Current Issues in Auditing*, 13(2), A19–A29.

Liu, R. X., Kuang, J., Gong, Q., & Hou, X. L. (2003). Principal component regression analysis with SPSS. *Computer Methods and Programs in Biomedicine*, 71(2), 141–147.

Melnychenko, O., & Hartinger, R. (2017). Role of blockchain technology in accounting and auditing. *European Cooperation*, 9(28), 27–34.

Moh'd Anwer, A. S. (2018). Towards better understanding of determinants logistical factors in SMEs for cloud ERP adoption in developing economies. *Business Process Management Journal*, 25(5), 887–907.

Nayak, G., & Dhaigude, A. S. (2019). A conceptual model of sustainable supply chain management in small and medium enterprises using blockchain technology. *Cogent Economics & Finance*, 7(1), 1667184.

Neagu, C. (2016). The importance and role of small and medium-sized businesses. *Theoretical and Applied Economics*, 23(3), 331–338.

Niranjanamurthy, M., Nithya, B. N., & Jagannatha, S. J. C. C. (2019). Analysis of Blockchain technology: pros, cons and SWOT. *Cluster Computing*, 22(6), 14743–14757.

Ooi, K. B., Lee, V. H., Tan, G. W. H., Hew, T. S., & Hew, J. J. (2018). Cloud computing in manufacturing: The next industrial revolution in Malaysia? *Expert Systems with Applications*, 93, 376–394.

Patil Vivek H., Singh, S. N., Mishra, S., & Donavan, D. T. (2017). *Parallel Analysis Engine to Aid in Determining Number of Factors to Retain using R [Computer software]*. https://analytics.gonzaga.edu/parallelengine/

Pham, C. T., & Thi Nguyet, T. T. (2023). Determinants of blockchain adoption in news media platforms: A perspective from the Vietnamese press industry. *Heliyon*, 9(1), e12747. https://doi.org/10.1016/j.heliyon.2022.e12747

Pilkington, M. (2016). Blockchain technology: principles and applications. In *Research handbook on digital transformations*. Edward Elgar Publishing.

Queiroz, M. M., & Wamba, S. F. (2019). Blockchain adoption challenges in supply chain: An empirical investigation of the main drivers in India and the USA. *International Journal of Information Management*, 46, 70–82.

Riemenschneider, C. K., Harrison, D. A., & Mykytyn Jr, P. P. (2003). Understanding IT adoption decisions in small business: Integrating current theories. *Information & Management*, 40(4), 269–285.

Russell, D. W. (2002). In search of underlying dimensions: The use (and abuse) of factor analysis in Personality and Social Psychology Bulletin. *Personality and Social Psychology Bulletin*, 28(12), 1629–1646. https://doi.org/10.1177/014616702237645

Salisbury, W. D., Pearson, R. A., Pearson, A. W., & Miller, D. W. (2001). Perceived security and World Wide Web purchase intention. *Industrial Management & Data Systems*, 101(4), 165–177.

Sarmah, S. S. (2018). Understanding blockchain technology. *Computer Science and Engineering*, 8(2), 23–29.

Swan, M. (2015). *Blockchain: Blueprint for a new economy.* O'Reilly Media, Inc.

Sumarliah, E., Li, T., Wang, B., Fauziyah, F., & Indriya, I. (2022). Blockchain-empowered halal fashion traceability system in Indonesia. *International Journal of Information Systems and Supply Chain Management (IJISSCM), 15*(2), 1–24.

Sun, W., Dedahanov, A. T., Shin, H. Y., & Li, W. P. (2021). Using extended complexity theory to test SMEs' adoption of Blockchain-based loan system. *PloS one, 16*(2), e0245964.

Tran Huyen (2022). Đại hội đại biểu Hiệp hội Doanh nghiệp nhỏ và vừa tỉnh Thái Nguyên lần thứ V, nhiệm kỳ 2021 – 2026. https://thainguyen.gov.vn/thoi-su/-/asset_publisher/L0n17VJXU23O/content/-ai-hoi-ai-bieu-hiep-hoi-doanh-nghiep-nho-va-vua-tinh-thai-nguyen-lan-thu-v-nhiem-ky-2021-2026/20181

Venkatesh, V., Morris, M. G., Davis, G. B., & Davis, F. D. (2003). User acceptance of information technology: Toward a unified view. *MIS Quarterly, 27,* 425–478.

Vidhyalakshmi, R., & Kumar, V. (2016). Determinants of cloud computing adoption by SMEs. *International Journal of Business Information Systems, 22*(3), 375–395.

Vincent, N. E. (2019). Blockchain research agenda for accounting. *Accounting and Finance Research, 8*(4), 1–93.

Wang, Y., & Qualls, W. (2007). Towards a theoretical model of technology adoption in hospitality organizations. *International Journal of Hospitality Management, 26*(3), 560–573.

Wang, H., Chen, K., & Xu, D. (2016). A maturity model for blockchain adoption. *Financial Innovation, 2*(1), 1–5.

Wang, Y., Han, J. H., & Beynon-Davies, P. (2018). Understanding blockchain technology for future supply chains: a systematic literature review and research agenda. *Supply Chain Management: An International Journal, 24*(1), 62–84.

Yamane, Taro. (1967). *Statistics: An Introductory Analysis,* 2nd Edition. New York: Harper and Row.

Yermack, D. (2017). Corporate governance and blockchains. *Review of Finance, 21*(1), 7–31.

Yli-Huumo, J., Ko, D., Choi, S., Park, S., & Smolander, K. (2016). Where is current research on blockchain technology?—A systematic review. *PloS one, 11*(10), e0163477.

Chapter 12

Deep learning techniques for the prediction of traffic jam management for smart city infrastructure

Arwa N. Aledaily and Kusum Yadav
University of Ha'il, Ha'il, Kingdom of Saudi Arabia

12.1 INTRODUCTION

Traffic jam or congestion is one of the most faced issues that is prevalent in all corners of the world; if we talk in terms of India, the population is 1,391,243,638 (as of world meters live data) [6], and according to a report of Statista, it shows that around 253 million vehicles were there in the year 2017 [7]. India has the world's second-longest road network spanning around 5.89 million kilometres [10]. However, the traffic on roads is mostly random and less predictable, which closes many avenues of efficient traffic management in cities and highways. In cities, one of the major problems people face due to traffic congestion includes losing their valuable time in commute and at intersections, and not being able to find parking spots efficiently. Eventually, all of this leads to them losing more money than they make from their purpose of moving around. The Internet of Things is a type of Internet-connected devices that have the ability of sensing, computing and communicating the parameters of the environment around them. These devices can be installed at multiple points in the physical world. They can compute upon the real-time sensory input and communicate the results over the network to centralised systems or networks anywhere in the world.

The huge amount of data collected can't be worked upon with normal tools like MS Excel and their equivalents. The huge amount of collected data can be stored in centralised data warehouses and further processed with tools like Apache Spark and Hadoop. This comes under Big Data. It can be used for drawing helpful insights like the time at which the traffic in particular areas of road systems is highest, the type of traffic each area experiences, the type of traffic that is most prominent during a particular time of the day or night and much more. These insights can be used to make data-based decisions about the most efficient ways of deploying resources such as traffic lights, traffic police, vehicle size limits, toll roads and traffic diversions and more. This can also be used to predict the traffic patterns and simulate them according to various factors like time, visibility, size and quality of roads and the type of establishments around the roads and can help us make more suitable decisions in other related fields like road constructions, assigning type of plots, and opening or closing roads for the type of traffic.

Our work in this research focuses on how the detection of the type of vehicles is possible with Deep Learning algorithms like YOLO v3 and how the results can be reported in the form of visualisations to clearly analyse the data patterns with respect to time. The rest of the chapter comprises of parts namely, "Literature Survey" section provides a review of relevant studies on the subject, "Methodology" describes the method by which we have collected imagery and video data and performed the detection on it to produce visualised results, "Results" section presents the results generated and discussion on them, and finally the chapter is concluded in "Conclusion" section.

12.2 LITERATURE SURVEY

In our research work, we have taken inspiration and suggestions from the following works on same or similar subjects. In this literature survey, we have tried summarising their works, identifying themes and outlining the structure of these papers/articles:

> Lu Sun et al. [1] have presented a paper on the estimation of expected travel time between two or more points using the method of moment. They start by defining how the travel time of vehicles is an important aspect of traffic management for both travellers and traffic management agencies as it provides essential information such as on selecting routes, congestion mitigation, automatic incident detection, intelligent freeway ramp control and congestion-based toll pricing. They later went on to describe the setup used for detecting the speed of traffic across the sample roads and defining the expected travel time (with a specified departure time) as an ensemble mean travel time over a number of days and comparing it to Monte Carlo simulation and direct sampling-based simulation of travel time. Neeraj K. Kanhere et al. [2] worked on Real-Time Detection and Tracking of Vehicle Base Fronts for Measuring Traffic Counts and Speeds on Highways. They started by describing the problem with real-time traffic monitoring cameras placed relatively low on the ground on the side of roads and freeways as the variable height of vehicles can obstruct the line of sight of cameras and produce erroneous results. They went on to try the newer approach of tracking the front base of vehicles on the road. Their system used Blepo Computer Vision Library developed at Clemson University in C++ language and the camera was mounted on a 26' tripod approximately 12' from the side of the road and the sequences were digitised at 320 × 240 resolution and 30 frames per second.
> Arunesh Roy et al. [3] wrote a research article on fusing the Doppler Radar and Video Information for Automated Traffic Surveillance.

They explained how the current Doppler-Radar-based traffic surveillance systems face the problem of not being able to differentiate the different vehicles on the road. They went on to separate the video into its components, identify unique tracks (multi-target tracking) and data association, then using the camera intrinsics and the camera extrinsics, found the location of these blobs in the image plane, on the ground (XY) plane, using the 3D positions of the tracked targets used as measurements in a Kalman filter to estimate the position and velocity of the vehicles. Sumeyye Cepni et al. [4] presented an article on Vehicle Detection Using Different Deep Learning Algorithms from Image Sequence, where they introduced the various Deep Learning Algorithms used for Object Detection in images and videos (since videos are effectively image sequences played at a high rate). They described the workings of CNN and YOLO algorithms, used them to detect vehicles in their dataset and presented the difference between their results.

Prayag Tiwari et al. [11] forecasted the traffic in a two-lane undivided highway using Artificial Neural Networks. They collected the short-term sample from the National Highway 58 (NH58) between Roorkee and Haridwar. They have trained it for the undivided two-lane highway in India with mixed traffic conditions because most datasets around the world vary greatly by their environmental factors and nature of the road. They compared their results with random forest, support vector machine, k-nearest neighbour classifier, regression tree and multiple regression models and found that their backpropagation artificial neural network performed better than other approaches. Luo-Wei Tsai et al. [12] presented a working approach to detecting vehicles from still images using colour and edges. Traditional approaches use motion features to detect vehicles in still images; however, this colour and edge-based approach filters out all the background pixels and analyses each pixel as a candidate for a possible vehicle; then matching it with the edge maps and coefficients of wavelet transform can produce extreme results under various lighting conditions with great speed and around 95% accuracy. This makes it a possible candidate to be used in multiple commercial applications where the lightning conditions do change a lot but not by a great margin. R. Cucchiara et al. [13] in their paper deployed two different techniques to detect vehicles during day and night. In their experiment, a supervising level detects the condition and selects the technique based on defined rules to be applied for the detection and tracking of vehicles. Also in their paper, they described the Vehicular Traffic Tracking System's (VTTS) architecture and have shown the two approaches it selects based on luminescence status in the environment. For tracking the vehicles, they have used a production rule system with forward chaining. Their system successfully detected 332 out of 343 vehicles, with an error rate of 3.1%.

12.3 METHODOLOGY

12.3.1 Data collection

The dataset used in this study was collected from multiple sources. About 190 images were clicked from various spots in multiple localities of Kanpur, Uttar Pradesh, India. The stated localities include Ganga Barrage, Jajmau and Barra mainly. As marked with circles in Figure 12.1, these areas lie in the Kanpur District of Uttar Pradesh and have typical moderate levels of traffic conditions. These images have been greatly compressed to only fill a few hundred kilobytes of space instead of MBs used by usual high-resolution pictures. Besides the images clicked first-hand from Kanpur, we took a similar number of images from the Indian Institute of Information Technology, Hyderabad's Driving Dataset developed to promote academic work on autonomous navigation. These pictures were clicked from parts of Hyderabad and Bangalore cities and their outskirts [5].

12.3.2 Data pipeline for captured images and videos

We have a rough idea of how busy Indian streets would be with vehicles all the time, especially in smaller unplanned cities, due to a smaller number of available parking slots and management, traffic jams mostly occur as spots are inversely proportional to the available traffic. If there is some predictive-based analysis or approach which is already equipped with our proposed

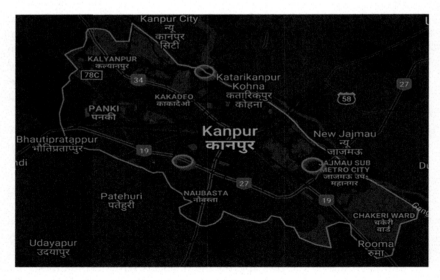

Figure 12.1 Map of Kanpur with areas of data collection marked.

method in surveillance cameras, then this will help us in approaching the traffic issue at different spots. Also, this will help the passengers help to understand the travel time or find a traffic spot based on its varied usage for this application. Let us look at the data pipeline in Figure 12.2, for the proposed system.

If we take inputs from Figure 12.2, we can see that it begins with surveillance cameras and other data producers; these are the vehicles which are being captured in the cameras at multiple points in a centralised traffic control system. The data is further categorised into two groups: one is the IDD image dataset [11] which is available for open use for analysis and other applications on Indian roads and traffic and the other source is a few collected images from the different sources collected locally which is loaded into the database. Now, this data is integrated together in layer 1 which is actually combining the available data altogether for a uniform view which can be further used for different analysis or computation; all these are monitored by a Hadoop developer; he/she writes a code in Java as Hadoop is written in Java.

This now goes to a data lake; a data lake is a cloud-based database which is very huge and data can be in the form of petabytes which is not possible to store locally on machines and is a raw form of data. Now, further, this data is again integrated based upon the interest of the user, based upon what type of data they want and by some codes written in Java or Python, the data is segregated in the form of data marts as we can see in the image; we have some data marts like Cars, Motorcycle, Truck and Buses. We can actually integrate more data marts as well, based on people and other vehicles but we here, took these four parameters only to show a glimpse of what we are trying to achieve. After differentiating the images, we process the data using some deep learning models; here, in our proposed system, we use YOLO v3 to identify the objects (objects here are cars, motorcycles, trucks and bus). But before we start detecting the images, we actually need to process a human practice, which is not perfect (similarly to the case for machines); that is why in machine learning, we train the machine using some sample images which is a huge data set used to train the model in order to give proper results. Often to do so, what we use is an approach called data augmentation; in this, in spite of using multiple datasets, we use the same dataset for training just by flipping it by 90 degrees, cropping it and padding it; these are some commonly used methods when we have to train neural networks on a large level. For our case, we used the COCO model (which is an abbreviation for common objects in context); it is a large image dataset for multiple objects spread across many categories, key points and instances; these are around 330,000 images and are labelled as less than 200,000 [8]. Thus, the advantage we got by training it with the COCO model was that we didn't have to work much for gathering many sources

168 Artificial Intelligence and Blockchain in Industry 4.0

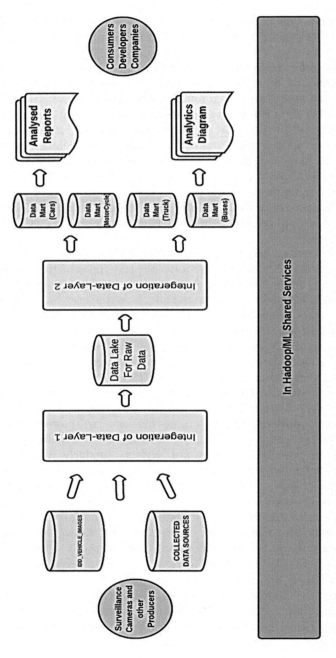

Figure 12.2 Data pipeline for intelligent traffic systems.

for training the image set, as COCO marks the objects when the code is run for detecting the images.

Also, we used Image AI for every instance which is a high-end API. This is mainly used for images where detection and extraction have to be performed. We used it with YOLO v3 to run the detection threads and analyse them better. We used Google Colab which is a wonderful tool for simulating deep learning models and machine learning. It is an Internet-based computing technology. After the object extraction and detection, it marks the object with a container, declaring the confidence percentage of detection, giving us the probability of exact detection of vehicles of different types; this can be used for detecting multiple objects of all types in COCO but here, we restricted it to the interest as discussed in data marts.

12.3.3 Algorithm

First, we declared the libraries required for the program such as OS, CSV, numpy, pandas, and imageai and then declared a variable and gave instructions to open a file and write inside it whatever is detected. Then we imported a class from image ai named detection. The execution path has to check the current working directory every time the process is repeated. The declared variable passes all the objects detected.

custom = detector. CustomObjects (bicycle = True, car = True, motorcycle = True, bus = True, truck = True)

A function named heatmap is defined and then arguments are passed, which is used to read the image set having the dimensions on x-, y- and z-axes. And whatever is detected is to be printed in the csv file. Then, the next is used for normalising the colour bar of the heatmap on the right in between 0 and 0.1. The next line looks for colours to be in order – basically a kind of check. Then it maps scalar data to RGBA; thus, we are able to see the heatmap on the right as a result. The next line adjusts the padding between and around the subplots. And the files having extension .csv are replaced with the .png extension. Heatmap images having the .png extension are then saved. In the next line of codes, we see that detectfromphoto function is detecting objects and returns an array along with the path to the original image here, the path is returned because it is used in the next function. Ind is a counter variable for how many images have been processed and also it looks for the logs in the folder where it is constantly searching for images. The image it gets is sends to the detectfromimage function and returns the output. And if the Boolean is true for it, then it is logged on terminal if there is any exception, it won't stop iterating and it will just skip that file and move to the next unless all files are checked.

170 Artificial Intelligence and Blockchain in Industry 4.0

12.4 RESULT

12.4.1 Detection report

We generate an analysed report in txt format with an identification container for each object detected. Also, we automated a script to generate individual csv files for each image detected. The image dataset we used created comprises 379 images after removing the ones illegible for our research. This gives us the locations or coordinates where the objects were discovered and what object it was, based on the masking it does. Thus, we can say that if the surveillance cameras data is equipped with these models, it will help us understand the patterns of traffic, and at what instant it is going to face traffic congestion can automatically alarm the department to arrive on the spot for helping the vehicles to move with a proper flow.

Figure 12.3 shows an excerpt of a matrix which shows the detected images with the positions for reference.

In Figure 12.3, a matrix for the detected image saved on our local machine, we can see the path first, then we can see what object was detected, which here is a truck with a percentage probability – a confidence percentage (CP) of the object which is 93.396 and box points which is the container coordinates (CC) in an array [170,94,228,162]. Similar are cases of other objects detected in the image such as motorcycles and cars. Likewise, in the other images detected with paths given test/165/663015_image.jpg and test/165/555079_image.jpg, we can see this time, it has detected a bus and a truck showing the CP and CC. Based on the requirement, we can change the sensitivity of the detection of images as well, to omit the chances of fine errors and making it too negligible so that it is not detected by the Python script.

12.4.2 Heatmaps of detected images

Below is a set of some results obtained from the Deep Learning model which we implemented for detecting the objects which here are the vehicles based upon the data marts. Showing the confidence percentage of each object

Figure 12.3 Snippet of generated report.

Deep learning techniques for traffic jam management 171

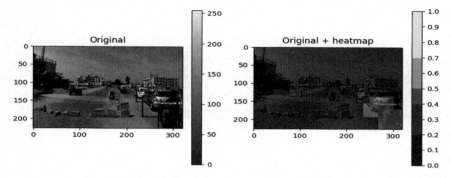

Figure 12.4 Original detected image with heatmap.

Table 12.1 Detected objects-1 matrix with CC and CP

Objects	x	y	x_1	Y_1	Probability value
Car	234	137	301	195	66.733
Car	221	124	248	154	82.83266
Motorcycle	268	172	318	236	64.82157
Motorcycle	79	120	103	136	73.05267

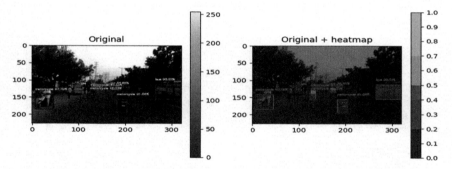

Figure 12.5 Original detected image with heatmap.

Table 12.2 Detected objects-2 matrix with CC and CP

Objects	x	y	x_1	y_1	Probability value
Bus	264	114	320	156	90.00018
Motorcycle	7	143	46	186	97.78981
Motorcycle	182	157	204	195	91.05596
Car	167	125	179	136	85.8622
Motorcycle	125	130	136	155	63.22891
Motorcycle	124	140	135	159	90.22681

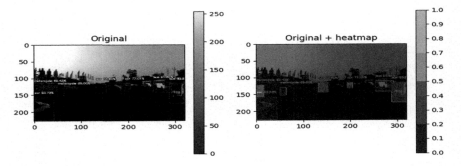

Figure 12.6 Original detected image with heatmap.

Table 12.3 Detected objects-3 matrix with CC and CP

Objects	x	y	x_1	y_1	Probability value
Bus	253	113	292	143	91.69381
Bus	287	118	319	175	85.83014
Motorcycle	0	123	17	163	69.43304
Car	1	159	43	225	93.72138
Car	193	117	205	133	77.00503
Car	132	118	153	145	75.79671
Motorcycle	51	131	64	153	88.99524

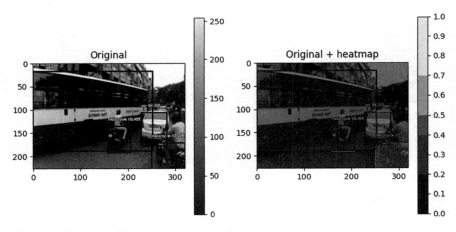

Figure 12.7 Original detected image with heatmap.

detected using the Deep Learning algorithm used with a heatmap showing the objects detected, the hotter the region, the more the number of occurrences of instances or objects or vehicles (marked with warm tones). The colder the region, the fewer the number of objects detected (marked with duller tones).

Table 12.4 Detected objects-4 matrix with CC and CP

Objects	x	y	x_1	y_1	Probability value
Bus	0	18	252	189	99.99149
Motorcycle	250	172	320	240	92.08702
Car	228	107	282	178	99.37789
Motorcycle	158	136	206	193	98.48207

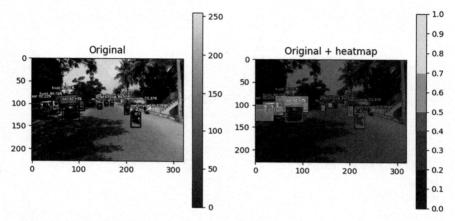

Figure 12.8 Original detected image with heatmap.

Table 12.5 Detected objects-5 matrix with CC and CP

Objects	x	y	x_1	y_1	Probability value
Truck	43	81	112	117	62.91277
Truck	17	94	60	124	66.78044
Car	0	100	39	137	62.23196
Car	65	105	101	139	93.42098
Motorcycle	210	119	230	153	88.92962
Car	154	96	162	105	65.43967
Motorcycle	140	100	148	113	69.43409
Motorcycle	196	103	206	125	73.37247

12.4.3 Count estimation in video

By using the above method, we can apply it to the instantaneous reports of vehicles present at that instance (which here can be referred to as in seconds or minutes) and create an analysis [9]. Now, we used the same approach for a video analysis of continuous streaming data from a surveillance camera installed and used our prediction model, which helps us to study much more

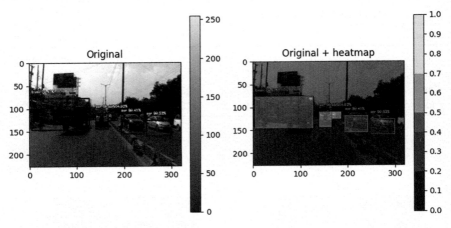

Figure 12.9 Original detected image with heatmap.

Figure 12.10 Image from the video file.

Table 12.6 Detected objects-6 matrix with CC and CP

Objects	x	y	x_1	y_1	Probability value
Truck	5	78	126	147	79.46603
Car	140	111	166	145	69.41035
Car	194	120	242	156	89.40948
Car	247	127	299	161	99.52367
Car	166	111	186	127	64.92664

effectively based on the similar approach we used here; in Figure 12.10 is an image of a frame from the video.

Thus, for our real-time analysis, we took a video of 34 minutes and sampled it into multiple frames in order to simplify it based on the GPU

Deep learning techniques for traffic jam management 175

and hardware. We reduced it to around 1/20th of the total frames produced for objects detected in a frame with accuracy for every second; this way, it was reduced to 2050 seconds and an equal number of frames detected. Below is a data sheet provided (Table 12.7) which was made after calculating a populated mean of the data after grouping the data in seconds to minutes as generating graph on the raw data was having some

Table 12.7 Mean of the vehicles detected in a particular minute

Minutes	Car	Truck	Bus	Motorcycle	Total vehicles
1	11.88333333	1.033333333	0.133333333	0	13.05
2	13.26666667	0.866666667	0.166666667	0	14.3
3	14.41666667	0.916666667	0.133333333	0.016666667	15.48333333
4	11.55	0.683333333	0.033333333	0.016666667	12.28333333
5	10.68333333	0.733333333	0.15	0.033333333	11.6
6	11.33333333	0.8	0.083333333	0.016666667	12.23333333
7	11.75	0.95	0.1	0	12.8
8	11.41666667	1.166666667	0.133333333	0	12.71666667
9	11.55	1.233333333	0.083333333	0	12.86666667
10	13.06666667	0.95	0.083333333	0	14.1
11	12.56666667	1.166666667	0.1	0	13.83333333
12	13.46666667	0.55	0.083333333	0.016666667	14.11666667
13	8.683333333	0.966666667	0.133333333	0	9.783333333
14	10.13333333	0.6	0.066666667	0	10.8
15	11.91666667	0.966666667	0.15	0	13.03333333
16	13.26666667	0.633333333	0.066666667	0	13.96666667
17	12.4	0.666666667	0.083333333	0.016666667	13.16666667
18	13.86666667	0.85	0.066666667	0	14.78333333
19	14.25	0.7	0.083333333	0	15.03333333
20	13.78333333	0.683333333	0.083333333	0	14.55
21	11.8	0.6	0.083333333	0	12.48333333
22	10.98333333	0.85	0.083333333	0	11.91666667
23	12.03333333	0.633333333	0.1	0	12.76666667
24	11.96666667	1.416666667	0.066666667	0	13.45
25	11.4	1.7	0.15	0	13.25
26	12.98333333	0.683333333	0.116666667	0	13.78333333
27	13.48333333	0.833333333	0.066666667	0	14.38333333
28	13.58333333	1.1	0.2	0	14.88333333
29	13.43333333	1.3	0.216666667	0	14.95
30	15.31666667	1.266666667	0.116666667	0	16.7
31	13.2	0.783333333	0.116666667	0	14.1
32	14.28333333	0.983333333	0.016666667	0	15.28333333
33	13.16666667	0.683333333	0.266666667	0	14.11666667
34	16.7	1.233333333	0.15	0	18.08333333

abnormality and was not apt for prediction; thus, populated mean of the data helped us make it more easily understandable. The formula used for the populated mean is

$$P.M. \text{ (Populated Mean)} = \left(\sum\nolimits^{*} Xi\right) / N$$

where
  $\sum$ means the submission or the total summation of the values in that column of the dataset,
  Xi means every data in that cohort or dataset,
  N means the number of values in the dataset.

Figures 12.11–12.15 are the graphs which are predicted over the above-calculated data which shows us the fluctuations at every minute; these are time series graphs for different parameters we used, such as cars, buses, motorcycles, trucks and overall vehicles.

Now, if we see Figure 12.11, the graph for the total vehicles detected, we can easily conclude saying that in minutes 3, 18, 19, 32, 33 and 34, there was a peak shoot in the graph; based on the values derived, we can keep track of when there is a probability of getting a traffic jam and this data can simultaneously be fetched and analysed using Hadoop and Spark framework in a centralised traffic control system and can alert the authorities to arrive on the spot and evacuate the space for smooth management of traffic and help in clearing the traffic jam.

12.4.4 Discussion

Thus, through the above traffic analysis using YOLO v3, we can conclude that we were able to distinguish the vehicles using a deep learning model by making a frame over the objects detected by the algorithm designed, which, in turn, gives us the log report of the detected objects in the form of an array which we can locate in the image. Also, the heatmap analysis showed us the detected object density at that region – the warmer the region, the higher the probability of an object being detected at that region, and the colder the region, lesser the probability of finding a vehicle. Using this method, we then underwent the analysis of a video file which was a surveillance camera which recorded streaming images; thus, we broke it down into frames of images detected for easier transformation which then again, with the above approach, gave us the number of objects detected and created a csv file of the logs detected, and then, in order to analyse and find the peak of vehicles for finding the probability of Traffic Jam, we calculated a populated mean of the dataset generated to normalise the abnormality and created graphs shown in Figures 12.11–12.15, where, in Figure 12.11, we can see that the peaks crossing over at certain points of normal density threshold which is

Deep learning techniques for traffic jam management 177

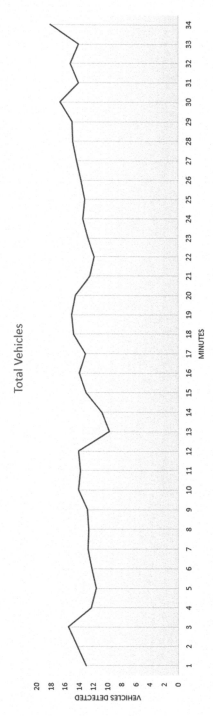

Figure 12.11 Minutes vs total vehicles detected.

178 Artificial Intelligence and Blockchain in Industry 4.0

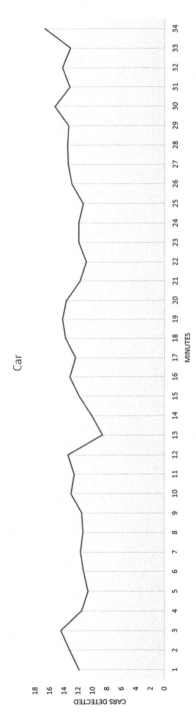

Figure 12.12 Minutes vs cars detected.

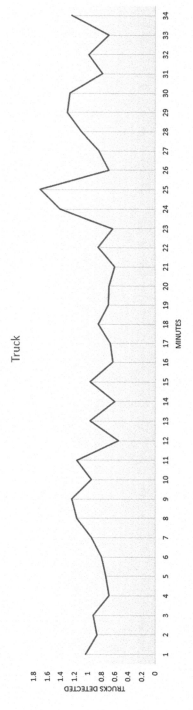

Figure 12.13 Minutes vs trucks detected.

180 Artificial Intelligence and Blockchain in Industry 4.0

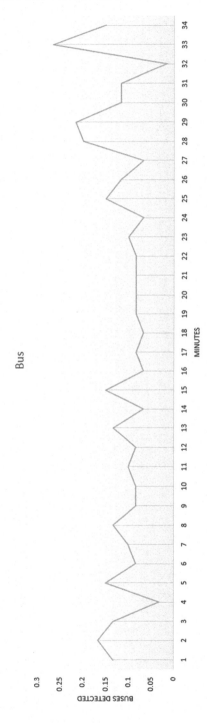

Figure 12.14 Minutes vs buses detected.

Deep learning techniques for traffic jam management 181

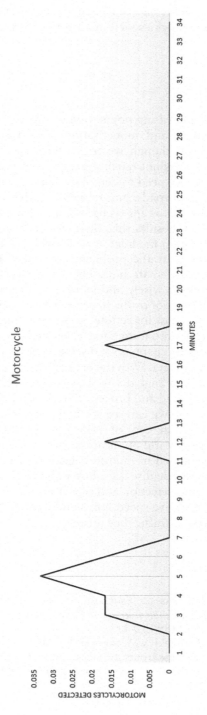

Figure 12.15 Minutes vs motorcycles detected.

constant for some time; these variations help us understand better the flow of traffic and alert the authorities well in advance to overcome such issues in future.

12.5 CONCLUSION

With the huge and ever-increasing population of private vehicles on Indian roads, traffic management and monetisation are becoming harder and harder to undertake. Even though we try to mitigate these problems by building more roads, there is not much space to keep buildings, especially in developed cities where we need to demolish major local commercial and residential spaces to rebuild roads. Not to mention, the cost of upkeep and regular maintenance of roads keeps rising with more roads, thus creating a self-perpetuating problem. Besides this, there are not many sources of revenue for agencies to maintain the huge road network of the country. Here, the alternative solutions to traffic problems come into the picture. With modern technology, we can create more efficient collection and reporting systems to use our resources wisely and make data-driven decisions about the management and policies of traffic in cities. This also, in turn, serves for less infrastructure and less cost for setting up the environment for detection (as we divided the Big Data into frames for faster analysis, instead of analysing the whole video). In this chapter, we discussed the data pipeline for the collection and reporting of data. With the help of a deep learning algorithm, YOLO v3 and different data analytics tools, we were able to show how the real-time data can be collected and processed to detect and analyse the traffic patterns in the freeways. We can see the highs and lows of traffic in the graphs and understand that the type of traffic on the road mostly comprises cars. In the long-term real-world application, this system with appropriate hardware can be implemented in multiple places on roads to collect real-time road vehicle data and visualise it for faster and proper decision-making with least hassle. The same detection and reporting system can also be used in custom toll pricing, adaptive speed limits and improving the road qualities according to the type of traffic they attract.

REFERENCES

1. L. Sun, W. Gu, H. Mahmassani. Estimation of expected travel time using the method of moment. *Canadian Journal of Civil Engineering*, 2011;38(2):154–165. https://doi.org/10.1139/L10-115
2. N.K. Kanhere, S.T. Birchfield, W.A. Sarasua, T.C. Whitney. Real-time detection and tracking of vehicle base fronts for measuring traffic counts and speeds on highways. *Transportation Research Record*, 2007;1993(1):155–164. https://doi.org/10.3141/1993-21.

3. A. Roy, N. Gale, L. Hong, Fusion of Doppler Radar and video information for automated traffic surveillance, *2009 12th International Conference on Information Fusion*, 2009, pp. 1989–1996.
4. S. Cepni, M. E. Atik, Z. Duran, Vehicle detection using different deep learning algorithms from image sequence. *Baltic Journal of Modern Computing*, 2020;8(2):347–358. https://doi.org/10.22364/bjmc.2020.8.2.1
5. IDD Images dataset, https://idd.insaan.iiit.ac.in/
6. India's population, https://www.worldometers.info/world-population/india-population/
7. Total vehicle population between 1951 and 2017, https://www.statista.com/statistics/664729/total-number-of-vehicles-india/
8. COCO data set, https://cocodataset.org/#home
9. Video file used for analysis, https://www.youtube.com/watch?v=wqctLW0Hb_0&t=162s
10. India Brand Equity Foundation. Road Infrastructure in India. *India Brand Equity Foundation*, 22 March 2021, https://www.ibef.org/industry/roads-india.aspx
11. B. Sharma, S. Kumar, P. Tiwari et al. ANN based short-term traffic flow forecasting in undivided two lane highway. *Journal of Big Data*, 2018;5:48. https://doi.org/10.1186/s40537-018-0157-0
12. L. Tsai, J. Hsieh, K. Fan, Vehicle detection using normalized color and edge map. *IEEE Transactions on Image Processing*, 2007;16(3):850–864. https://doi.org/10.1109/TIP.2007.891147
13. R. Cucchiara, M. Piccardi, Vehicle Detection under Day and Night Illumination, *Proceedings of 3rd International ICSC Symposium on Intelligent Industrial Automation (IIA 99)*.

Chapter 13

The role of ethical chatbots for enhancing customer experience

An interdisciplinary perspective

Priyanka Tyagi and Narendra Mohan Mishra
SRM Institute of Science and Technology, Delhi, India

13.1 INTRODUCTION

Artificial intelligence (AI) is a revolutionary technology emerging from the fourth Industrial Revolution (Kumar et al., 2022). Artificial intelligence is a term that is frequently used, often with a degree of confusion about what it means (Ryan, 2022). Since artificial intelligence is a broad category of technologies with several benefits for businesses, now organizations are increasingly turning to AI to generate business value (Enholm et al., n.d.).

The present study is crucial as emerging intelligent environments are assumed to introduce great opportunities to significantly improve human life, both on an individual and societal level. They are considered to provide practical ways to assist people in their daily lives (Burzagli et al., 2022).

In recent years, ethical issues raised by the development of artificial intelligence have drawn increased attention (Sekiguchi & Hori, 2021). Several studies emphasized the merits of artificial intelligence for our society (Ferrario et al., 2020) (Isensee et al., 2021). A closer investigation of how AI-based ethical chatbots impact customer experiences has been neglected in line with the current emphasis placed on ethical chatbot's impact on customer experiences. In the public domain as of mid-2019, there were more than 80 AI ethics standards. Despite this, news reports of unethical usage of AI were very common in 2020. This is due to the fact that the theory of AI ethics is abstract and has only a limited number of real-world applications for those in charge of building algorithms and AI systems (Morley et al., 2021).

The emergence and expansion of artificial intelligence pose fresh and crucial questions to technologists, to mankind, and sentient life in general (Gabriel, 2020), and unethical use of AI could threaten privacy, equality, and general well-being (Zhao et al., n.d.). Therefore, several ethics directives have been publicly disclosed (Hagendorff, 2022). Besides, AI can offer incredibly accurate insights that will help an organization base its decisions on gathered data more effectively. But the usage of AI can be viewed differently by society as moral and ethical questions may be raised, particularly about obtaining and utilizing public data (Anshari et al., 2022). This study intends to determine whether ethical chatbots affect customer experience. By knowing the role of ethical chatbots on customer experiences, marketing managers would be able to make better decisions. The remainder of this research paper is organized as follows: researchers begin with an overview of the previous literature on the importance of ethical chatbot agents, the gaps in the literature, the methodology employed, and the in-depth analysis followed by a conclusion and limitations.

13.2 LITERATURE REVIEW

A growing body of knowledge has emerged regarding the dangers AI poses to human rights in terms of issues with data security and privacy. A rising application will have an impact on how society gathers, uses effectively, and safeguards the data (Cebulla et al., 2022).

Across the world, researchers, decision-makers, and the general public have been debating on the societal and ethical consequences of artificial intelligence (Mao & Shi-Kupfer, 2021). Artificial intelligence operates using sensors, data, algorithms, and computing power (Henz, 2021). It is vital for AI systems to keep track of their realized values and engage in ongoing redesign operations (van de Poel, 2020). To elevate individual and service ethics, rapid advances also call for technical expertise (P. Kumar et al., 2021). Today, media covers ethical issues surrounding artificial intelligence technology as they become more prevalent in our daily lives. The selection and preparation of the data, the choice and configuration of algorithmic tools, and fine-tuning of various parameters based on interim results are the three iterative steps that commonly accompany the implementation of an AI system and are under the control of AI engineers (Ouchchy et al., 2020).

Moreover, modern AI is regularly attacked for its "black box" features despite having humans involved at every level of the development process, from conception and design to implementation. Sometimes, users are unsure of what is happening internally or how and why certain outcomes occur (Bickley & Torgler, 2022). Issues about integration of artificial intelligence into social life have spread around the world, and there are now more efforts than ever to reduce the harm that occurs from applying such systems (Foffano et al., 2022). The concept of using artificial intelligence for social good is becoming more popular among members of the AI community. But the idea behind what makes AI socially beneficial is still only partially understood (Floridi et al., 2020). However, artificial intelligence offers a wide range of possible applications and an understanding of ethical concerns. This awareness has led to a rise in the number of frameworks, codes, and guidelines for ethics that are founded on principles (Morley et al., 2021). Several initiatives are aimed at developing guidelines and rules to assure that the development, implementation, and application of AI are morally acceptable (Stahl et al., 2022). The ethical implications of AI systems such as robots, chatbots, avatars, and other intelligent agents are a key area of recent research and development (Dignum, 2018). Recently revolution driven by artificial intelligence has an ever-greater impact on prosperity, the planet, and the people (Goh & Vinuesa, 2021). Hence, it is crucial that autonomous systems, whether software or hardware, that communicate with customers behave ethically (Anderson & Anderson, 2021).

Today, it is increasingly challenging to define AI's precise boundaries and what it specifically includes in the field of study and applications has grown tremendously (Devedzic, 2022). The advancements in artificial intelligence enable automation of some consumer tasks and micro-targeting marketing strategies (André et al., 2018). Moreover, artificial intelligence is altering the consumer behavior and the way customers purchase. The use of AI in businesses offers incredible benefits (Giroux et al., 2022). Sinceunderstanding consumer satisfaction in marketing is a crucial aspect, customer service representatives offer a variety of information for assessing customer satisfaction (Ahmed et al., 2022). Chatbot agents communicate with customers in businesses, improving customer experiences (Yen & Chiang, 2021). Chatbots are employed for marketing and customer care will depend on customer experience. For these chatbots, the customer experience is especially focused on whether the customer receives suitable answers to their inquiries and whether their interaction with the chatbot promotes their search for a solution (Følstad & Taylor, 2021). Previous research emphasizes the chatbot's capacity to impact society while assuring adherence to moral and ethical guidelines for artificial intelligence (Ng et al., 2022). The following hypotheses are put forth in this study based on earlier research studies.

H_1: Relationship commitment of ethical chatbots positively affects the customer experience.
H_2: Perceived benefits of ethical chatbots have a positive impact on customer experience.
H_3: Safety positively influences the customer experience.

13.3 RESEARCH GAP

The influence of ethical chatbots (relationship commitment, perceived benefits, and safety) on customer experience needs to be understood, but no research has yet attempted to theoretically and empirically investigate the phenomenon. Understanding what affects consumers' brand experiences is crucial, whereas studies specifically linking ethical chatbots to customer experiences are lacking. To advance the state of AI technology's current uses, research on ethical chatbots powered by AI in marketing is crucial. The purpose of this present study is to assess the impact of ethical chatbots on customer experience to fill the gap in the existing literature.

13.4 METHODS

The current study aims to investigate the impact of ethical chatbots (relationship commitment, perceived benefits, safety) on customer experiences. A thoughtful study is designed to accomplish these objectives.

13.4.1 Sampling framework and questionnaire design

The sample for this current study consists of Indians specifically from Delhi NCR region customers. Respondents from various demographic profiles participated in an online survey. The participants aged between 18 and 51 and above and represented a wide range of backgrounds and qualifications in terms of education and employment. In the present study, 271 respondents participated in an electronic survey that collected the data for this study. The survey had two sections: the first section of the questionnaire included questions about the respondent's demographic profile. While the second section comprises questions, to know how independent variables influenced the dependent variable. In this study, customer experience is the dependent variable, whereas ethical chatbots (relationship commitment, perceived benefits, safety) are independent variables (Figure 13.1).

13.4.2 Measures

The measurement of constructs by a measurement scale is an important aspect of research (Ringle & Sarstedt, 2016). Based on validity and reliability,

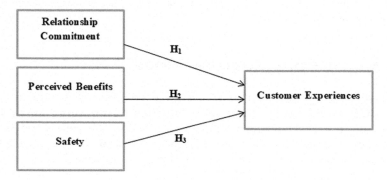

Figure 13.1 Research framework developed by researchers.

a measurement scale's importance can be assessed. Based on previous literature, researchers developed 16 items to measure relationship commitment, perceived benefits, and safety of ethical chatbots (Balakrishnan & Dwivedi, 2021), (Ameen et al., 2021). To measure customer experience, researchers developed six items based on previous literature (Cachero-Martínez &Vázquez-Casielles, 2021). Researchers measured these items using a 5-point Likert scale (1 = "strongly disagree," 5 = "strongly agree").

13.5 RESEARCH TOOLS AND TECHNIQUES

The Construct Reliability was measured by using the Fornell and Larcker (1981) criterion, Moreover, the Construct Reliability value should be 0.7 or more for an acceptable level of reliability. Further, AVE (average variance extracted) was used in the present study to know the convergent validity, the AVE value should be above 0.50. The cross-loading values of the latent variables were used to obtain discriminant validity. The cross-loading values in corresponding variables should be more in comparison to the loadings of other constructs. To determine the suitability of the scale items, a measurement model has been employed. To know the model's prediction accuracy, R^2 (the coefficient of determination) is used (Hair et al., 2019).

13.6 ANALYSIS OF THE MEASUREMENT MODEL

The overall results for the measurement model were acceptable, with a NFI = 0.80 and SRMR = 0.06. To determine the convergent validity and discriminant validity of the observed variables, researchers examined the measurement model. Accordingly, the loading values, the AVE values, and the CR values must all be above 0.6, 0.5, and 0.7 as described in previous studies (Hair et al., 2019). The fulfillment of these above-mentioned requirements assures a model's fitness. In Table 13.1, the measuring features of the constructs are shown in detail.

13.7 ANALYSIS OF STRUCTURAL MODEL

The results of the structural model showed goodness of fit and predictive validity of the model. The results show that relationship commitment (β = 0.560, p < .001), perceived benefits (β = 0.650, p < .001), and safety (β = 0.970, p < .001) have a positive impact on customer experience. Since NFI = 0.861, R^2 = 96.6 %. Thus, H_1, H_2, H_3 were supported.

Table 13.1 Constructs, factor loadings, Cronbach's alpha, composite reliability, and average variance extracted

Constructs	Factor loadings	Cronbach's α	CR	AVE
Relationship Commitment				
I have an emotional attachment to ethical virtual assistants.	0.795			
It would be difficult for me to switch away from ethical chatbots.	0.765			
My life will be disturbed if I switch away from ethical chatbots.	0.736	0.919	0.919	0.658
It would be costlier for me to switch away from the usage of ethical chatbots.	0.962			
I believe chatbot agents are trying to make long term commitment.	0.859			
I believe whenever chatbot agents make an important decision, I know it will be concerned about me.	0.724			
Perceived Benefits				
The ethical chatbots always meet myexpectations.	0.949			
The ethical chatbots has good features.	0.824			
The ethical chatbots shows interest in me as acustomer.	0.804	0.934	0.932	0.735
The ethical chatbots allow me to use theirService whenever I want.	0.809			
I value ethical chatbots that acquire my personal information and personalizethe products for me.	0.890			
SS Safety				
The ethical chatbots that are being used are reliable.	0.788			
Ethical chatbots are honest.	0.948			
I believe ethical chatbots are safe to use.	0.947	0.959	0.959	0.826
I believe ethical chatbots are highly ethical.	0.911			
I am not concerned about the loss of personal information when using these ethical chatbots.	0.941			
Cu Customer Experience				
I feel usage of ethical chatbots a memorable experience.	0.929			
I feel entertained by using ethical chatbots.	0.941			
I feel excited to use ethical chatbots.	0.951	0.956	0.957	0.790
I can use ethical chatbots comfortably.	0.840			
I feel welcomed whenever, I use ethical chatbots.	0.899			
My visual sense and other senses are greatly, affected by ethical chatbots.	0.759			

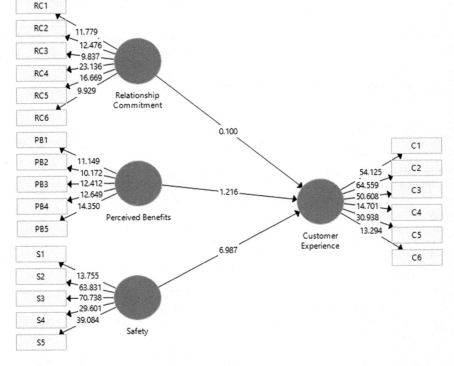

Figure 13.2 Measurement model outcome.

13.8 CONCLUSION

Businesses utilize chatbots more frequently as virtual customer care agents on their websites and mobile applications, but there are a few studies on the effects of ethical chatbots on customer experiences. This study has strengthened the theory by revealing how customer experiences are influenced by deploying ethical chatbots, specifically (relationship commitment, perceived benefits, and safety) ethical chatbots. Since the use of ethical chatbots enhances customer experience, marketing managers and chatbot developers must take special care and continually work to develop better ethical chatbots. The continuous development and deployment of ethical chatbots will give customers a positive brand experience.

Researchers urge marketing managers to increase the usage of ethical chatbots in marketing and customer service to enhance the customer experience. Researchers suggest employing artificial intelligence not only to improve internal procedures but also as a tool (for example, ethical chatbots) to enhance customer service for low-complexity tasks, shifting human resources to other business activities. This research provides information that will be useful to practitioners and marketers worldwide. The empirical results of the current study have given several recommendations that could help practitioners, developers, and marketers. Every study has limits, which can lead to some compelling new research questions. Similarly, artificial intelligence not only offers numerous chances to improve people's lives and the development of economies and communities, but also raises several technological issues. Despite substantial discoveries and insights offered by the present study, this study was time and budget limited.

REFERENCES

Ahmed, C., ElKorany, A., & ElSayed, E. (2022). Prediction of customer's perception in social networks by integrating sentiment analysis and machine learning. *Journal of Intelligent Information Systems*. https://doi.org/10.1007/s10844-022-00756-y

Ameen, N., Tarhini, A., Reppel, A., & Anand, A. (2021). Customer experiences in the age of artificial intelligence. *Computers in Human Behavior*, 114. https://doi.org/10.1016/j.chb.2020.106548

Anderson, S. L., & Anderson, M. (2021). AI and ethics. *AI and Ethics*, 1(1), 27–31. https://doi.org/10.1007/s43681-020-00003-6

André, Q., Carmon, Z., Wertenbroch, K., Crum, A., Frank, D., Goldstein, W., Huber, J., van Boven, L., Weber, B., & Yang, H. (2018). Consumer choice and autonomy in the age of artificial intelligence and big data. *Customer Needs and Solutions*, 5(1–2), 28–37. https://doi.org/10.1007/s40547-017-0085-8

Anshari, M., Hamdan, M., Ahmad, N., Ali, E., & Haidi, H. (2022). COVID-19, artificial intelligence, ethical challenges and policy implications. *AI and Society*. https://doi.org/10.1007/s00146-022-01471-6

Balakrishnan, J., & Dwivedi, Y. K. (2021). Conversational commerce: Entering the next stage of AI-powered digital assistants. *Annals of Operations Research*. https://doi.org/10.1007/s10479-021-04049-5

Bickley, S. J., & Torgler, B. (2022). Cognitive architectures for artificial intelligence ethics. *AI and Society*. https://doi.org/10.1007/s00146-022-01452-9

Burzagli, L., Emiliani, P. L., Antona, M., & Stephanidis, C. (2022). Intelligent environments for all: A path towards technology-enhanced human well-being. *Universal Access in the Information Society*, 21(2), 437–456. https://doi.org/10.1007/s10209-021-00797-0

Cachero-Martínez, S., & Vázquez-Casielles, R. (2021). Building consumer loyalty through e-shopping experiences: The mediating role of emotions. *Journal of Retailing and Consumer Services*, 60. https://doi.org/10.1016/j.jretconser.2021.102481

Cebulla, A., Szpak, Z., Howell, C., Knight, G., & Hussain, S. (2022). Applying ethics to AI in the workplace: The design of a scorecard for Australian workplace health and safety. *AI and Society*. https://doi.org/10.1007/s00146-022-01460-9

Devedzic, V. (2022). Identity of AI. *Discover Artificial Intelligence*, 2(1). https://doi.org/10.1007/s44163-022-00038-0

Dignum, V. (2018). Ethics in artificial intelligence: Introduction to the special issue. In *Ethics and Information Technology* (Vol. 20, Issue 1). Springer, Netherlands. https://doi.org/10.1007/s10676-018-9450-z

Enholm, I. M., Papagiannidis, E., Mikalef, P., & Krogstie, J.(n.d.). *Artificial intelligence and business value: A literature review*. https://doi.org/10.1007/s10796-021-10186-w/Published

Ferrario, A., Loi, M., & Viganò, E. (2020). In AI we trust incrementally: A multi-layer model of trust to analyze human-artificial intelligence interactions. *Philosophy and Technology*, *33*(3), 523–539. https://doi.org/10.1007/s13347-019-00378-3

Floridi, L., Cowls, J., King, T. C., & Taddeo, M. (2020). How to design AI for social good: Seven essential factors. *Science and Engineering Ethics*, *26*(3), 1771–1796. https://doi.org/10.1007/s11948-020-00213-5

Foffano, F., Scantamburlo, T., & Cortés, A. (2022). Investing in AI for social good: An analysis of European national strategies. *AI and Society*. https://doi.org/10.1007/s00146-022-01445-8

Følstad, A., & Taylor, C. (2021). Investigating the user experience of customer service chatbot interaction: A framework for qualitative analysis of chatbot dialogues. *Quality and User Experience*, *6*(1). https://doi.org/10.1007/s41233-021-00046-5

Fornell, C., & Larcker, D. F. (1981). Evaluating structural equation models with unobservable variables and measurement error. *Journal of Marketing Research*, *18*(1), 39. https://doi.org/10.2307/3151312

Gabriel, I. (2020). Artificial intelligence, values, and alignment. *Minds and Machines*, *30*(3), 411–437. https://doi.org/10.1007/s11023-020-09539-2

Giroux, M., Kim, J., Lee, J. C., & Park, J. (2022). Artificial intelligence and declined guilt: Retailing morality comparison between human and AI. *Journal of Business Ethics*, *178*(4), 1027–1041. https://doi.org/10.1007/s10551-022-05056-7

Goh, H.-H., & Vinuesa, R. (2021). Regulating artificial-intelligence applications to achieve the sustainable development goals. *Discover Sustainability*, *2*(1). https://doi.org/10.1007/s43621-021-00064-5

Hagendorff, T. (2022). Blind spots in AI ethics. *AI and Ethics*, *2*(4), 851–867. https://doi.org/10.1007/s43681-021-00122-8

Hair, J. F., Risher, J. J., Sarstedt, M., & Ringle, C. M. (2019). When to use and how to reportthe results of PLS-SEM. *European Business Review*, *31*(1), 2–24. https://doi.org/10.1108/EBR-11-2018-0203

Henz, P. (2021). Ethical and legal responsibility for Artificial Intelligence. *Discover Artificial Intelligence*, *1*(1), 2. https://doi.org/10.1007/s44163-021-00002-4

Isensee, C., Griese, K.-M., & Teuteberg, F. (2021). Sustainable artificial intelligence: A corporate culture perspective. *Sustainability Management Forum | Nachhaltigkeits Management Forum*, *29*(3–4), 217–230. https://doi.org/10.1007/s00550-021-00524-6

Kumar, P., Dwivedi, Y. K., & Anand, A. (2021). Responsible artificial intelligence (AI) for value formation and market performance in healthcare: The mediating role of patient's cognitive engagement. *Information Systems Frontiers*. https://doi.org/10.1007/s10796-021-10136-6

Kumar, S., Lim, W. M., Sivarajah, U., & Kaur, J. (2022). Artificial intelligence and blockchain integration in business: Trends from a bibliometric-content analysis. *Information Systems Frontiers*. https://doi.org/10.1007/s10796-022-10279-0

Mao, Y., & Shi-Kupfer, K. (2021). Online public discourse on artificial intelligence and ethics in China: Context, content, and implications. *AI and Society*. https://doi.org/10.1007/s00146-021-01309-7

Morley, J., Elhalal, A., Garcia, F., Kinsey, L., Mökander, J., & Floridi, L.(2021). Ethics as a service: A pragmatic operationalisation of AI ethics. *Minds and Machines, 31*(2), 239–256. https://doi.org/10.1007/s11023-021-09563-w

Ng, J., Haller, E., & Murray, A. (2022). The ethical chatbot: A viable solution to socio-legal issues. *Alternative Law Journal, 47*(4), 308–313. https://doi.org/10.1177/1037969X221113598

Ouchchy, L., Coin, A., & Dubljević, V. (2020). AI in the headlines: The portrayal of the ethical issues of artificial intelligence in the media. *AI and Society, 35*(4), 927–936. https://doi.org/10.1007/s00146-020-00965-5

Ringle, Christian M., & Sarstedt, Marko (2016). Gain more insight from your PLS-SEM results: The importance-performance map analysis (October 31, 2015). *Industrial Management & Data Systems, 116*(9), 1865–1886.

Ryan, M. (2022). The social and ethical impacts of artificial intelligence in agriculture: Mapping the agricultural AI literature. *AI and Society.* https://doi.org/10.1007/s00146-021-01377-9

Sekiguchi, K., & Hori, K. (2021). Designing ethical artifacts has resulted in creative design: Empirical studies on the effect of an ethical design support tool. *AI and Society, 36*(1), 101–148. https://doi.org/10.1007/s00146-020-01043-6

Stahl, B. C., Antoniou, J., Ryan, M., Macnish, K., & Jiya, T. (2022). Organisational responses to the ethical issues of artificial intelligence. *AI and Society, 37*(1), 23–37. https://doi.org/10.1007/s00146-021-01148-6

van de Poel, I. (2020). Embedding values in artificial intelligence (AI) systems. *Minds and Machines, 30*(3), 385–409. https://doi.org/10.1007/s11023-020-09537-4

Yen, C., & Chiang, M. C. (2021). Trust me, if you can: A study on the factors that influence consumers' purchase intention triggered by chatbots based on brain image evidence and self-reported assessments. *Behaviour and Information Technology, 40*(11), 1177–1194. https://doi.org/10.1080/0144929X.2020.1743362

Zhao, Y., Zhang, Y., Feng, J. (Wang), Schrock, W. A., & Calantone, R. J. (n.d.). *Brand relevance and the effects of product proliferation across product categories.* https://doi.org/10.1007/s11747-020-00727-1

Chapter 14

An efficient gas leakage detection and smart alerting system using IoT

K. Muthumanickam
Kongunadu College of Engineering and Technology, Trichy, India

P. Vijayalakshmi and S. Kumarganesh
Knowledge Institute of Technology, Salem, India

T. Kumaravel
Kongu Engineering college, Erode, India

K. Martin Sagayam
Karunya Institute of Technology and Sciences, Coimbatore, India

Lulwah M. Alkwai
University of Ha'il, Ha'il, Kingdom of Saudi Arabia

14.1 OVERVIEW

14.1.1 Internet of Things

If we can reach its full potential, the Internet of Things is poised to rank among the most important technological advancements of our time. IoT is a worldwide infrastructure that uses cutting-edge services to connect things relied mainly upon exploiting present or new interoperable infrastructure and also communication technologies. Without human interaction, IoT links electrical devices to a distributed server and helps to exchange information [1–2]. The IoT is considered to be a grid of interconnected smart gadgets that can gather data, distribute information, and frequently respond appropriately on their own. These internet-connected devices, which include sensors, software, and embedded electronics, collect and exchange data. These are typically self-configuring and made for clever communication with nearby nodes.

14.1.2 Layers in IoT

The IoT architecture that serves as an entry point for many applications was designed to connect devices and bring IoT-reliant smart facilities to every home. To transmit/receive diverse information, several communication protocols such as WiFi, narrowband/wideband, ZigBee, and LPWAN are adopted

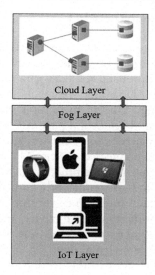

Figure 14.1 Various layers involved in fog computing.

in different layers of the architecture [3]. According to Sadique et al. [38], the typical IoT architecture is made up of three different layers (1) the physical layer, (2) the network layer, and (3) the application layer (Figure 14.1).

Application layer: The application layer in the IoT system serves consumers through web-based and mobile software. According to current usages and developments, IoT offers a wide range of uses in today's high-tech society. Through the good fortune of the IoT system and its untold services, dwellings, homes, and structures, as well as areas for education, transportation, agriculture, health, and businesses, among other things, have become intelligent.

Network layer: This layer is more significant in the IoT system because it serves as an intermediate for information transmission and redirection using a variety of connection protocols like GSM, IPv6, 5G, etc., that link devices offering smart services [4]. Local servers and clouds at the network layer store and process data, acting as a middleman between the network layer and the subsequent layers.

Perception layer: Physical and medium access control layers make up the perception layer and is the initial layer of the Internet of Things architecture. The PHY layer is primarily concerned with hardware, such as smart devices and sensors utilized for transmitting and receiving data utilizing various communication protocols, such as Bluetooth, Zigbee, RFID, etc. [5]. To enable proper communication between physical devices and networks, the MAC layer creates that connection. The MAC connects to network layers using several protocols, including cellular networks (EC-GSM/LTE-M). A significant percentage of large data is produced by plug-and-play devices, which make up the majority of IoT layer devices [5].

14.2 MACHINE LEARNING IN IoT

The goals of IoT are to comprehend human behavior and thought processes, forecast desired and undesirable outcomes, and develop situational management skills. IoT must comprehend the data generated by masses of objects to accomplish all of this. Algorithms for machine learning can be used to achieve this understanding (MLAs). In the IoT pattern, machine learning (ML) can show a crucial and important part. IoT is omnipresent by nature, so being accessible everywhere is one of its main objectives. By mining the data created by millions of connected devices, ML will play a crucial part in this. IoT devices will become more helpful thanks to ML, and only pervasive IoT can exist. To enable IoT to significantly contribute to the accomplishment of its goals, embedded intelligence (EmInt) will be at the center. To improve connectivity, automation, productivity, and efficiency, EmInt fuses product and intelligence [6]. Learning is the only way to become intelligent, whether in the real world or a virtual one.

The basis for intelligence comparable to that of humans may lie in ML's propensity to detect patterns. Enhanced comprehension of the environment around us is provided by a further generalization of these forms into more insightful insights and trends. The real goal of ML in IoT is to provide mechanization by improving learning to support intellect through means of smarter items [7]. After learning from the data, ML offers IoT-reliant devices the capacity to emulate human-like choices and progressively develop their awareness of our surroundings. The humanoid visual system is greatly influenced by information visualization, which helps systems comprehend data and insights.

Information visualization offers its consumers several benefits, including (1) improved understanding without requiring extensive data analysis and (2) the application of cognitive talents to improve data comprehension in humans. IoT will substitute various, presently in use, expensive devices and also preserve technologies, with low-cost sensor-reliant ML structures. For instance, more people perished in poor nations as a result of extreme weather. Radar-based weather monitoring systems (WMSs) are typically used to monitor the weather. Radar WMSs, however, are expensive and not available everywhere throughout the world. In economically underdeveloped nations like Guinea and Haiti, an ML-reliant IoT scheme with a low-cost network that relies on sensors and analysis cloud patterns to forecast different weather conditions has been effectively implemented [8].

IoT will have an impact on how we perceive technology as well as how it advances and improves the state of the planet [9]. Every day, the IoT makes numerous parts of our lives easier and more connected. IoT becomes intelligent and omnipresent thanks to ML. Given that the IoT is made up of a variety of devices, protocols, network-reliant technologies, data kinds, applications, and users, it can serve as a good synonym for the word "heterogeneous." IoT's diverse nature presents ML with several difficulties.

While MLAs won't be able to learn fast from all IoT applications' copious data, a huge number of small-sized data are also generated by this kind of new type of algorithms that are required to study from sparse data [10].

Sensor-reliant technology can't always be precise and trustworthy [11]. Some of the operations that must be carried out on the data before ML commences include data imputation and outlier recognition. To understand the perceptions of data produced by millions of IoT objects, ML provides an intellect to IoT-reliant systems. IoT uses several MLAs that learn from a variety of data which distinguishes ML from IoT. But we still require conventional or we essentially want to have a distinct MLA. MLAs can be classified into several groups, and a few of them won't rely on elaborate mathematical proofs to function.

Early in 2001 [12], Eric Brill et al. produced an intriguing work demonstrating that increasing the amount of training data improves learning instead of improving and developing new-fangled MLAs. IoT will not ever have a big data issue. Massive amounts of data will be produced by the internet's billions of linked IoT devices [13]. As a result, algorithms used in IoT-based ML will be able to acquire from the huge dataset. The IoT area is only moderately accurate because IoT also involves tiny data. Minimal attributes are present in small datasets. Small data can be created by combining big data and can be utilized to define the present situation, start events, and produce small data.

There are several IoT-enabled applications available to governments, businesses, and individuals that make use of ML. MLAs and IoT application areas were discussed by Shanthamallu et al. [14]. On MLAs, which combine ML with IoT infrastructure, Sharma and Nandal concentrated at the same time [15]. IoT viewpoints on ML can be summarized as (1) pattern recognition and (2) data quality. MLAs can be utilized to improve data superiority and eventually lead to greater learning in addition to making predictions from a wealth of data. Before train the MLAs for prediction, for instance, MLAs can also be utilized to detect ascribed data.

14.3 VIEW OF IoT FROM DIFFERENT PERSPECTIVES

14.3.1 Data perspective

Data are gathered by using a range of sensors and offer worth to the IoT paradigm. IoT's sensor arsenal includes both affordable and pricey models. The sensor for temperature finding, for instance, is less expensive than an effective detection system, which is excessively expensive. Perception about the environs and sending that information to smart appliances that will be exercised to anticipate and predict using ML algorithm is one of the main uses of IoT. The learning outcomes are also utilized to create AI that can make decisions. Later, utilizing actuators, the choice is converted into mechanical output [16]. Today, our daily lives are surrounded by billions of sensor-equipped devices. IoT generates and will continue to generate a vast quantity of information that must be analyzed, archived, and saved for use in the future. Even in wealthy economies, IoT networks have not yet been completely integrated. Megasmart-city projects utilizing IoT infrastructure are progressively being developed in emerging economies like Malaysia, India, and others.

IoT refers to a vast volume of live data. For instance, autonomous vehicles by themselves can generate enormous volumes of data. For storing the data for autonomous vehicles, Wang et al. suggested HydraSpace [17] multilevel storage architecture. The cloud might be more scalable, adaptable, and widespread. However, data stored in clouds cannot be used for real-time data analytics. IoT data storage that is based on the edge and fog is therefore crucial [18]. Real-time insights from the observed data can be obtained using AI and ML offered to edge devices. The collected information can be kept on the cloud for later use. An IoT landscape that is more useful and realistic will result from data being transferred on the edge first.

IoT has made the world more interconnected, and this combination of inexpensive sensors and dispersed intelligence will change the way we see the environment. The result of this merger will be huge data that contain important information. Since sensing instruments are not always accurate and dependable, sensing data has serious quality problems. Before supplying IoT data to MLAs, pre-processing is necessary to obtain crucial insights. The data generated in IoT, especially in setups with a lot of sensors [19], are often large and has several dimensions. As a result, it is necessary to use FS approaches to obtain more precise data. The literature has many thorough studies of numerous dimensionality-reduction and feature-selection techniques [20]. The most important variables in the datasets are determined by using MLAs like KNN [21], SVM [22], AdaBoost [24], decision tree [23], random forest [25–26], entropy assessment criteria [27], etc.

14.3.2 Application perspective

Recently, governments all over the world, particularly in the design of [28] and rising economies [29], have given the "smart city" concept a lot of attention. Urban development planning must now include smart cities. A "smart city" has no official definition. However, it may be described as a result of rapid development and cutting-edge information technology that attempts to better the socioeconomic standing of residents and raise the standard of living in general. IoT is about utilizing the internet to connect physical objects in order to promote the efficient exchange of information. Deprived of the practical assistance of IoT that is essential to achieving smart city goals, the idea of a IoT-reliant smart city would not be attainable. Smart grids, smart traffic, smart transportation, smart homes, smart logistics, smart supply chains, smart environments, etc., are only a few examples of real-time applications. These applications are all dependent on IoT and ML ideas. Let's take a smart healthcare application as an example.

14.3.2.1 Smart healthcare

By introducing new, sophisticated sensors that are online and produce crucial data in real time, IoT is transforming the healthcare sector. Intelligent healthcare real-time applications aim to: (1) improve the accessibility of getting smart services; (2) raise the standard of care; and (3) lower the cost of care. Perceiving patterns and important perceptions from healthcare records is the key to attaining the aforementioned goals [30–31]. The automated assessment of a person's welfare and the notification of others of certain health risks for the patient have both been extensively studied. Sharma et al. [32] describes the creation of an intelligent system to keep tabs on people's health in their homes. Solutions for managing weight, monitoring cardiovascular diseases, and monitoring physiology are provided in [33–34]. An abiliment armband multisensor system called Body Media FIT uses machine learning to continuously measure physiological data and regulate body weight.

Moveable-based ML supporting methods for monitoring cardiovascular diseases (M-CVD) are discussed in [35]. Mobile devices are used to track heart conditions. By contextualizing vital health sign patterns with clinical datasets, M-CVD analyzes them locally. SVM has demonstrated great accuracy in identifying individuals at risk for cardiovascular diseases and is used to analyze features taken from wearable sensors and medical datasets to diagnose a patient with cardiovascular diseases [32]. A comprehensive IoT-enabled perceptive healthcare solution that caters to a wider range of patients is offered by IBM Watson. It blends the strength of MLAs with the capacity of healthcare records to produce fresh understandings. While cognitive care offers up-to-date ways for medical professionals to communicate with patients, enhancing investigative inevitability and lowering error rates, the ML-reliant IoT healthcare mechanisms improve people's proactive and preventative healthcare interventions and lower healthcare expenses. IoT-based healthcare solutions can aid in the discovery of insights that can assist improve the standard of healthcare globally.

14.3.3 Industry perspective

IoT is still primarily in its infancy as it develops and is being adopted by the information technology sector. The implementation of Industry 4.0 which will transform industry practices will be considerably improved and accelerated by expanding the usage of IoT infrastructure. As ITI gathers additional information for its IoT-reliant systems, they begin by adding MLAs. Popular IoT-enabled ML systems covered in this chapter include Microsoft Azure, Google TensorFlow, and IBM Watson. Microsoft built the cloud computing system known as Azure [33]. According to Joseph Sirosh, who is working as an ML expert at Microsoft, businesses cannot wait months or weeks for insights produced by data in this brand-new, quickly evolving world of cloud and devices. His remarks are reflected in the most recent changes to Azure. To enhance big data capabilities and prepare for IoT, the Azure cloud added ML through advanced analytics. Customers can use services like Azure Event Hubs and Stream Analytics to process information from IoT ecosystem-supported devices. Microsoft Azure's Scott Hanselman, principal program manager, provided an example of how such a platform participates in different domains and supports ML for IoT.

14.3.4 Evolving trends

In the recent past, intriguing new developments in IoT have emerged, including connected autonomous vehicles, fog computing, edge computing, and deep learning. Additionally, in the recent past, we have witnessed the successful application of IoT in the management and containment of the COVID-19 epidemic. Let's use IoT-based fog computing as an example. As opposed to transmitting sensing data to a distant cloud, edge computing carries computational capability closer to the data [34]. As a result, devices and applications operate more quickly and perform better while transporting data. An emergent architecture known as fog computing might be considered a subgroup of the edge computing pattern. Fog computing allows for the placement of the cloud nearer to the actuators which act on recognized data and the smart objects that produce data. It outlines the requirements for edge computing, networking, data handover, storing, and processing [35]. Fog and edge computing become emerging technologies which will support smart applications on IoT infrastructure and further the global goal of smart cities (Figure 14.2).

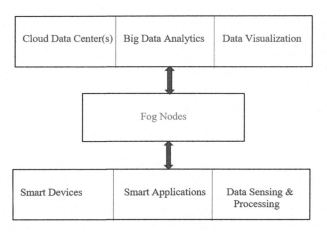

Figure 14.2 Environment of smart computing.

14.4 CHALLENGES IN IoT PARADIGM

By creating a hyperconnected environment, the IoT archetype is the ideal option to fill the space that exists between the physical world and the digital age. However, it has significant technological and nontechnological problems that must be overcome and managed.

14.4.1 Technical challenges

Four key technologies—sensing, networking, actuators, and security—are combined to create the Internet of Things (IoT). There are still connectivity problems in the world. Internet accessibility and mobile connectivity are challenges, especially in developing nations. Another issue is that the current IoT platform has little cross-platform capacity, which slows down adoption. The safety and security of such "things" is a major issue that poses danger to the widespread adoption of IoT systems [36]. Decentralized edge devices, which are more vulnerable to cyberattacks, are what the IoT is all about. A few additional security concerns also arise due to the lack of universal standards for authorization and authentication for IoT devices. One of the most important concerns that IoT must answer in order to become a successful and accepted paradigm is security.

14.4.2 Business challenges

A huge economic potential exists in services, applications, and manufacturing, thanks to IoT. But it is unable to match the publicity which was generated. Businesses are struggling with obstacles such as the absence of worldwide platforms, the absence of business standards, data collecting, and safety concerns, among others [37]. One significant challenge is the lack of IoT experts with the necessary skills. Aside from this, international market glitches such as COVID-19, the Russia vs Ukraine crisis, and lacking global chips have slackened down IoT development and decreased industry and government enthusiasm as priorities have changed.

14.5 CASE STUDY

14.5.1 Objective of the case study

When the data from the gas sensor is submitted to prediction, the location authority receives a prompt notification. To forecast how violent the accident would be, a dataset that is retrieved based on the previous accident is employed. The data from the gas sensor are analyzed using predictive analysis utilizing a dataset made up of accident reports from the past. The prediction study of the liquefied petroleum gas (LPG) accident leakage was done using the Random Forest method. As a result, proper action that should be taken following the accident is anticipated beforehand. When the data from the gas sensor is submitted to prediction, the location authority receives a prompt notification. To forecast how violent the accident would be, a dataset that is retrieved based on the previous accident is employed. The data from the gas sensor is analyzed using predictive analysis utilizing a dataset made up of accident reports from the past. The Random Forest algorithm is used for the prediction analysis of the LPG accident leakage. As a result, proper action that should be taken following the accident is anticipated beforehand.

14.5.2 Review of existing solutions

Recent technologies like convolutional neural networks and machine learning can be effectively used in the medical field to optimize the prediction of many diseases [39]. In addition, in the field of disaster management, several academics have been working in IoT and predictive-reliant algorithms in recent years to foresee, stop, and steer clear of big accidents globally. It helps to reduce the loss of life and financial resources. This is taken directly from [40], a review report on IoT-based gas leak detection. This automatic detection and alerting system has an advantage over the manual approach. There exist one goal gas and continuous data transfer, and lack of M2M are drawbacks of this. IoT-based LPG Gas Detection is found in [41]. The benefit is that it can locate the leak and transmit the info to a website. Additionally, temperature and humidity have a role in this system. As a result, this component would alter the system's sensitivity. The benefit of [41]'s automatic gas leak recognition and deterrence system is that it prevents fire disasters in buildings and saves lives.

Algorithms were trained using supervised machine learning techniques on temperature datasets to categorize a region as "Gas leaking" or "Normal" [42]. The best prediction performance came from the regression logistic approach, demonstrating that "Gas leakage" areas can be found in programmed decision-making system in an industrial environment. The method presented in [43] utilizing multimodal AI fusion approaches provides a novel method to sense and detect gaseous emissions. Since most of the gases and its vapors lack any discernible color, odor, or taste, they present a challenge to our five senses. Sensor fusion is necessary for robust and consistent detection in a number of real-world applications since sensing reliant on one sensor may not be precise.

Another difficult task that calls for technological attention is finding a specific gas or several different gases in a mixture of gases. A Colorimetric Tape can be used as one of the existing approaches for mixed gas detection [44]. With this technique, a dry piece of tape reacts with the gas being released and creates a unique stain for each gas being tested. Adbul Majeed [45] proposed a strategy for choosing the topmost weighted vital features from large datasets, which increased the accuracy and time complexity of machine learning models.

In order to locate the point of leakage in the ground utilizing six separate gas sensors, Bilgera et al. [46] suggested a fusion of multiple AI models for Gas Source Localization. The system utilizing a thermal camera to detect methane gas and ethane gas leaks is presented in [47]. A technique for identifying gas leaks by exploiting the properties of infrared image investigation was presented by Jadin and Ghazali [48–51]. The technique of image processing, which includes data collection, image pre-processing, processing, vital feature identification, and classification, was used to design the system.

Due to their extremely complicated architectural frameworks, advanced frameworks like AlexNet will increase the computational difficulty of the system. Utilizing CNN enables quicker processing and is appropriate for use in real-time systems. The findings demonstrate that the fused output has lower false positive and false negative rates than the distinct modalities. Typically, a gas sensor collection together with a thermal camera will be used in the experimental setup to gather real-time data, which will be used to pre-process the data and evaluate the framework that has been created.

14.5.3 Architecture of the case study

The design, manufacture, and testing of the prototype under actual working circumstances add yet another novel element to this study. As shown in Figure 14.3, the suggested approach would be able to predict the conditions

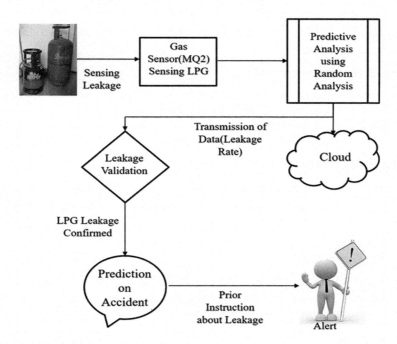

Figure 14.3 Prototype of the case study.

of distributed generators accurately and only plan trips to those that are malfunctioning. The wire and the energy meter are both monitored by sensors in the suggested system.

14.5.3.1 IoT connection module

- The Internet of Things is utilized to collect data from real-world physical objects and communicate them to the virtual world for analysis.
- The sensitive material is inexpensive and has reduced conduction in unsoiled air for the MQ2 gas sensor.
- The Arduino board is linked to the MQ2 Gas sensor (Arduino UNO).
- The MQ2 Gas sensor's input is read using a microcontroller.
- The Wi-Fi unit is coupled because it establishes a physical link to the internet and allows for the transmission of data to the cloud over that link.

14.5.3.2 Detection module

- The MQ2 sensor is utilized to find gas molecules like LPG, propane, and methane that are present in the environment between 200 and 10,000 parts per million (ppm).
- The MQ2 gas sensor measures the rate of LPG leakage, and the microcontroller reads the data.
- The NodeMCU transmits the data via the internet (Wi-fi Module).

14.5.3.3 Integration module

The prediction analysis will function with the sensor data that is implemented in the Arduino IDE and is integrated with the hardware-developed code for the accident prediction.

14.5.3.4 NodeMCU

NodeMCU is basically a microcontroller panel that is used to read sensor data and transmit it as an output, as well as operate and monitor the operations of a particular system.

14.5.3.5 Gas sensor

- The MQ2 gas sensor type is employed, and each device covers 112 m^2, which is roughly equivalent to 100 m^2.
- The area of coverage for heat detectors is 56 m^2, which is rounded down to 50 m^2.
- Thus, the gas level in the ambient air is detected by this MQ2 gas sensor.

14.5.3.6 Prediction module

- This module is responsible for prediction analysis which is employed to receive notification of the impending accident.
- With the aid of the data (LPG leakage) obtained from the sensor, it will be possible to prevent it from happening.
- The most often used predictive analysis is linear regression.
- In particular, the dataset used during the analysis phase is compared to the present LPG leakage rate using the Random Forest algorithm.

14.5.4 Implementation

Gas leaks in smart homes are a severe problem which leads to loss of lives and property. Solutions for prevention and warning are currently available. In smart homes, this system will help to improve safety and protect property. To concentrate on the development of important trends in gas drip and fire accidents in smart homes, following are the data acquisition step's criteria: The Random Forest classification algorithm is associated to a gas sensor (MQ2sensor) through a NodeMCU. The LPG level is sensed using a gas sensor, and data transmission is done using a NodeMCU (Figure 14.4).

The data are combined and sent to the PC for processing via an RF module by the aggregation node. Additionally, data pre-processing is done to

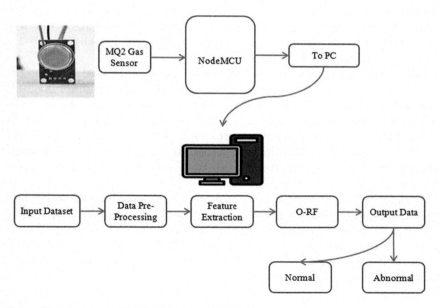

Figure 14.4 System with IOT hardware with RF.

raise the caliber of the data. Using predetermined threshold values for the parameters, the approach for detecting the incidents is used. Predicting unknown information and determining the degree of risk are the major goals. If the sensor data is higher than the threshold value, the alert level grew more significant (based on location). The system will send out mail notifications to the relevant individuals at appropriate times (Location Incharge). Additionally, the information is kept in the cloud (ThingSpeak) for later use.

14.5.4.1 Data processing

Information was gathered from the Environmental Defense Fund (EDF) which contains information on LPG leakage accidents. Data pre-processing works as the process of transforming raw information into something that is utilized by a machine learning prototype. The data which will be put into the prototype is contained using the training dataset. Simply by utilizing this data, our model would learn. It contains 70% of the data. The data that can be taken for testing the trained and approved prototype belongs to the testing process. The data make up 30% of it. Diagrammatically it is shown in Figure 14.5.

Artificial intelligence's machine learning subfield uses intelligent software to enable machines to carry out tasks with proficiency [9]. It makes use of information derived from real-world features which can then be handled by either a model or an algorithm capable of performing forecasting tasks, including classification, estimation value, and forecasting. The model or algorithm can be thought of as an approximate representation of the procedure which we need computers to mimic. It employs various algorithms like classification-based or supervised/unsupervised learning techniques, reinforcement learning, and unclassified data [9].

Algorithms for supervised machine learning are constructed based on the entry attributes and organization of the output-related examples. Regions classified as "Leakage" and "Normal" were employed in this investigation. The previous experiment yielded data for training as well as testing various algorithms that mapped input features such as temperature, product, and level of the leak over time (t0,2), (t0,4), and (t0,6) (t20). Table 14.1 displays the overall training dataset used to validate the method; a total of 32 cases were used.

In order to assess the efficiency of supervised machine learning techniques in classifying and predicting the "Leakage" zone as True Positives, namely TP, and the zone indicating "Normal," is treated as True Negative, specifically TN, four different algorithms, including Decision Tree, Naive Bayes, Logistic Regression, and Support Vector Machine, were used for training and testing.

An efficient gas leakage detection and smart alerting system 207

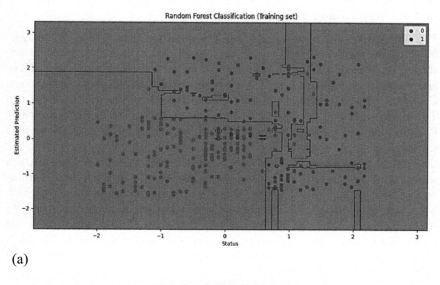

(a)

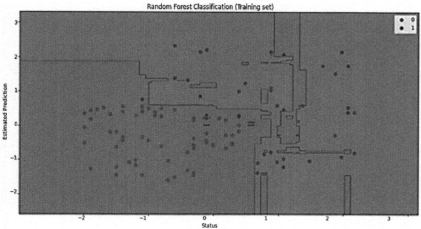

(b)

Figure 14.5 Data training and testing. (a) Data training and (b) Data testing.

Table 14.1 Training and testing data

	Training data		Testing data		Total
No. of Instances	25	10	25	10	70

14.5.4.2 Results and discussion

The simulation testbed of the case study is shown in Figure 14.6. The majority of the methods utilized to train the model using supervised learning are shown in Table 14.2 which compares their performance outcomes. Here are the most popular metrics for evaluating classification task proficiency. Comparing the performance of the four algorithms reveals that the Decision Tree performed well in terms of features like the precision of the system and Recall value, next, Naive Bayes incurs a high false positive, signifying that the method stays "Normal."

Better results were obtained using logistic regression: TP classification accuracy of 100% and FP classification accuracy of 0%. With the FN influencing 50% of accurate TP, there is room for enhancing the value of recall. More manufacturing process information for air conditioners, primarily from "Leakage," is required to increase the accuracy evaluation utilizing more evenly distributed TP and TN instances and boost trust in the forecasts. We also send an alert message to the owner of a house through mail service as shown in Figure 14.7.

Figure 14.6 Simulation of the proposed work.

Table 14.2 Algorithm's assessment on classification

Algorithm	Accuracy	Recall	Precision	FP	FN
Naïve bayes	48	53	18	6	6
Decision Tree	86	0	0	3	13
Support Machine Vector	93	28	100	0	9
Logistic Regression	96	53	100	0	5

An efficient gas leakage detection and smart alerting system 209

Figure 14.7 Alert message about gas leakage through mail.

14.6 CONCLUSIONS

IoT is now much more developed. IoT applications are being used in more settings now. Leveraging IoT's prospects has attracted a lot of interest from people, organizations, and governments. How IoT will learn to give a high level of automation is still a crucial unanswered subject. The solution is found in other areas of computer science especially the use of ML to comprehend and behave like humans. Instead of conducting a traditional literature review in this study, we tried emphasizing the significance of ML for the development of numerous IoT applications. We categorize ML advancements in the IoT based on three factors: industry, application, and data. The IoT ecosystem can fully or partially benefit from the research we have done. Although the Internet of Things uses smart devices to connect physical entities with virtual ones and make life easier, more pleasant, and smoother, security has become a top concern for IoT systems in order to keep their services running.

REFERENCES

1. A. Abane, M. Daoui, S. Bouzefrane, P. Muhlethaler A lightweight forwarding strategy for named data networking in low-end IoT. *Journal of Network and Computer Applications*, vol. 148, 1–12, 2019.
2. H. Haddad Pajouh, et al. *A survey on Internet of Things security: Requirements, challenges, and solutions.* Internet of Things, 2019.
3. M. Saadeh, A. Sleit, K. E. Sabri, W. Almobaideen Hierarchical architecture and protocol for mobile object authentication in the context of IoT smart cities. *Journal of Network and Computer Applications*, vol. 121, 1–19, 2018.
4. A. Singh, A. Payal, S. Bharti. A walkthrough of the emerging IoT paradigm: Visualizing inside functionalities, key features, and open issues. *Journal of Network and Computer Applications*, vol. 143, 111–151, 2019.
5. M. K. Saggi, S. Jain A survey towards an integration of big data analytics to big insights for value-creation. *Information Processing & Management*, vol. 54, no. 5, 758–790, 2018.
6. B. Guo, D. Zhang, Z. Yu, Y. Liang, Z. Wang, X. Zhou From the Internet of Things to embedded intelligence. *World Wide Web*, vol. 16, 399–420, 2013.
7. R. Jain Recent machine learning applications to Internet of Things (IoT). abstract: Table of contents: 1–19, 2020. Available online: https://www.cs.wustl.edu/~{}jain/cse570-15/ftp/iot_ml/#sec7 (accessed on 5 August 2022)
8. B. L. Davarzani, M. Purdy The Internet of Things is now a thing 2015. Available online: https://ssir.org/articles/entry/the_internet_of_things_is_now_a_thing (accessed on 5 August 2022)
9. M. Purdy, D. Ladan The growth game-changer: How the industrial internet of things can drive progress and prosperity, 2015. Available online: https://www.bl.uk/business-and-management/collection-items/suppressed-by-publisher/accenture/growth-gamechanger-how-the-industrial-internet-of-things-can-drive-progress-and-prosperity (accessed on 5 August 2022)
10. D. Dansana, R. Kumar, A. Parida, R. Sharma, J. D. Adhikari et al., Using susceptible-exposed-infectious-recovered model to forecast coronavirus outbreak. *Computers, Materials & Continua*, vol. 67, no. 2, 1595–1612, 2021.
11. M. T. Vo, A. H. Vo, T. Nguyen, R. Sharma, T. Le, Dealing with the class imbalance problem in the detection of fake job descriptions, *Computers, Materials & Continua*, vol. 68, no. 1, pp. 521–535, 2021.
12. Smriti Sachan, Rohit Sharma, Amit Sehgal Energy efficient scheme for better connectivity in sustainable mobile wireless sensor networks, *Sustainable Computing: Informatics and Systems*, vol. 30, 100504, 2021.
13. S. Ghanem, P. Kanungo, G. Panda et al. Lane detection under artificial colored light in tunnels and on highways: an IoT-based framework for smart city infrastructure. *Complex & Intelligent Systems*, 2021. https://doi.org/10.1007/s40747-021-00381-2

14. S. Sachan, R. Sharma, A. Sehgal SINR based energy optimization schemes for 5G vehicular sensor networks. *Wireless Personal Communications*, 2021. https://doi.org/10.1007/s11277-021-08561-6.
15. I. Priyadarshini, P. Mohanty, R. Kumar et al. A study on the sentiments and psychology of twitter users during COVID-19 lockdown period. *Multimedia Tools and Applications*, 2021. https://doi.org/10.1007/s11042-021-11004-w
16. C. Azad, B. Bhushan, R. Sharma et al. Prediction model using SMOTE, genetic algorithm and decision tree (PMSGD) for classification of diabetes mellitus. *Multimedia Systems*, 2021. https://doi.org/10.1007/s00530-021-00817-2
17. I. Priyadarshini, R. Kumar, L.M. Tuan et al. A new enhanced cyber security framework for medical cyber physical systems. *SICS Software-Intensive Cyber-Physical Systems.*, 2021. https://doi.org/10.1007/s00450-021-00427-3
18. Ishaani Priyadarshini, Raghvendra Kumar, Rohit Sharma, Pradeep Kumar Singh, Suresh Chandra Satapathy Identifying cyber insecurities in trustworthy space and energy sector for smart grids, *Computers and Electrical Engineering*, vol. 93, 107204, 2021.
19. Rajesh Singh, Rohit Sharma, Shaik Vaseem Akram, Anita Gehlot, Dharam Buddhi, Praveen Kumar Malik, Rajeev Arya Highway 4.0: Digitalization of highways for vulnerable road safety development with intelligent IoT sensors and machine learning, *Safety Science*, vol. 143, 105407, 2021.
20. L. Sahu, R. Sharma, I. Sahu, M. Das, B. Sahu, R. Kumar Efficient detection of Parkinson's disease using deep learning techniques over medical data. *Expert Systems*, e12787, 2021. https://doi.org/10.1111/exsy.12787
21. R. Sharma, R. Kumar, D. K. Sharma et al. Water pollution examination through quality analysis of different rivers: A case study in India. *Environment, Development and Sustainability*, 2021. https://doi.org/10.1007/s10668-021-01777-3
22. D. H. Ha, P. T. Nguyen, R. Costache et al. Quadratic discriminant analysis based ensemble machine learning models for groundwater potential modeling and mapping. *Water Resources Management*, 2021. https://doi.org/10.1007/s11269-021-02957-6
23. G. Dhiman, R. Sharma SHANN: An IoT and machine-learning-assisted edge cross-layered routing protocol using spotted hyena optimizer. *Complex & Intelligent Systems*, 2021. https://doi.org/10.1007/s40747-021-00578-5
24. Rohit Sharma, Deepak Gupta, Zdzislaw Polkowski, Sheng-Lung Peng Introduction to the special section on big data analytics and deep learning approaches for 5G and 6G communication networks (VSI-5g6g). *Computers and Electrical Engineering*, vol. 95, 107507, 2021. https://doi.org/10.1016/j.compeleceng.2021.107507
25. Prabh Deep Singh, Gaurav Dhiman, Rohit Sharma, Internet of Things for sustaining a smart and secure healthcare system. *Sustainable Computing: Informatics and Systems*, vol. 33, 100622, 2022. https://doi.org/10.1016/j.suscom.2021.100622
26. R. Sharma, R. Arya A secure authentication technique for connecting different IoT devices in the smart city infrastructure. *Cluster Computing*, 2021. https://doi.org/10.1007/s10586-021-03444-8
27. R. Sharma, R. Arya Secure transmission technique for data in IoT edge computing infrastructure. *Complex & Intelligent Systems*, 2021. https://doi.org/10.1007/s40747-021-00576-7

28. M. Rai, R. Sharma, S. C. Satapathy et al. An improved statistical approach for moving object detection in thermal video frames. *Multimedia Tools and Applications*, 2022. https://doi.org/10.1007/s11042-021-11548-x
29. Rasika Verma, Rohit Sharma Dual notched conformal patch fed 3-D printed two-port MIMO DRA for ISM band applications. *Frequenz*, 2022. https://doi.org/10.1515/freq-2021-0242
30. N. Sharma, R. Sharma Real-time monitoring of physicochemical parameters in water using big data and smart IoT sensors. *Environment, Development and Sustainability*, 2022. https://doi.org/10.1007/s10668-022-02142-8
31. R. Anandkumar, K. Dinesh, Ahmed J. Obaid, Praveen Malik, Rohit Sharma, Ankur Dumka, Rajesh Singh, Satish Khatak, Securing e-Health application of cloud computing using hyperchaotic image encryption framework, *Computers and Electrical Engineering*, vol. 100, 107860, 2022. https://doi.org/10.1016/j.compeleceng.2022.107860
32. R. Sharma, Q. Xin, P. Siarry W.-C. Hong, Guest editorial: Deep learning-based intelligent communication systems: Using big data analytics. *IET Communications*, 2022. https://doi.org/10.1049/cmu2.12374
33. Rohit Sharma, Rajeev Arya, UAV based long range environment monitoring system with Industry 5.0 perspectives for smart city infrastructure, *Computers & Industrial Engineering*, vol. 168, 2022, 108066, ISSN 0360-8352, https://doi.org/10.1016/j.cie.2022.108066
34. M. Rai, T. Maity, R. Sharma et al. Early detection of foot ulceration in type II diabetic patient using registration method in infrared images and descriptive comparison with deep learning methods. *The Journal of Supercomputing*, 2022. https://doi.org/10.1007/s11227-022-04380-z
35. R. Sharma, D. Gupta, A. Maseleno S.-L. Peng Introduction to the special issue on big data analytics with internet of things-oriented infrastructures for future smart cities. *Expert Systems*, vol. 39, e12969, 2022. https://doi.org/10.1111/exsy.12969
36. R. Sharma, D. Gavalas, S. -L. Peng Smart and future applications of Internet of Multimedia Things (IoMT) using big data analytics. *Sensors*, vol. 22, 4146, 2022. https://doi.org/10.3390/s22114146
37. Sharma, R, Arya, R. Security threats and measures in the Internet of Things for smart city infrastructure: A state of art. *Transactions on Emerging Telecommunications Technologies*, e4571, 2022. https://doi.org/10.1002/ett.4571
38. K. M. Sadique, R. Rahmani, P. Johannesson Towards security on Internet of Things: Applications and challenges in technology. *Procedia Computer Science*, vol. 141, 199–206, 2018.
39. Y. Patil 6 key challenges to consider for successful IoT implementation. Available online: https://www.saviantconsulting.com/blog/iot-implementation-challenges-enterprises.aspx (accessed on 2 June 2022).
40. R. A. Sadek Hybrid energy aware clustered protocol for IoT heterogeneous network. *Future Computing and Informatics Journal*, vol. 3, no. 2, 166–177, 2018.
41. P. Pandiaraja, K. Muthumanickam Convolutional neural network-based approach to detect covid-19 from chest X-ray images. In D. P. Agrawal, N. Nedjah, B. B. Gupta, G. Martinez Perez (eds) *Cyber security, privacy and networking. Lecture Notes in Networks and Systems*, vol 370, pp 231–245. Springer, Singapore, 2022.

42. Mithun Mukherjee Leishu, Xu Xioaling, Kun Wang.: A survey on gas leakage source detection and boundary tracking with wireless sensor network. *IEEE Access*, vol. 4, 1700–1715, 2016.
43. Muhammad Benny Chaniago, Revy CahyaAlamsyah: Design of cloud computing based gas detection systems. *International Journal on Informatics for development*, vol. 8, no. 2, 69–73, 2019.
44. S. F. Moreira, V. Shah, M. L. R. Varela, A. C. Monteiro, G. D. Putnik Supervised machine learning applied to gas leak detection in air conditioner cooling system. In: *14th International Conference on Axiomatic Design (ICAD 2021)*, pp. 1–6, 2021.
45. Parag Narkhede, Rahee Walambe, Shruti Mandaokar, Pulkit Chandel, Ketan Kotecha, George Ghinea Gas detection and identification using multimodal artificial intelligence based sensor fusion. *Applied System Innovations*, vol. 4, no. 1, 1–14, 2021.
46. MDC Systems Inc. Detection Methods, https://mdcsystemsinc.com/detection-methods/ accessed on 19 March 2022.
47. A. Majeed Improving time complexity and accuracy of the machine learning algorithms through selection of highly weighted top k features from complex datasets. *Annals of Data Science*, vol. 6, 599–621(2019).
48. C. Bilgera, A. Yamamoto, M. Sawano, H. Matsukura, H. Ishida Application of convolutional long short-term memory neural networks to signals collected from a sensor network for autonomous gas source localization in outdoor environments. *Sensors*, vol. 18, 4484, 2018.
49. X. Pan, H. Zhang, W. Ye, A. Bermak, X. Zhao A fast and robust gas recognition algorithm based on hybrid convolutional and recurrent neural network. *IEEE Access*, vol. 7, 100954–100963, 2019.
50. B. Thiyaneswaran, K. Anguraj, S. Kumarganesh, K. Martin Sagayam, S. Ghosh IOT based smart cold chain temperature monitoring and alert system for vaccination container. *International Journal of Przegląd Elektrotechniczny*, vol. 2022, no. 8, pp. 206–208, 2022.
51. Marathe, S. Leveraging drone based imaging technology for pipeline and RoU monitoring survey. In: *Proceedings of the Asia Pacific Health, Safety, Security, Environment and Social Responsibility*, Kuala Lumpur, Malaysia, 2019.
52. B. Thiyaneswaran, P. Elayaraja, P. Srinivasan, S. Kumarganesh, K. Anguraj IOT based air quality measurement and alert system for steel, metal and copper processing industries. *Materials Today: Proceedings Journal*, Elsevier https://doi.org/10.1016/j.matpr.2021.02.696
53. P. Srinivasan, S. Anthoniraj, K. Anguraj, S. Kumarganesh, B. Thiyaneswaran, Development of embedded based biometric authenticated vehicles ignition system. *Materials Today: Proceedings Journal*, Elsevier. https://doi.org/10.1016/j.matpr.2021.03.632
54. B. Thiyaneswaran, S. Kumarganesh, K. Martin Sagayam, D. Hien An effective model for the iris regional characteristics and classification using deep learning alex network. *IET Image Processing*, vol. 17, no. 1, 227–238, 2023. https://doi.org/10.1049/ipr2.12630
55. S. Kumarganesh, S. Anthoniraj, T. Senthil Kumar, P. Elayaraja, et al. A novel analytical framework is developed for wireless heterogeneous networks for video streaming applications. *Journal of Mathematics*, vol. 2022, no. 1, 1–7, 2022. https://doi.org/10.1155/2022/2100883

Chapter 15

Principles and goals of Industry 4.0

K. Umapathy, D. Muthukumaran, G. Poojitha, A. Sai Samvida and S. Prabakaran
SCSVMV Deemed University, Kanchipuram, India

Safia Yasmeen
Alfaisal University, Saudi Arabia

15.1 INTRODUCTION

The problems associated with Industry 4.0 are analyzed among researchers, business holders and general institutions. There is a modulation in the environment of universal manufacturing lines recently as an outcome of promotions along with creations. Industry 4.0 can be correlated with certain past uprisings and illustrate the best remarkable troubleshooting changes in the industrial area with correspondence to technological developments [1]. The arrival of the steam engine hastened the initial revolution with industry that got initiated in England in the 1850s. The next revolution with industry rolled out in the European countries and the US in the 1950s. This mutiny was characterized by large-scale production and replacement of condensation power both chemically and electrically [2]. The third revolution was provoked by the design of microchips (IC). Each Industrial Revolution concentrated on increasing the level of production. The above revolutions had a noteworthy effect on commercial performances, thus enhancing productivity and adaptability by making use of creative technological developments which include digitization, power and types of engines [3]. Industry 4.0 can be conclusively quoted as the fourth revolution which is an extremely compounded arrangement frequently argued with determination. It includes revealing the effect on the organizational field as it establishes applicable advancements linked with appropriate industries. This evolving concept is a sunshade for the latest industrial standard that comprises Big Data, CPS (Cyber-Physical Systems), Cloud Computing, Augmented reality, Robotics, etc. [4]. Acquisition of the above-mentioned automation combines digital and substantial worlds by accepting a group of future organizational advancements and is vital in the expansion of additional inventive manufacturing activities. This assumption involves devices and appliances which interchange messages separately, which ensures an intelligent domain for production [5]. Figure 15.1 shows the principles of Industry 4.0.

Figure 15.1 Principles of Industry 4.0.

15.2 AN OVERVIEW OF INDUSTRY 4.0

The above-discussed revolutions with industry have been derived from variations in industrial designs: automaton through means of power and water, large-scale manufacturing in suitable patterns and motorization by means of computers [6]. The initial revolution evolved in the UK later mostly with the approach of power, water and collectivization of production. Then the second revolution initiated devices and machines based on electrical power. The railway establishment into a manufacturing scheme was aided by the mass production of steel, which consecutively improved huge manufacturing [7]. The third revolution started by acquiring various types of devices within the equipment. This led to completely automated implementation inside the production activities. The fourth revolution is currently under execution and is called "Industry 4.0." This can be determined in any organization by employing appropriate techniques of communication and information. Manufacturing systems using automated technology are increased by grid links connected on the web. These systems facilitate appropriate devices and the manufacturing of information with respect to them [8]. All the above are interconnected to form a production system in a cyber-physical format and correspondingly industries with manufacturing devices and relevant persons become approximately independent. Machines can anticipate errors and take action on conserving activities all alone [9]. Figure 15.2 shows the generations of revolution.

Machines also have the power to vary the working modes. Because of Industry 4.0, persons can be brought into intelligent networks that will tend toward systematic functioning. The persons involved in maintenance undergo appliance regulation and would rather spend their time directing the matters

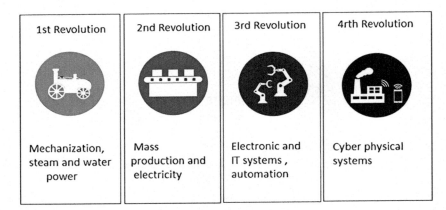

Figure 15.2 Generations of revolution.

Figure 15.3 Industrial revolution.

in place of wasting time searching for technical information [10]. Industry 4.0 is revolutionary with respect to industries. Industry 4.0 is currently a bewitching model and its assisting technologies such as Cloud Manufacturing (CM) and the Internet of Things (IoT) are however imperfectly determined and understudied. Figure 15.3 shows the Industrial Revolution. Figure 15.4 shows a picture of an old steam engine used in the past.

15.3 DEFINITION OF INDUSTRY 4.0

Industry 4.0 is the current revolution that determines the forthcoming manufacturing system's perception. The proposal of Industrial 4.0 was entrenched by a class of executives from various careers as a fundamental

Principles and goals of Industry 4.0 217

Figure 15.4 Old steam engine.

attempt to fuse all production organizations to attain longevity. Basically, as an outcome of Industry 4.0, functioning and productivity will in turn become economic and systematic. These are achieved by the easy exchange of data and the functioning of industries that operate collectively in a logical manner [11, 12].

The Fourth Industrial Revolution gestates quick substitutes to knowledge, companies and methodologies currently due to rising affinity and brilliant techniques [13, 14]. Schwabk et al. distinguished a small number of the latest technologies; the enhancement in these has consequences on factories, administration plans and financial systems [15]. Industry 4.0 also indicates communal, governmental and financial progress from the computer era of the 20th century to a period of interrelation noted by extensive utility [16]. Wang et al. described Industry 4.0 to be contemporary and more enlightened equipment and devices with modern software and wired sensors that can be implemented to design, foresee and command public outcomes and entrepreneur replicas [17, 18]. Subsequently, Industry 4.0 is applicable to denote three eventual subjects relevant to it, for instance, handling with difficulty, ability for creation and malleability [19]. Figure 15.5 shows the Fourth Revolution.

15.4 COMPONENTS OF INDUSTRY 4.0

Industry 4.0 is a tangled scientific standard marked basically by linking, incorporation and computerization, spotlighting the chances for integrating complete elements in an enriched structure and advancements are disturbing the contents between electronic and real-time worlds by combining individuals, things, products and procedures [20]. The enabling technologies of IoT are shown in Figure 15.6.

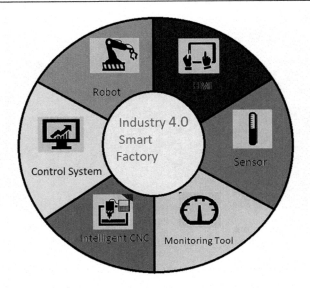

Figure 15.5 Fourth industrial revolution.

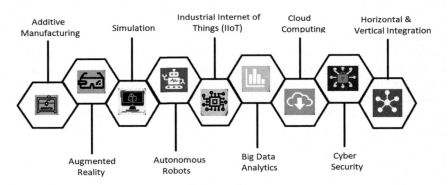

Figure 15.6 Enabling technologies of the IoT.

15.4.1 Cyber-physical system (CPS)

CPS is an amalgamation of manipulated and actual procedures that become important elements of Industry 4.0 execution. Bergera et al. interpreted sensor-based systems and unique varieties of embedded systems that rely on controlling other systems [21]. Subsequently, CPS can uplift users' consciousness and on the other hand, it has a few security issues, i.e., additional consumption surely out-turns into increased dangers.

15.4.2 Cloud systems (CS)

Cloud is a word that is availed against operations. The typical examples are services at remote and quality standard applications. This indicates that cloud has a significant importance in the IT sector but is a bit apart in various business fields. It ameliorates capability by approving those input providers, staff and customers having admittance to identical information during the same time period [22].

15.5 MACHINE-TO-MACHINE (M2M) COMMUNICATION

M2M is a methodology which permits immediate communication between gadgets by implementing any medium in any mode of operation. It can comprehend technical implementation and any form of communication [23]. It is also contemplated to be an important fundamental of Industry 4.0. It is also regarded as an indispensable element of Industry 4.0. The programs are regulated toward the upgradation of endeavors by initiating substitute revenue paths and minimizing expenses and outgoings [24]. Figures 15.7 and 15.8 illustrate the projects and communication of M2M, respectively.

15.6 THE INTERNET OF THINGS

The IoT is an urging idea for integrating different methods and techniques, contingent on communication between actual things and the Internet. Enhancement in this area is broadly considered the most basic driver of Industry 4.0 and has generated the upgradation of various visions and

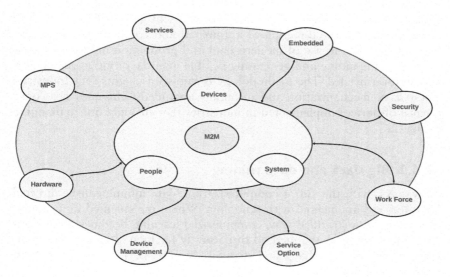

Figure 15.7 Machine-to-machine projects.

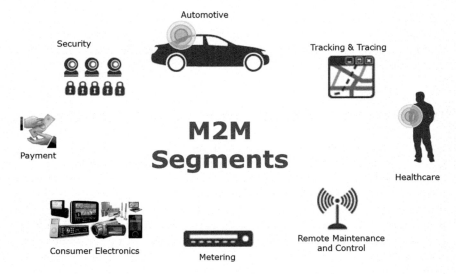

Figure 15.8 Machine-to-machine communications.

meanings for the term "IoT." The IoT relates to the interlink between cars, physical devices, buildings and other institutions armed with IT, electronics and interconnection to combine and contribute information for a brilliant production environment, which can be called a smart factory [25].

15.6.1 Smart factories

A smart factory is a method of assembling to focus on developing idea generation, manufacturing and product interlinks by forwarding away from classical ways in the direction of automated and computerized networks. It points to making use of modern data and production industries for the sake of commanding and generating flexible manufacturing at the greatest acceleration needed. The methodology is known as Lights Out Industries. These days, mechanization and autonomy upgraded techniques as a consequence of devices implemented in industries that strongly utilize manufacturing [26].

15.6.2 Big data and data mining

Nowadays, with the aid of computerization, data mining can be directed, liberated and augmented with schooling. When implemented in a number of sections in a stratified way, computerized learning becomes meticulous. Machine learning is the method of naturally fetching attributes via monitored or liberated studying in a step-wise manner [27]. Figure 15.9 shows the five Vs of Big Data.

Principles and goals of Industry 4.0 221

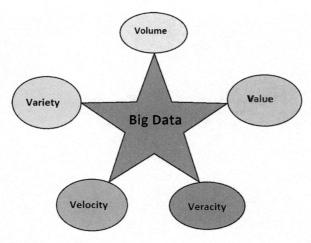

Figure 15.9 Five Vs of Big Data.

15.6.3 Intelligent robotics

In day-to-day life, the latest commodities and systems appear as a consequence of technological upgrades. Android robots will be an element of everyday life quite soon. Modern implementations have led to art that authorizes robots to deal with their surroundings. AI will put up to the improvement of undergoing robot groups collaborating and contributing in attaining various duties determined for a particular purpose [28]. Executing a synergetic robot in an industry will furnish many favors to the company, counting:

- Averting people from doing monotonous and risky work
- Manufacturing the finest goods with praising profitable proportions and also improving yield
- Enhancing motivation when compared with nations with low-wage employment [29]

On the other hand, the profits of the robot implementation are interlinked with the endeavor of a mechanic. In the beginning, there is no coordination between the operator and the machine. But in the end, the workshop is completely divided between the robot and the man [30]. Figure 15.10 shows the interlink between AI and robotics.

15.6.4 Augmented reality

Information retrieved from the cloud is exploited as an input to specific patterns for assessing attainable outlines associated with the product pattern, improvement and manufacturing. Process examining and hiking through the simulation allows individuals to minimize business alteration, danger

222 Artificial Intelligence and Blockchain in Industry 4.0

Figure 15.10 Interlink between AI and robotics.

Figure 15.11 Impact of AR on healthcare.

and adjustment time and improve quality assurance for further procedures and facilities, just prior to the enactment of accustomed in the real corporal world. Augmented reality blended with simulation is an improved methodology for realizing corporal actual surroundings integrated with computerized graphics and illustrations predicted on top of them. Figure 15.11 illustrates the impact of AR on healthcare.

15.6.5 Smart virtual product development (SVPD)

It is a system of merchandise agreement that reinforces methodology which rescues, utilizes and supplies practical study of past supporting incidents in the pattern of SOEs. It was generated to acknowledge the need for the computerized understanding grabbed in manufacturing ideas and supervision. Due to this, the grade of the product and growth span will be enhanced, as needed by the notion of Industry 4.0 [31].

15.7 CHARACTERISTICS OF INDUSTRY 4.0

The key improvement from the classical productivity with regard to Industry 4.0 resulted in four main attributes and qualities [32]. This is shown in Figure 15.12.

- Vertical networking of intelligent productivity schemes
- Horizontal amalgamation through the latest formation of worldwide worth conglomerate connections
- Persisting engineering across the complete value link
- The effect of increasing instrumentation and technologies

15.7.1 Vertical networking

An important feature of Industry 4.0 is vertical networking applied for intelligent productivity in organizations. It set ups an interlink between the various measures of the industry, from the mass production basement, through production observing, command and direction, standard supervision, product oversight, working and so on. This connection to overall collaborative standards supplies a fluid, crystalline information flow which permits data-driven deliberate and prudent choices. Vertical networking upgrades an institution's ability to appropriately acclimate to modify to advertise preconditions and profits from new chances.

15.7.2 Horizontal integration

In Industry 4.0 conviction, horizontal combination determines a collection of miscellaneous procedures and industries in order to achieve globally accepted standards. This is projected at manufacturing intensity as a complete reinforcement for correlated productivity procedures. On the other side, Vertical integration determines a great level of bringing together between manufacturing and highly authorized layers like management of quality and manufacturing jurisdiction [33].

224 Artificial Intelligence and Blockchain in Industry 4.0

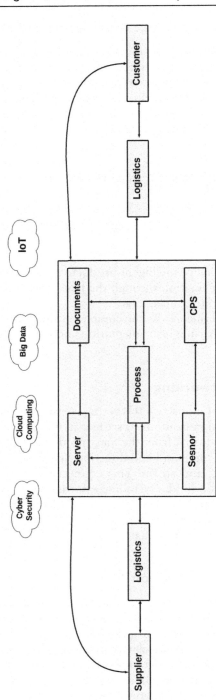

Figure 15.12 Key features and benefits of Industry 4.0.

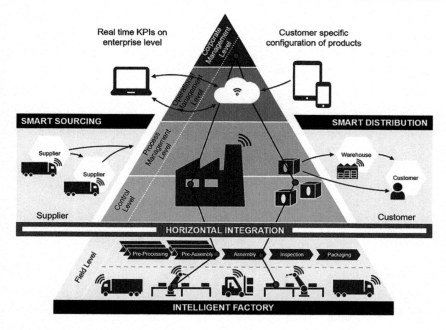

Figure 15.13 Horizontal combination of Industry 4.0.

15.7.3 Through-life engineering

From the discussion, the constituents of the fourth Industrial Revolution are "ten methods of creation," effective method of creation and totally systematic supervision. The computerized modification to Industry 4.0 brings about the feasibility to enhance the adding capability of transformation governance in all the constituents of Industry 4.0. Communicative and mapped-out modules prepare personalized learning attainable, resulting in boosting deliberate implementation and institutional improvement [34]. Figure 15.13 shows the horizontal integration of Industry 4.0 architecture.

15.8 EFFECTS OF INDUSTRY 4.0

Innovation and technological evolutions execute a crucial role in professions, nations and zones. Industry 4.0 paves the way to prospective large-scale commutations in diversity in spheres of the organizational portion. Its impact can be categorized into the following:

- Industrial sector
- Services and products
- Models of business

226 Artificial Intelligence and Blockchain in Industry 4.0

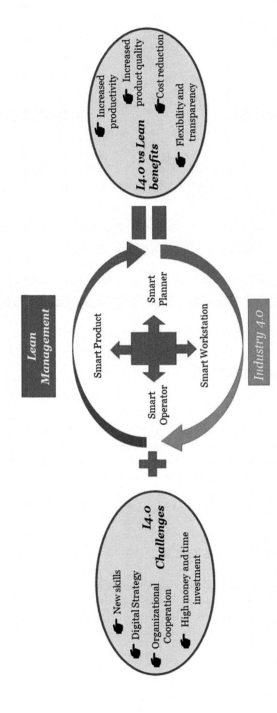

Figure 15.14 Bilateral effects of Industry 4.0.

- Economies of the nation
- Work environment
- Skills improvement

The organizational sector will be the primary to experience the consequence of Industry 4.0. The evolving prototype suggests completely integrating products and procedures, substituting commercial perception from large-scale manufacturing to great personalization and ensuing in enhanced difficulty [35]. Productivity growth is the fundamental individual revolution. In contrast, the fourth Industrial Revolution will affect the whole supply chain, from manufactured formation and production to withdrawing planning, thus further improving productivity and yield [36, 37]. As an outcome, foreseen improvements in estimated production signs show that the adverse impact of mechanization on the new era was only short-termed when this Industry 4.0 age has been started; nevertheless, adding strategies are required to switch on the industry development [38–40].

15.9 CONCLUSION

This chapter commits to associating the censorious space by talking about the main elements, traits and fallouts on several proportions, drivers, barricades and other executive provocations of Industry 4.0 – the fourth industrial revolution which reports subsequent perception for manufacturing. It also emphasizes the characteristics and effects of Industry 4.0. The main characteristics of Industry 4.0 are synergy and amalgamation of strategies in various forms of combination. The horizontal and vertical strategies authorize the utility of elements of Industry 4.0 for manufacturing to acknowledge appealing variations in stock levels. Innovation and scientific upgradations bring out a vital role in institutions, sectors and nations. There are various solicitations for Industry 4.0, enforced by the KUKA organization which initiates various domains such as smart industries, cloud computing, smart machines and online business. These advancements aid Industry 4.0 to distinguish quickly from other standards in constructing intelligent productive systems for the physical world.

REFERENCES

1. Pereira AC and Romero F, "A review of the meanings and the implications of the industry 4.0 concepts", *Procedia Manufacturing*, 2017, Volume 13, pp. 1206–1214, https://doi.org/10.1016/j.promfg.2017.09.032
2. Mukha D, "Impact of industry 4.0 on global value chains, business models and foreign direct investment", 2021, Volume 13, pp. 75–84, https://doi.org/10.21122/2309-6667-2021-13-75-84

3. Mhlanga D, "Artificial intelligence in the industry 4.0 and its impact on poverty, innovation, infrastructure development and the sustainable development goals: Lessons from emerging economies", *Sustainability* (Switzerland), 2021, Volume 13, Issue 11, pp. 1–16, https://doi.org/10.3390/su13115788
4. Oztemel E and Gursev S, "Literature review of industry 4.0 and related technologies", *Journal of Intelligent Manufacturing*, 2020, Volume 31, Issue 1, pp. 127–182, https://doi.org/10.1007/s10845-018-1433-8
5. Deloitte AG, "Industry 4.0. Challenges and solutions for the digital transformation and use of exponential technologies", 2015, pp. 1–30.
6. Tay SI, Lee TC, Hamid NZA and Ahmad ANA, "An overview of industry 4.0: Definition, components and government initiatives", *Journal of Advanced Research in Dynamical and Control Systems*, 2018, Volume 10, Issue 14, pp. 1379–1387.
7. Ojra A, "Revisiting Industry 4.0: A new definition", *Advances in Intelligent Systems and Computing*, 2019, 858, pp. 1156–1162, https://doi.org/10.1007/978-3-030-01174-1_88
8. Beier G, Ullrich A, Niehoff S, Reibig M and Habich M, "Industry 4.0: How it is defined from a socio-technical perspective and how much sustainability it includes – A literature review", *Journal of Cleaner Production*, 2020, 259, pp. 1–13, https://doi.org/10.1016/j.jclepro.2020.120856
9. Culot G, Nassimbeni G, Orzes G and Sartor M, "Behind the definition of industry 4.0: Analysis and open questions", *International Journal of Production Economics*, 2020, 226, 107617, https://doi.org/10.1016/j.ijpe.2020.107617
10. Nosalska K, Piątek ZM, Mazurek G and Rządca R, "Industry 4.0: Coherent definition framework with technological and organizational interdependencies", *Journal of Manufacturing Technology Management*, 2020, Volume 31, Issue 5), pp. 837–862, https://doi.org/10.1108/JMTM-08-2018-0238
11. Pereira AC and Romero F, "A review of the meanings and the implications of the industry 4.0 concept", *Procedia Manufacturing*, 2017, Volume 13, pp. 1206–1214, https://doi.org/10.1016/j.promfg.2017.09.032
12. Wahlster H, Helbig J, Hellinger A and Wahlster W, "Recommendations for implementing the strategic initiative INDUSTRIE 4.0. Securing the future of German manufacturing industry; final report of the Industrie 4.0", Working Group, Forschungsunion, 2013, München Retrieved from: http://forschungsunion.de/pdf/industrie_4_0_final_report.pdf
13. Qin J, Liu Y and Grosvenor R, "A categorical framework of manufacturing for industry 4.0 and beyond", *Procedia CIRP*, 2016, Volume 52, pp. 173–178, https://doi.org/10.1016/j.procir.2016.08.005
14. Bai C, Dallasega P, Orzes G and Sarkis J, "Industry 4.0 technologies assessment: A sustainability perspective", *International Journal of Production Economics*, 2020, 229, https://doi.org/10.1016/j.ijpe.2020.107776
15. Schwab K, "The fourth industrial revolution: What it means and how to respond", *World Economic Forum*, 2016, Volume 21, Retrieved from: https://www.weforum.org/agenda/2016/01/thefourth-industrial-revolution-what-it-meansand-how-to-respond
16. Philbeck T and Davis N, "The fourth industrial revolution: Shaping a new era", *Journal of International Affairs*, 2019, Volume 72, pp. 35–40.

17. Wang S, Wan J, Zhang D, Li D and Zhang C, "Towards smart factory for industry 4.0: A self-organized multi-agent system with big data based feedback and coordination", *Computer Networks*, 2016, Volume 101, https://doi.org/10.1016/j.comnet.2015.12.017
18. Mrugalska B and Wyrwicka MK, "Towards lean production in industry 4.0", *Procedia Engineering*, 2017, 182, https://doi.org/10.1016/j.proeng.2017.03.135
19. Bauernhansl T, Schatz A and Jager J, "Complexity management – industry 4.0 and the consequences: New challenges for socio-technical production systems", *ZWF Zeitschrift fuer Wirtschaftlichen Fabrikbetrieb*, 2014, Volume 109, Issue 5, pp. 347–350.
20. Santos L, Brittes G, Fabián N and Germán A, "The expected contribution of industry 4.0 technologies for industrial performance", *International Journal of Production Economics*, 2018, Volume 204, pp. 383–394.
21. Berger C, Hees A, Braunreuther S and Reinhart G, "Characterization of cyber-physical sensor systems", *Procedia CIRP*, 2016, Volume 41, https://doi.org/10.1016/j.procir.2015.12.019
22. Ghadge A, Er Kara M, Moradlou H and Goswami M, "The impact of industry 4.0 implementation on supply chains", *Journal of Manufacturing Technology Management*, 2020, Volume 31, Issue 4, https://doi.org/10.1108/JMTM-10-2019-0368
23. Chen M, "Towards smart city: M2M communications with software agent intelligence", *Multimedia Tools and Applications*, 2013, Volume 67, Issue 1, https://doi.org/10.1007/s11042-012-1013-4
24. Biral A, Centenaro M, Zanella A, Vangelista L and Zorzi M, "The challenges of M2M massive access in wireless cellular networks", *Digital Communications and Networks*, 2015, Volume 1, Issue 1, https://doi.org/10.1016/j.dcan.2015.02.001
25. Karnik N, Bora U, Bhadri K, Kadambi P and Dhatrak P, "A comprehensive study on current and future trends towards the characteristics and enablers of industry 4.0", *Journal of Industrial Information Integration*, 2021, 100294, https://doi.org/10.1016/j.jii.2021.100294
26. Mohammed A and Wang L, "Brainwaves driven human-robot collaborative assembly", *CIRP Annals*, 2018, Volume 67, Issue 1, https://doi.org/10.1016/j.cirp.2018.04.048
27. Sunhare P, Chowdhary RR and Chattopadhyay MK, "Internet of things and data mining: An application oriented survey", *Journal of King Saud University – Computer and Information Sciences*, 2020, In press (1), pp. 1–22, https://doi.org/10.1016/j.jksuci.2020.07.002
28. Wang L, Liu S, Cooper C, Wang XV and Gao RX, "Function block-based human-robot collaborative assembly driven by brainwaves", *CIRP Annals*, 2021, Volume 70, Issue 1, https://doi.org/10.1016/j.cirp.2021.04.091
29. Lima F et al., "Digital manufacturing tools in the simulation of collaborative robots: Towards industry 4.0", *Brazilian Journal of Operations and Production Management*, 2019, Volume 16, Issue 2, https://doi.org/10.14488/bjopm.2019.v16.n2.a8
30. Sharma N and Sharma R, "Real-time monitoring of physicochemical parameters in water using big data and smart IoT sensors", *Environment, Development and Sustainability*, 2022. https://doi.org/10.1007/s10668-022-02142-8

31. Anandkumar R, Dinesh K, Obaid AJ, Malik P, Sharma R, Dumka A, Singh R and Khatak S, "Securing e-Health application of cloud computing using hyperchaotic image encryption framework", *Computers & Electrical Engineering*, 2022, Volume 100, 107860, ISSN 0045-7906, https://doi.org/10.1016/j.compeleceng.2022.107860
32. Sharma R, Xin Q, Siarry P and Hong W-C, "Guest editorial: Deep learning-based intelligent communication systems: Using big data analytics", *IET Communications*, 2022. https://doi.org/10.1049/cmu2.12374
33. Sharma R and Arya R, "UAV based long range environment monitoring system with Industry 5.0 perspectives for smart city infrastructure", *Computers & Industrial Engineering*, Volume 168, 2022, 108066, ISSN 0360-8352, https://doi.org/10.1016/j.cie.2022.108066
34. Rai M, Maity T, Sharma R et al. "Early detection of foot ulceration in type II diabetic patient using registration method in infrared images and descriptive comparison with deep learning methods", *The Journal of Supercomputing*, 2022. https://doi.org/10.1007/s11227-022-04380-z
35. Sharma R, Gupta D, Maseleno A and Peng S-L, "Introduction to the special issue on big data analytics with internet of things-oriented infrastructures for future smart cities", *Expert Systems*, 2022, Volume 39, e12969. https://doi.org/10.1111/exsy.12969
36. Sharma R, Gavalas D and Peng S-L, "Smart and future applications of Internet of Multimedia Things (IoMT) using big data analytics", *Sensors*, 2022, Volume 22, 4146. https://doi.org/10.3390/s22114146
37. Zheng J, Wu Z, Sharma R and Haibin LV, "Adaptive decision model of product team organization pattern for extracting new energy from agricultural waste", *Sustainable Energy Technologies and Assessments*, 2022, Volume 53, Part A, 102352, ISSN 2213-1388, https://doi.org/10.1016/j.seta.2022.102352
38. Sharma R and Arya R, "Security threats and measures in the Internet of Things for smart city infrastructure: A state of art," *Transactions on Emerging Telecommunications Technologies*, 2022, e4571. https://doi.org/10.1002/ett.4571
39. Mou J, Gao K, Duan P, Li J, Garg A and Sharma R, "A machine learning approach for energy-efficient intelligent transportation scheduling problem in a real-world dynamic circumstances", *IEEE Transactions on Intelligent Transportation Systems*, 2022, https://doi.org/10.1109/TITS.2022.3183215
40. Priyadarshini I, Sharma R, Bhatt D et al. "Human activity recognition in cyber-physical systems using optimized machine learning techniques", *Cluster Computing*, 2022. https://doi.org/10.1007/s10586-022-03662-8

Chapter 16

The positionality of culture in teaching EFL in technology-supported classrooms

Teachers' perceptions and practices

Chau Thi Hoang Hoa and Lien Bao Tran
Tra Vinh University, Tra Vinh city, Vietnam

16.1 INTRODUCTION

Teaching English as a foreign language (EFL) in Vietnam has been gradually changing to meet a special mission: to contribute to interculturality so that Vietnamese people can be competent global citizens. This change has been prompted by the trend of integration within the Association of Southeast Asian Nations (ASEAN) as well as globally. Also, the redesigned general curriculum for teaching English in upper secondary school makes explicit that interculturality is the learner's outcome. The new secondary school curriculum's specific aims are to (1) increase learners' intercultural awareness (2) foster positive attitudes about one's and others' cultures, and (3) reflect the values of Vietnamese culture in English (MOET, 2018). As well as this, the new EFL coursebook editions have been updated to include a variety of intercultural knowledge on English-speaking culture, Vietnamese or native culture, and other cultures with a view to meeting these objectives. Teachers are believed to play active contributors to the success of this educational reform as their strategies to exploit intercultural content in the coursebooks decide the degree of intercultural awareness and understanding that learners should achieve. Notwithstanding, some EFL instructors favour emphasising language skill development and tend to ignore intercultural teaching (Fandio, 2013; Nguyen et al., 2016; Rajabi & Ketabi, 2012). The gap between the stated outcomes and current teaching practices could be accused by the lack of official guidance from MOET for intercultural teaching towards intercultural objectives (Chau & Truong, 2019).

The risks of ignoring intercultural objectives became bigger during emergency remote teaching (so-called online teaching in this study). In fact, during the COVID-19 pandemic, the turn of teaching mode from offline to online entailed the shortening of class meeting time and reduction of intercultural content, which is focused in "Culture and Communication" sections across the units. That is to say, intercultural content was no longer focused in a separate section, but it is scattered in other sections of each unit. The flexibility and discrepancy of MOET about intercultural integration or

DOI: 10.1201/9781003452591-16

intercultural language teaching (ILT) before and while online teaching due to COVID-19 could be proved. Hence, it is worth studying if there exists any gaps between the perceptions and practices of teachers regarding online ILT and what strategies teachers used to deal with online ILT. Specifically, the issues are stated in three research questions (RQ) as follows:

RQ1. How do EFL teachers perceive the importance of ILT?
RQ2. How do EFL teachers perceive the feasibility of online ILT?
RQ3. How do EFL teachers practise online ILT?

16.2 LITERATURE REVIEW

16.2.1 Defining culture

Culture is a term that lacks a unified definition for its being defined in various ways by various fields. Hofstede (1984) defines culture as the shared mental programming that sets apart individuals belonging to different categories or groups. Another definition by Brooks (1997, as cited in Choudhury, 2013) distinguishes culture between "big C" Culture and "small c" culture. "Big C" Culture refers to the formal and visible aspects of a country's literature or civilization, whereas "small c" culture is more subtle. Liddicoat (1997) takes a dynamic view of culture, arguing that culture encompasses not only facts and artefacts but also includes actions and comprehension. According to Liddicoat et al. (2003, p. 45), culture is "a complex system of concepts, attitudes, values, beliefs, conventions, behaviours, practices, rituals, and lifestyle of the people who make up a cultural group, as well as the artefacts they produce and the institutions they create". Browett (2003) and Sewell (2005) support the dynamic nature of culture and the constant evolution of all cultural aspects of cultural groups. Consistent with Liddicoat et al.'s (2003), Browett's (2003), and Sewell's (2005) perspectives, this study defines culture as a multifaceted and dynamic structure that comprises a group's ideas, viewpoints, principles, traditions, customs, way of life, objects, and organisations.

16.2.2 Culture and language connection

The relationship between language and culture has been viewed differently. Hymes (1972) has an opposing view from Sapir (1929) about the relationship between language and culture. Hymes (1972) suggests that language is a crucial part of a culture, whilst Sapir (1929) argues that language shapes culture. Kramsch (1998) views this relationship as bilateral, meaning that language and culture are interwoven since people use language to express their cultural reality. Similarly, Liddicoat (2002) acknowledges the mutual connection between the two components and highlights the role of language in conveying one's cultural meaning. In addition, Risager (2007) suggests that language and culture can be separated in language education as culture can be treated as either context or content in language teaching. It is worth noting that this understanding does not mean to separate culture from language in education, but rather suggests two primary strategies to deal with them effectively. In conclusion, the findings of Kramsch (1998), Liddicoat (2002), and Risager (2007) all support the relationship between language and culture, and highlight the significance of incorporating culture into language instruction or ILT.

16.2.3 Framework and strategies for online ILT

A variety of strategies for introducing local culture via online applications or integrating cultures into foreign language teaching could be found in related literature. However, integrating culture into EFL teaching via online applications or platforms is rare. The context of teaching foreign languages, detached from the target language-speaking environment, favours the conducting of intercultural teaching strategies using the internet or online platforms. Hence, this study embraces the framework Chau and Truong (2019) for ILT. Adopting the intercultural development model of Fantini (2000), the ILT model aims to build learners' intercultural competence (IC) at four levels of "intercultural knowledge, intercultural skills, intercultural attitude, and intercultural awareness". This model can be applied as a guideline for ILT offline and online teaching contexts.

As cited in Chau and Truong (2019), common ILT strategies and activities are based on theories and principles of language learning (Vygotsky, 1978) and intercultural language learning (Newton et al., 2010), and proposed from intercultural teaching practices (see Liu & Zhang, 2014; Stern, 1992; Lien & Chau, 2022). The suggested ILT strategies in practice are described as follows:

- Enabling learners to access a variety of intercultural resources available online (videos, audio, pictures, movies, songs, etc.) (adopted from Lien & Chau, 2022, p. 131).
- Exposing learners to a wide range of cultures via multimedia, social networking sites, internet apps, having online friends from other countries, etc. (as adapted from Stern, 1992).
- Utilising resources from either the local cultural community or learner variations in cultural identification, such as comparing cultural practices and experiences, discussing holidays and celebrations, rules to follow, customs, and traditions (as proposed by Newton et al., 2010).
- Engaging learners into intercultural interactions, such as discovering, presenting, comparing, reflecting on cultures, role-playing to solve intercultural conflicts, and doing projects (as suggested by Newton et al., 2010; Newton, 2016; and Vygotsky, 1978).

- Instructing students on what should be done and how things should be done in intercultural contexts (as proposed by Lo Bianco, 1999, and Liu & Zhang, 2014).

16.2.4 Previous studies on teachers' perceptions and practices of ILT

Teachers' perceptions and practices of ILT have been well-researched in recent times according to the spread of English among people of different cultures. Among them, Chau and Truong (2019), Nguyen (2013), Oranje and Smith (2018), and Stapleton (2000) shed light on the importance and objectives of intercultural teaching and specified strategies for ILT in practices.

In respect of teachers' perceptions of ILT, the most common finding is that teachers had the propensity to ignore culture in their language lessons (Czura, 2016; Maedeh, 2020; Nguyen et al., 2016; Stapleton, 2000). Stapleton (2000) indicated that Japanese teachers integrated cultural elements randomly in spite of the fact that they acknowledged the significance of ILT. Similarly, Czura (2016) also pointed out that teachers tended to underestimate ILT due to the lack of time. In the local context, Nguyen et al. (2016) confirmed that a number of EFL teachers in Vietnam hardly devote much effort to teaching culture because they believed that building communicative language skills for students was a big task for teachers already.

Various strategies for integrating culture into foreign language teaching both online and offline were reported in the following studies: Chau (2020), Nguyen (2013), Jata (2015), Ho (2011), Sugianto and Ulfah (2020), and Trede et al. (2013). Jata (2015) and Trede et al. (2013) recommended that having learners write reflections on or keep journals about their intercultural issues or experiences and discovering other cultures and discussing what they had learned from the new cultures could be effective strategies for intercultural teaching. The recommended activities could be implemented in online teaching. In Vietnam, according to Ho (2011) and Nguyen (2013), the most commonly reported ILT practice was treating cultural aspects as themes or content in EFL lessons. In the upper-secondary education context, Chau and Truong (2019) found that activities teaching intercultural knowledge are less frequently conducted than ones for building intercultural attitudes. Furthermore, making use of the internet to enrich intercultural resources was the most common activity to build learners' intercultural knowledge and attitudes. Sugianto and Ulfah (2020) proved that ILT could be more feasible if conducted online, thanks to prompt teachers' responsive feedback, more classroom interactions, abundant online materials (videos, pictures, etc.) for teaching cultures, the uses of social media, applications, platforms namely Zoom, Skype, Google Meet, Cisco Webex, WhatsApp, and Google Form. However, some internal and external challenges could be named as limited internet quota, the dearth of internet connection, and students' psychological health.

What teachers believe and do is defined as their perceptions and practices (Borg, 2003). Surely, belief and actions are related because belief drives one's actions and the actions adjust one's belief. In most studies of teachers' perceptions and practices of ILT, it was found that teachers' perceptions and practices were correlated but the level of perceptions was much higher than that of practices (Chau & Truong, 2018, 2019; Gönen & Sağlam, 2012; Hoang, 2015; Sercu et al., 2005; Zhou, 2011). Simply stated, teachers had positive perceptions of ILT, but they rarely included it in their practices on a daily basis. In light of ILT, the question of teachers' perceptions and onsite and online teaching are in line with teachers' practices has not been investigated. Those are the gaps in the literature this study aims to bridge.

16.3 METHODOLOGY

This study follows a descriptive mixed-method research design with the use of quantitative and qualitative data collected from a questionnaire and semi-structured interview.

The questionnaire is a Likert five-point-scale with 23 items, focusing on teachers' perceptions of and practices of ILT. Teachers' perceptions are measured according to opinion-based scales, from *strongly disagree* to *strongly agree*, and their practices are frequency scale-based, from *never* to *always*. The interpretation of the data follows the scale of agreement (Oxford, 1990) (1.0–2.4 for low; 2.5–3.5 for medium; 3.6–4.4 for high; 4.5–5.0 for very high).

The ideas for the items in the questionnaire are adapted from Sercu (2006) and Chau and Truong (2019) with adjustments focusing on ILT online. The 23-item questionnaire is divided into three main clusters. The first addresses (1) teachers' perceptions of the importance of ILT (items 1–7, coded as Cluster P1) and the second is about (2) teachers' perceptions of the feasibility of ILT online (items 8–12, Cluster P2). The third cluster focuses on (3) teachers' practices of ILT online (items 13–23, Cluster Pr).

To ensure the validity and reliability of the questionnaire, the researchers added one reversed item (Item 7, P1), and coded with reversed value from 5 to 1 instead of from 1 to 5. Besides, the questionnaire was piloted to 25 non-participants to improve the clarity in understanding and translating the statements. The final questionnaire was delivered to 64 teacher participants via Google Forms with 100% of positive results for all the required responses. The Cronbach's alphas of the three clusters were .742, .684, and .718, which are acceptable levels of reliability.

In supporting quantitative data and understanding the gaps (if any) among the results in cross-comparison, the semi-structured interviews were conducted with up to 10% of the participants (six subjects). Each interview happened from 45 minutes to an hour. The themes of the interviews were relevant to the research issues in the questionnaire but open to teachers' more ideas, suggestions, clarifications, justifications, and confirmation. Some of the questions in the interviews were adapted from (Ghavamnia, 2020). The language of the interviews was Vietnamese to create comfortable conditions for interviewees to answer the questions. The qualitative data collected from the interviews were analysed through the content analysis approach, starting with reading for overall understanding, and then comparing, and figuring out the correlation among and across the transcripts. For the process of data analysis and interpretation, the framework of Deductive Category Assignment (Mayring, 2014) has been specifically used.

16.4 RESULTS AND DISCUSSIONS

This part presents and discusses quantitative and qualitative research results in general and in relevance to the three research questions. The mean scores of the three clusters P1, P2, and Pr are presented in Table 16.1.

Table 16.1 demonstrates a downward trend from (1) teachers' perceptions of the importance of ILT, (2) teachers' perceptions of the feasibility of ILT online, to (3) teachers' practices of ILT online. Teachers' perceptions of ILT in terms of importance and feasibility and teachers' practice gain high mean scores (MP1 = 4.4576; MP2 = 4.45; MPr = 3.6719 > 3.5). The gap between teachers' perceptions of the importance of ILT and teachers' perceptions of the feasibility of ILT online was not statistically significant (Sig. = .855 > .005). The gap between teachers' perceptions and practices of ILT online was confirmed (Sig. = .000 < .005).

Pair correlations between (1) teachers' perceptions of the importance of ILT and the feasibility of online ILT and (2) teachers' perceptions and practices of ILT online are shown in Table 16.2.

As reported in Table 16.2, the two pairs P1 & P2, and P2 & Pr are strongly correlated (r = .680, p = .000; r = .680, p = .000). This means the teachers who have good perceptions of intercultural integration are likely to have good perceptions towards online intercultural integration and implement it in their practice frequently.

Table 16.1 Mean scores of (P1) teachers' perceptions of ILT (P2) teachers' perceptions of the feasibility of online ILT (Pr) teachers' practices of online ILT

	N	Min	Max	Mean	STD. deviation
MEANP1	64	3.14	5	4.45	0.4001
MEANP2	64	3	5	4.45	0.42464
MEANPR	64	2.8	4.5	3.67	0.39729
VALID N	64				

Table 16.2 Correlations between teachers' perceptions of ILT and online ILT and teachers' perceptions and practices of online ILT

Correlations	N	Pearson correlation*	Sig. (2-tailed)
P1 and P2	64	.680	.000
P2 and Pr	64	.669	.000

Note:
* Correlation is significant at the level of 0.01 (2-tailed).

The gap between and correlations of teachers' perceptions and practice has been confirmed. This result is in line with many studies in the field (see Chau & Truong, 2018, 2019; Gönen & Sağlam, 2012; Hoang, 2015; Sercu et al., 2005; Zhou, 2011). Explaining the sources of difference, Chau and Truong (2019) and Gönen and Sağlam (2012) agreed that teachers faced many obstacles in implementing ILT. However, Nguyen (2013) believed that it was caused by the lack of teachers' responsibility.

RQ 1 How do EFL teachers perceive the importance of ILT?

Teachers' perceptions of the importance of ILT are shown in Table 16.3.

The high level of mean score of the whole cluster (M1 = 4.4567) shows that teacher participants agree that integrating into EFL teaching is important. All of the seven items have Standard Deviation (SD) values smaller than 1, which means that the distributions of choices are clustered around the mean score of each item. Gaining the highest mean scores of 4.70 and 4.52, items 7 and 2 confirm the steady position of intercultural integration

Table 16.3 Teachers' perceptions of the importance of ILT

Items	Mean	SD
P1. Teachers' perceptions of the importance of ILT	**4.4567**	**.40018**
1. Teaching culture should be an important part of foreign language teaching.	4.45	.641
2. Integrating culture into language teaching facilitates students to learn English better.	4.52	.534
3. Language and culture should be intertwined in ELF teaching.	4.28	.723
4. Integrating culture into language teaching enables students to be more tolerant of new cultures.	4.39	.748
5. Integrating culture into language teaching helps students improve their understanding of foreign cultures.	4.45	.754
6. Integrating culture into language teaching enables students to communicate with people from different cultures effectively.	4.41	.660
7. Integrating culture into language teaching is questionable. (reversed version)	4.70	.494

in the EFL teaching context. Giving a smaller score for item 1 *"Teaching culture should be an important part of foreign language teaching"* (Mi 1 = 4.45) is not a coincidence. The gaps between the mean scores of items 1 and those of items 7 and 2 prove that teachers gave higher approval to the integration of culture than the equivalence of positionality of culture in comparison to language. Items 4, 6, and 5 focus on the benefits of intercultural integration: building students' intercultural attitudes (Mi 4 = 4.39), intercultural skills (Mi 6 = 4.41), and intercultural knowledge (Mi 5 = 4.45). The upward trend of the three mean scores is concurrent with the spiral dimensions of intercultural competence, moving from knowledge, skills, and attitudes, to awareness (Fantini, 2006).

Giving answers to the questionnaire, the participant teachers experienced an ambivalent attitude towards the positionality of culture in comparison to language. However, in the interview, all six teachers confirmed that teaching culture was as essential as teaching language (language areas and language skills) and that teaching language and culture should be integrated. For example, three teachers contented that the intercultural content embedded in the coursebooks was updated, diverse, interesting, and logically arranged, so the reduction of sections focusing on cultures was not necessary. It was proven by the fact that two teachers admitted that they still insisted on teaching the omitted sections focusing on culture. The other two senior teachers with more than 20-year teaching experience had more insights into the role of ILT. They admitted that ILT was thought to enable learners' communicative ability in an intercultural communication context. Noticeably, all six teachers focused on the importance of teaching intercultural knowledge. They agreed that teaching cultural content was interesting and that it facilitated their language teaching. Though the findings were drawn from a small number of participants, it is strongly believed that the participant teachers had positive perceptions of the importance of culture and integration of culture into language education.

From the findings, it is safe to confirm that participant teachers had a positive attitude towards the importance of culture and ILT. The findings were relevant to those of Chau and Truong (2019), Gönen and Sağlam (2012), Hoang (2015), and Zhou (2011).

RQ 2 How do EFL teachers perceive the importance of online ILT?

Teachers' perceptions of the feasibility of integrating culture into online teaching are investigated based on qualitative data presented in Table 16.4 and qualitative data collected from the interviews with participant teachers.

Table 16.4 certifies that teachers had good perceptions of the feasibility of integrating cultures into EFL teaching online (MP2 = 4.45). Of the five items, item 8 addressing the reduction or omission of teaching intercultural contents during online teaching attained the highest score of 4.81. In reverse, item 11, relevant to item 8 got the lowest mean score of 3.97. The gap between the two scores implies that teachers appreciated ILT but the

Table 16.4 Teachers' perceptions of the feasibility of online ILT

Items	Mean	SD
P2. Teachers' perceptions of the feasibility of ILT online	**4.4500**	**.42464**
8. Intercultural integration should not be put aside from language teaching during online teaching due to Covid-19 pandemic.	4.81	.432
9. ILT could be conducted online.	4.42	.686
10. ILT could be implemented even better in the context of online teaching thanks to a variety of online resources and applications.	4.55	.589
11. I still find teaching culture as important as teaching language when teaching online.	3.97	.872
12. The activities to integrate culture into language teaching were more diverse thanks to online teaching.	4.50	.563

equivalence of the positionality of culture in comparison to language in EFL education is not that stably confirmed. The three items 9, 10, and 12 achieved high scores of 4.42, 4.55, and 4.50, respectively, which confirms that the teachers approved of the integration of culture and agreed that when teaching online, they could deal with culture better thanks to the variety of activities and diversity of accessible teaching and learning resources.

The results from teacher interviews were consistent with those of the questionnaire regarding teachers' feasibility of teaching culture online. In fact, Teachers One and Six affirmed that teaching culture was vital in teaching language even though the mode of teaching was switched to online. Also, the confirmation of teaching culture online was proven by the fact that two teachers admitted that they integrated cultures during online teaching though it could be omitted or optional in online teaching during COVID-19.

The positive perceptions of online ILT were proven by Sugianto and Ulfah (2020). Sugianto and Ulfah (2020) pinpointed that ILT could be more feasible if conducted online thanks to abundant online materials (videos, pictures, etc.) on the internet and the uses of applications and platforms.

RQ 3 How do EFL teachers practise online ILT?

Teachers' practices of online ILT are reported from the combination of teacher questionnaires (Table 16.5) and semi-structured interviews.

Though teachers gave a rather high mean score to the general description of their practice of online ILT (M = 3.75), their responses to detailed descriptions of intercultural language strategies were much higher (Mi 14 = 3.75; Mi 15 =. 4.45; and Mi 16 = 4.12). The highest score given to item 14 indicates the most frequent activity teachers did to integrate culture into their teaching was extending intercultural knowledge from online resources like artefacts or cultural products of different cultures. In alignment with Chau and Truong (2019), making use of online resources is one of teachers'

Table 16.5 Teachers' practices of online ILT

Items	Mean	SD
Pr. Teachers' practices of online ILT	**3.6719**	**.46021**
13. I embedded intercultural knowledge into my online language lessons (e.g. vocabulary, grammar, structure, idioms)	3.75	.816
14. I illustrated foreign cultural contents in English textbooks (e.g. music, fashion, festivals,) from online resources (namely videos, audio, pictures, movies, songs, etc.) for references.	4.73	.479
15. I referred to the similarities and differences between local and foreign cultures when integrating cultural contents in the textbooks.	4.45	.561
16. When teaching online, I encouraged my students to justify the comparison of cultures presented in the coursebooks.	4.12	.807
17. Besides the synchronous mode of teaching, I asked my students to write their reflections related to the cultural content they had learned via available learning management systems (LMS) such as Google Classroom and MODDLE	2.89	1.114
18. I enriched intercultural materials relevant to the intercultural aspects of my teaching by creating group works with my colleagues via social networks.	1.08	.522
19. I created a chat group with foreigners so that students can exchange their intercultural knowledge and experiences.	1.81	1.153
20. I created an online multicultural exchange environment for my students to learn languages and cultures.	1.51	1.053
21. I asked students to do small projects to present their native culture to their friends on social media or LMS.	4.44	.588
22. I asked students to discover an aspect of a new culture to share with their friends via online interaction.	4.52	.776

favourite activities. The condition of online teaching with internet access and availability of necessary gadgets together with the enjoyment that cultural products could bring to the students originates from the commonness of this activity. The second frequent strategy for integrating cultures was comparing cultures. Acknowledging and justifying the differences and similarities, which are the two different levels of intercultural language teaching, were mentioned in items 14 and 15 (TMi 15 = 4.45 and Mi 16 = 4.12). It is shown that building intercultural knowledge was more frequently conducted than building intercultural attitudes, a higher level of intercultural competence, which reconfirms a similar finding of Chau and Truong (2019). The level of frequency of activities building intercultural knowledge and attitudes in offline teaching is parallel to that of online teaching.

Doing projects related to local cultures and introducing them on LMS or social media to catch the attention of a wider audience and discovering a foreign aspect of culture and presenting it to their friends are of high (Mi 21 = 4.44) and very high frequency (Mi 22 = 4.52) mean scores. These findings are contrary to those of Chau and Truong (2019) for the same strategies for ILT less happened offline (M = 2.58 and 2.68) than in online (Mi 21 = 4.44 and Mi 42 = 4.52) teaching mode. It is concluded that online teaching creates better conditions for intercultural integration thanks to the use of social media and the convenience of online interaction.

Of the ten items, items 17, 19, and 20 got SD values of bigger than 1. It means that there was a rather big gap among the participants in the frequency of choosing those ILT practices activities. Concurrently, two (items 19 and 20) of the three items were among the groups with the lowest mean scores (Mi 18 = 1.08; Mi 19 = 1.81; Mi 20 = 1.51) and both focus on intercultural practice encouraging students to participate on genuine intercultural communication via social media platforms. With the medium mean score (M = 2.89), item 19 relating an activity building intercultural attitude by asking students to write reflections on intercultural issues was not often conducted. The sparsity of ILT activities at the level of building intercultural practices and attitudes was also relevant to Chau and Truong (2019), Ho (2011), and Nguyen (2013) and irrelevant to Jata (2015) and Trede et al. (2013). The discrepancy could rely on the nature of students and the tests prescribed in the coursebooks applied.

When being asked about intercultural language activities that teachers most often applied to teaching, most of the teachers stated that they often collected online cultural products as an add-in to their lessons. For clarification, teachers stated that they made use of multimedia resources from the internet (such as short films, YouTube videos, or animated pictures) to make their lessons more lively and interesting. The second frequent strategy for intercultural teaching was fun activities such as games and puzzles on websites for foreign languages and cultures teaching like Bamboozle, Kahoot, and Quizzes. The teachers further explained that they had known these websites and applications but they rarely took their advantages when teaching offline because it entailed much proper preparation. Making use of internet applications and resources to integrate cultures is in line with the findings of Sugianto and Ulfah (2020). The other two activities mentioned by teachers were asking students to write their reflections on intercultural issues and sharing their experiences with their friends online. Of the two strategies, teachers admitted that writing reflection less frequently happened because they avoided using asynchronous activities to reduce the burden for their students during the difficult time of COVID-19. The teacher's explanation gave an account for the lower mean score of item 17 (Mi 17 = 2.89).

16.5 CONCLUSION, RECOMMENDATION, AND LIMITATION

The main quantitative and qualitative findings from the questionnaire and semi-structured interview dealing with the research issues can be summarised as follows. First, the participating teachers had equally high levels of (1) perceptions of the importance of culture and the integration of culture in language teaching and (2) perceptions of the feasibility of the integration of culture in online language teaching. However, the frequency of applying ILT activities to integrate culture was not as high as the level of their perceptions. One of the reasons for the existence of the gaps was the reduction of intercultural content during COVID-19. Teachers' most frequent activities to deal with cultures were illustrating intercultural content in the coursebook with additional materials like videos, audio, pictures, movies, and songs to make the lessons more vivid and interesting. Teachers conducted asynchronous activities less often than synchronous activities to lessen students' stress during online learning. Activities involving teachers' group work for sharing materials and connection with foreigners to create genuine interaction in intercultural communication were rarely conducted. Most of the findings of teachers' practices of intercultural integration into EFL teaching online were parallel with those of offline teaching mode reported in related research. All of the interviewed teachers agreed that online teaching is more inclined to intercultural integration than offline teaching, thanks to the availability of online resources, websites, and applications, which are continuously developed for educational purposes.

Based on the main findings, this study has three recommendations. Firstly, teachers should establish professional groups sharing experiences by applying technology, making use of online learning software such as Quizziz, Bamboozle, Azota, Zalo, Facebook, or Kahoot to increase interaction, and introducing useful websites and social platforms to enrich online materials. Or else, they are at risk of ignoring their good teaching habits or diminishing their skills using online technologies to facilitate their intercultural integration when they are back to their normal teaching and studying practices, with more face-to-face interaction, and not relying much on technology. Secondly, teachers can set up intercultural communication on an online basis if they are keen on it. It is admitted that having reliable foreign teaching mates to conduct virtual meetings or an environment to exchange cultures or form intercultural communication seems impossible for teachers. For that reason, the teachers can create educational friend circles with foreigners and start with simple activities at first, like encouraging students to post their presentations of introducing local cultures on social media and call for notice, and comments from the foreigners and keep following it interactively. Thirdly, it is recommended that the educational managers should encourage a variety of action research on integrating culture into EFL teaching in general education. This engagement or action will raise teachers' awareness and skills of teaching language and culture in integration.

Two major limitations of this study can be listed as follows. Firstly, the modest number of participants can negatively affect the reliability of the quantitative data reports. As a remedy for this shortcoming, the researchers combine qualitative data collected from semi-structured interviews to compensate for the reliability of the results. In fact, the gaps of data collected from the questionnaire were bridged and well-explained in teachers' interviews. However, a bigger data size will surely contribute to the quality of the findings. Secondly, from a broader view, this research study seems not a complete one. The teachers' perceptions and practices of online ILT were reported. The gaps in teachers' perceptions and practices of online and offline ILT were reported within data collected during COVID-19 and implications drawn by the researchers in comparison with related studies. The findings would gain better validity if a similar research is conducted with the same participants focusing on teachers' perceptions and practices of their intercultural integration in offline teaching after COVID-19. This limitation can be a recommendation for further research.

REFERENCES

Borg, S. (2003). Teacher cognition in language teaching: A review of research on what language teachers think, know, believe, and do. *Language Teaching*, 36(2), 81–109. DOI:10.1017/S0261444803001903

Browett, J. (2003). Culture: Are we speaking the same language. *Babel*, 38(2), 18–38.

Chau, T. H. H., & Truong, V. (2018). Integrating cultures into teaching EFL in Vietnam: Teachers' perceptions. *LEARN Journal: Language Education and Acquisition Research Network*, 11(2), 103–115.

Chau, T. H. H., & Truong, V. (2019). The integration of intercultural education into teaching English: What Vietnamese teachers do and say. *International Journal of Instruction*, 12(1), 441–456.

Chau, T. H. H. (2020) Integrating culture into teaching EFL in general education: A context of Vietnam. *Language Related Research*, 11(5), 227–252. DOI: 10.21859/LRR.11.5.227; http://lrr.modares.ac.ir/article-14-46663-en.html

Choudhury, M. H. (2013). Teaching culture in EFL: Implications, challenges and strategies faculty of languages and translation. *Journal of Humanities and Social Science*, 13(1), 20–24.

Czura, A. (2016). Major field of study and student teachers' views on intercultural communicative competence. *Language and Intercultural Communication*, 16(1), 83–98.

Fandiño, Y. (2013). Knowledge base and EFL teacher education programs: A Colombian perspective. *Íkala, revista de lenguaje y cultura*, 18(1), 83–95.

Fantini, A. (2000). A central concern: Developing intercultural competence. *SIT Occasional Papers Series: Addressing Intercultural Education, Training and Service* (pp. 25–33). Brattleboro, VT: School for International Training.

Fantini, A. (2006). Exploring and assessing intercultural competence. Retrieved from http://www.sit.edu/publications/docs/feil_research_report.pdf

Ghavamnia, M. (2020). Iranian EFL teachers' beliefs and perceptions on incorporating culture in EFL classes. *Intercultural Education*, 31(3), 314–329.

Gönen, S., & Sağlam, S. (2012). Teaching culture in the EFL classroom: Teachers' perceptions. *International Journal of Global Education*, 1(3), 2196–9296.

Hoang, V. (2015). The development of the ten-year English textbook series for Vietnamese schools under the National Foreign Language 2020 Project: A cross-cultural collaborative experience. *VNU Journal of Foreign Studies*, 31(3), 1–17.

Hofstede, G. (1984). National cultures and corporate cultures. In L. A. Samovar & R. E. Porter (Eds.) *Communication between cultures* (pp. 51–63). Belmont, CA: Wadsworth.

Ho, S. T. K. (2011). *An investigation of intercultural teaching and learning in tertiary EFL classrooms in Vietnam* (Doctoral dissertation, Victoria University of Wellington, Wellington, New Zealand).

Hymes, D. H. (1972). On communicative competence. In J. B. Pride & J. Holmes (Eds.), *Sociolinguistics: Selected readings* (pp. 269–293). Harmondsworth, UK: Penguin.

Jata, E. (2015). Perception of lecturer on intercultural competence and culture teaching time (Case Study). *European Journal of Interdisciplinary Studies*, 1(3), 176–180.

Kramsch, C. (1998). *Language and culture*. Oxford, UK: Oxford University Press.

Lien, B. T., & Chau, T. H. H. (2022). Vietnamese EFL teachers' perceptions towards and strategies for teaching culture during the COVID-19 pandemic. *Ho Chi Minh City Open University Journal of Science-Social Sciences*, 12(2), 3–18

Liddicoat, A. J. (1997). Everyday speech as culture: Implications for language teaching. In A. J. Liddicoat & C. Crozet (Eds.), *Teaching Language, Teaching Culture* (pp. 55–70). Canberra: Applied Linguistics Association of Australia.

Liddicoat, A. J. (2002). Static and dynamic views of culture and intercultural language acquisition. *Babel*, 36(3), 4–11.

Liddicoat, A. J., Papademetre, L., Scarino, A., & Kohler, M. (2003). *Report on intercultural language learning*. Canberra, ACT: Commonwealth of Australia.

Liu, Y. & Zhang, M. (2014). The application of Constructivism to the teaching of intercultural communication, *English Language Teaching* 7(5), 136–141. DOI:10.5539/elt.v7n5p136

Lo Bianco, J. (1999). *Training teachers of language and culture: Presentation to the national conference on teacher education*, Melbourne: Language Australia.

Mayring, P. (2014). *Qualitative content analysis: Theoretical foundation, basic procedures, and software solution*. Klagenfurt.

Maedeh, G. (2020). Iranian EFL teachers' beliefs and perspectives on incorporating culture in EFL classes. *Intercultural Education*, 31(3), 314–329. DOI: 10.1080/14675986.2020.1733862

MOET. (2018). *Overall Curriculum for General Education*. https://data.moet.gov.vn/index.php/s/CXTqAkDQNTwyEPt#pdfviewer

Nguyen, L., Harvey, S., & Grant, L. (2016). What teachers say about addressing culture in their EFL teaching practices: the Vietnamese context. *Intercultural Education*, 27(2), 165–178.

Nguyen T. L. (2013). *Integrating culture into Vietnamese University EFL teaching: A critical ethnographic study* (Doctoral dissertation, AUT University, School of Language and Culture, New Zealand).

Newton, J., Yates, E., Shearn, S., & Nowitzki, W. (2010). *Intercultural communicative language teaching: Implications for effective teaching and learning*. Report to the Ministry of Education.

Newton, J. (2016). Cultivating intercultural competence in tertiary EFL programs. *Crossing Borders in Language Teaching and Business Communication: Proceedings* of the 11th ELT conference at AE CYUT (pp. 1-22). Chaoyang University of Technology, Chaoyang, Taiwan, 27 May 2016. ISBN 978-986-5631-24-6

Oranje, J., & Smith, L. F. (2018). Language teacher cognition and intercultural language teaching: The New Zealand perspective. *Language Teaching Research*, 22(3), 310–329.

Oxford, R. (1990). *Language learning strategies: What every teacher should know*. New York: Newbury House.

Rajabi, S., & Ketabi, S. (2012). Aspects of cultural elements in prominent English textbooks for EFL setting. *Theory & Practice in Language Studies*, 2(4), 705–712.

Risager, K. (2007). *Language and culture pedagogy*. Clevedon Multilingual Matters.

Sapir, E. (1929). The status of linguistics as a science. In E. Sapir (1958): *Culture, Language and Personality* (ed. D. G. Mandelbaum). Berkeley, CA: University of California Press.

Sercu, L. (2006). The foreign language and intercultural competence teacher: The acquisition of a new professional identity. *Intercultural Education*, 17(1), 55–72. DOI:10.1080/14675980500502321

Sercu, L., Méndez García, M., & Prieto, P. C. (2005). Culture learning from a Constructivist perspective. An Investigation of Spanish foreign language teachers' views. *Language and Education*, 19(6), 483–495.

Sewell, W. H. (2005). *Logics of history: Social theory and social transformation*. University of Chicago Press.

Stapleton, P. (2000). Culture's role in TEFL: An attitude survey in Japan. *Language Culture and Curriculum*, 13(3), 291–305.

Stern, H. (1992). *Issues and options in language teaching*. Oxford: Oxford University Press.

Sugianto, A., & Ulfah, N. (2020). Construing the challenges and opportunities of intercultural language teaching amid Covid-19 Pandemic: English teachers' voices. *Journal of English Language Teaching and Linguistics*, 5(4), 363–381.

Trede, F., Bowles, W., & Bridges, D. (2013). Developing intercultural competence and global citizenship through international experiences: Academics' perceptions. *Intercultural Education*, 24(5), 442–455.

Vygotsky, L. S. (1978). *Mind in society: The development of higher psychological processes*. Cambridge, MA: Harvard University Press.

Zhou, Y. (2011). *A study of Chinese university EFL teachers and their intercultural competence teaching* (Doctoral dissertation, University of Windsor), [https://scholar.uwindsor.ca/etd/428/] (Retrieved May 12, 2021).

Chapter 17

Algorithm for secured energy-efficient routing in wireless sensor networks

A review

Jyoti Srivastava and Jay Prakash

Madan Mohan Malviya University of Technology, Gorakhpur, India

17.1 INTRODUCTION

Spam is not only a prevalent attack in non-second-generation (non-2G) networks, but it has also become standard practice to inject misleading information or propagate unneeded promotional information. IP-based statement in third-generation (3G) wireless networks facilitates the transfer of Internet-based vulnerabilities and challenges to mobile devices. The need for high-speed Internet Protocol (IP) connections is on the rise in the mobile industry, and fourth-generation (4G) cellular networks will help meet this demand. This new twist makes the threat more complex or ever-changing. With the introduction of the fifth-generation (5G) network, security risks will increase, or users will have more discretionary space. Because transmission is inherently unstable and there are an infinite number of other ways to communicate, it can be expensive to implement authentication, authenticity, and privacy. Both authenticity and privacy can also be broken. Today's mobile networks are especially vulnerable to attacks on network security and privacy concerns at the media access control (MAC) and physical layer (PHY). Using techniques like data monitoring, network authentication, and user device (EU) authentication, traditional security architectures can protect both text and audio communications. With this technology, customers and network operators get the highest level of security and dependability for conventional Long Term Evolution (LTE) mobile networks [1]. Users and services use standard authentication, but EU and the foundation database can also use their own independent authentication methods. Using a huge key management system can help keep the security of LTE access and traffic control [2]. Some research has been done into how secure LTE-related technologies are, but new kinds of security are needed to accommodate developing network designs and usage types.

There is a chance that some innovative use cases will need to meet certain constraints, such as a minimum amount of time that must pass between conversations between users. The discovery of new information could show security holes or force the installation of extra security measures in 5G [3], and it will also make the services offered more complicated. When setting up a HetNet, it's important to remember that, due to the different security measures of the different access technologies and the possibility of a wide range of chaotic events, regular authentication and strict latency limits may be needed. This is something you should always keep in mind. Massive MIMO is an important part of 5G technology because it lets you use airwaves more efficiently and uses less power overall. It's also a good way to keep someone else from taking over your thoughts. In addition, 5G's support for SDN or NFV in new service rescue models necessitates the introduction of novel security capabilities [4]. With the advent of the 5G network paradigm, new safety planning is needed to solve these problems; safety must be measured as an integral part of large architecture or included in the structure of the system from the outset. In order to effectively support a selection of use cases or new reliability models, a secure security system is needed. The reliability models of traditional communication networks or 5G wireless systems are shown in Figure 17.1. Verification of identity is not only required by the customer and the applicant, but also between them.

17.1.1 Challenges in 5G networks

Every new technology brings with it systemic challenges or challenges posed by new operations. Not all complications are those that cause unnecessary damage to the system. Some are just a few elements. If you don't listen, it may cause the system to malfunction or cause other problems. Identifying these problems in advance can have a significant impact on the response of the technology introduced. In the case of 5G, some issues that need to be addressed are highlighted, and then the underlying system can be properly explored. Because so many devices can connect to 5G networks, there is an urgent need to be able to prevent attackers who try to circumvent the authentication system and result in denial of service (DoS).

17.1.2 Attacks and security services in 5G wireless networks

Because broadcast media is open to everyone, wireless information programs can be abused in many different ways. In this section of the article, we'll talk about four different ways that 5G wireless networks can be attacked: traffic analysis, interference, DoS, and man-in-the-middle attacks. The security measures we give will cover four different areas: authentication, privacy, usability, and safety.

    i. **Attacks on 5G wireless networks:** Figure 17.2 shows the types of attacks: each type of attack is discussed in the following three areas: nature of the attack (passive or active), safety services supply to protect touching the attack, or appropriate technique used to evade or to

248 Artificial Intelligence and Blockchain in Industry 4.0

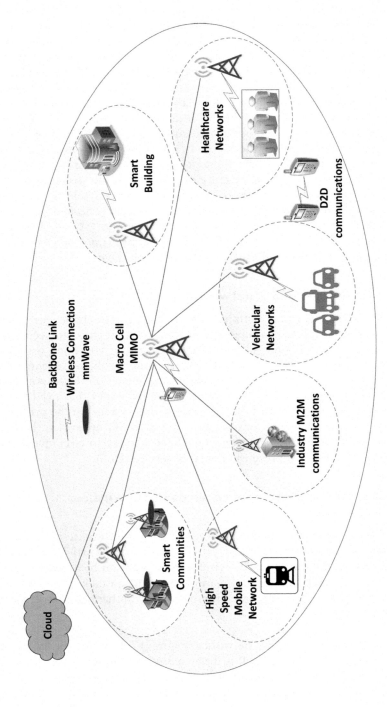

Figure 17.1 Structure for fifth-generation wireless networks.

Algorithm for secured energy-efficient routing 249

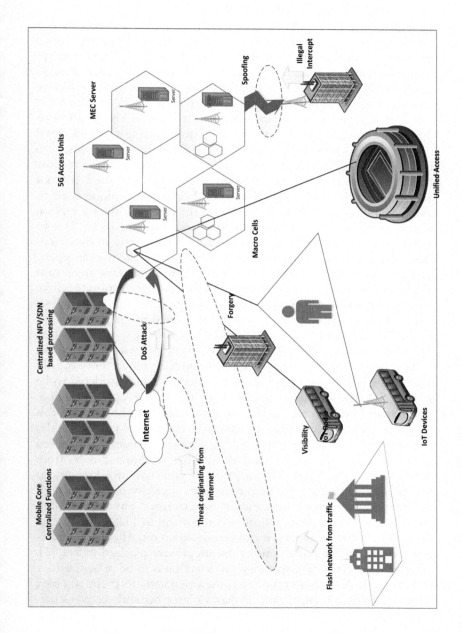

Figure 17.2 Attacks on 5G wireless networks.

prevent it. We focus on attackers targeting safety attacks at the PHY layer or MAC layer, which are significant variations in safety levels between wireless and wired networks.

ii. **Eavesdropping and traffic analysis**: Eavesdropping is a bias assault because a regular relationship has no effect on eavesdropping. Due to its transient nature, it is difficult to detect its opening. Signal encryption over wireless links is primarily used to protect against malicious attacks. Due to encryption, the listener cannot honestly receive received signals. Traffic examination is another attack in which the recipient inadvertently obtains information such as the location or information of communicating party by investigating received signal traffic without knowing the content of the signal. In other words, even if an indication is deleted, traffic examination can still be used to show how the parties communicate. The traffic examination attack will not affect the legal relationship. The eavesdropping technique used to avoid eavesdropping depends largely on the power of the encryption algorithm or the power of the eavesdropping machine meter. Due to rapid growth in accounting power or rapid development of data analysis techniques, new methods can be used to attack eavesdroppers. The current mechanisms to deal with eavesdropping face significant challenges, as many of them assume that the number of collectors here and there is small, but the ability to compute data and data analysis is very low. In addition, some technologies used in 5G wireless networks (such as HetNet) may amplify the complexity of combating eavesdroppers. In universal, new operations on 5G wireless networks make it more difficult to maintain the monitoring mode. Since the old writing methods have been thoroughly studied and considered to be serious, PLC research has attracted more and more people's attention.

iii. **Interference**: Interference, unlike monitoring and traffic analysis, can completely disrupt communication between real users. Malicious nodes can cause intentional intrusion, halting data communication between legitimate users. Interference can also prevent authorized users from accessing radio property. The majority of proactive attack solutions are based on what the attacker is doing. For cognitive users, a pseudo-random, time-hopping, anti-jamming mechanism is envisaged in Ref. [5] that can restore presentation in comparison to FHSS. Because of the way interference works, it can be picked up on. Allocation strategies for fusion center and jammer resources were planned in Ref. [6]. To reduce the error rate, spending on detection is to be prioritized.

iv. **DoS and DDoS**: DoS attacks can cause opponents to drain complex wealth. DoS is a security attack that violates network convenience. Interference can be used to launch DoS attacks. When there are several distributed opponents, DDoS can be formed. There is a lot of energy and versatility in both DoS and DDoS attacks. DoS and DDoS attack identification is currently the primary application of recognition technology. DoS and DDoS can pose a serious danger to operators due to the widespread adoption of massive equipment in the 5G wireless system. Many interconnected gadgets in the 5G wireless complex make it vulnerable to DoS and DDoS assaults. A DoS attack can be heralded as either an attack on the transport layer of a system or an attack on a device or user. Network communications are vulnerable to DoS assaults on several levels, including the signal plane, user plane, control plane, maintenance system, radio goods, logical property, and animal property. Battery life, data storage, central processing units, radios, actuators, and sensors are all vulnerable to DoS assaults.

17.1.3 Security services in 5G wireless networks

5G wireless networks have introduced a new skin texture and new rations for safety services as a result of new architectural features, new knowledge, and new utilities. We focus on four types of security services in this section: authentication (confirmation of devices and messages), confidentiality (data privacy and secrecy), accessibility, and truthfulness.

- **Authentication**: There are two types of authentication, namely device confirmation and message authentication. In 5G wireless networks, device authentication or messaging authentication are both important in resolving the above attacks. Authentication of the device is done to make sure that it is exactly what it says it is. User equipment (UE) and the mobility management unit (MMU) need to be set up before traditional mobile networks can be used for communication (Mobile Management Entity [MME]). For a cellular network to work, both the UE and the MME need to agree on the same security model. In a wireless network that works with 4G LTE, symmetric keys are used for both authentication and key agreement (AKA). Nonetheless, 5G calls for clearance from parties outside of the UE and MME (such as service providers).

5G necessitates a hybrid or flexible verification organization since the trust model is different from the trust model employed in the conventional mobile complex. There are three distinct approaches to carrying out hybrid confirmation and flexible UE authentication: exclusively through the identity verification network, exclusively through the identity verification service provider, and either through the network or the identity verification service provider [7]. 5G identity verification is expected to be quicker than ever before because of the high data rates or extremely low latency needs of 5G wireless networks. It's also possible that regular transfers or approvals between 5G layers will crop up in multilayer planning. An SDN-enabled technique using weighted security context information transmission was presented as a means to circumvent the complexity of key organization in

HetNets and reduce the waiting time brought on by repeated switching or authentication between different layers. enhance effectiveness. Integrate identity verification into the transmission process while still maintaining 5G latency. A public key-based AKA is envisaged in Ref. [4] to enable the provision of additional safety services in 5G wireless networks.

17.2 RELATED WORK ON 5G TECHNOLOGY

Bendale et al. (2018): Compared to the traditional mobile network (4G), the new 5G mobile wireless network will be thoroughly researched. It starts with the specificity and the new requirements for 5G networks and the motivation for 5G mobile safety. Threats and security issues will be investigated. Device authentication, network access in restricted areas, data confidentiality, privacy, and intrusion detection are examples of security services. This security research or advancement is used to explain new 5G mobile wireless security measures. New interference recognition technologies will also be required for the next generation of mobile wireless networks. They concluded by summarizing the scope of 5G security management and potential future directions [5].

Ilgaz et al. (2018): They used their knowledge of millimeter waves to describe and look into the possible uses of optoelectronic oscillators in the coming development of 5G mobile or wireless complexes. There is a chance of making a 39 GHz single-ring optoelectronic oscillator, which would be the heart of any future 5G wireless or mobile networks. This setup is meant to make it easier to organize base stations in the future or to save money for the upcoming 5G system. The scheme design avoids power failure due to chromatic scattering and is temperature stable for long-term stability or has low secondary state effects to ensure low-phase noise [6].

Wang et al. (2019): Still, these systems lose a lot of signal strength or heat when they are close to solid objects. This makes the cell radius smaller or the network architecture more complicated. Millimeter wave communication, also called mmWave, will be used in 5G mobile cellular systems. It's been shown that an ultra-dense small cellular network can be combined with a large-scale antenna array (LSAA) coverage expansion method without adding too much to the cost. There is a chance that the importance of wireless network architecture will be changed by factors like the rate of proliferation. In the last step of this research project, the effects of the coverage behavior of a system with a lot of antenna arrays on the development of mm-wave network architecture and oscillations at multiple cellular levels are studied [8].

Zhang et al. (2018): As the next phase in the production of ultra-high-speed wireless networks, the 5G mobile wireless complex has recently garnered a lot of investigative focus or attempts from academia or industry. The unparalleled delay limited quality of service (QoS) promises of 5G mobile wireless networks will likely benefit a wide range of multimedia services, apps, or consumers with drastically different ratios. However, many new, severe challenges have emerged that were not present in 4G wireless networks, such as how to provide sufficient multimedia services on 5G wireless systems. Here, a unique software-defined network architecture (SDN) for 5G multimedia wireless networks with heterogeneous statistical QoS settings is introduced, which uses mobile traffic scanning to address these new difficulties. Device authentication, network access in restricted areas, data confidentiality, privacy, and intrusion detection are examples of security services. Because 5G uses cutting-edge technologies and methodologies such as the Internet of Things (IoT), Massive Multiple Input Multiple Output (mMIMO), Device-to-Device Communication (D2D), and SDNs, its mobile wireless networks must also implement cutting-edge security measures (SDN). This security research or advancement is used to explain new 5G mobile wireless security measures. New interference recognition technologies will also be required for the next generation of mobile wireless networks. They concluded by summarizing the scope of 5G security management and potential future directions [7].

Kim et al. (2019): The 5G stand-alone (SA) classification is a real 5G classification, which uses 5G radio among base stations, and the base network also uses the 5G network. In the core 5G network of the SA 5G system, the service is transmitted to the user's base through different techniques such as switching mode or separation mode. These methods combine wireless and 5G, wireless Wi-Fi used. However, the operator of a 5G system may not be aware of how often traffic transmission techniques such as switching and sharing are achieved. Therefore, this paper presents a 5G network traffic monitoring system that allocates operators to simply identify the exchanges and distribution of traffic through traffic monitoring messages sent from operating components of the 5G network [9].

Muhammad Farhan Khan et al. (2020): The interest in using heterogeneous networks (HetNets) based on 5G to aid manufacturing processes in smart plant use cases has recently increased. Supporting mobile robots and other autonomous vehicles, for example, requires a high-performance wireless system with high data rates or increased spectrum competency. Multilayer, femto, Pico, or macrocellular networks are all being considered as potential options to fulfill required data rate ratios for low-latency, high-coverage 5G implementations. For low-mobility robots or autonomous cars with high-mobility and fixed wireless sensors actuator equipment, this work analyzes a variety of base station connection options on 5G HetNet, including the highest received authority and the highest signal-to-interference-and-noise ratio (SINR). The experimental results favor the maximum received power cell correlation method used in their collective implementation scheme over the largest signal-to-noise ratio (SNR) cell correlation strategy [10].

Sung et al. (2020): In this study, the effectiveness of 5G telephony communication based on optical input frequency (IFWF) between radio input (IFoF) (IFANF) was reported. An IFOF-based cyclic antenna system (CAS) is offered to make it easier for mmWave-based 5G networks to meet reporting requirements. During the 2018 Winter Olympics, a distributed antenna system (DAS) based on an IFOF and equipped with a 22-integrated circuit (MIMO) was set up in the Pyeongchang area. This will allow 5G test displays. The IFoF-based DAS makes it possible for anyone to test 5G technology. One of these abilities is the ability to send and receive data quickly. Within the boundaries of the DAS application, the throughput is about 1. A link with a speed of 1 Gb/s can send about 200 Mb/s per second. Their company has also made available a 5G mobile endpoint that works with IFoF and can get around the bandwidth limits of RAN. They will be using IFoF-based mobile fronthaul to connect to mm-based 5G networks. When using a 4 × 4 MIMO interface, it is estimated that the downstream link development is reached in a short time by each 5G terminal. An outdoor demonstration was conducted at 9 Gb/s to verify the capability of the 5G front-end network technology based on IFoF equipment. When moving connecting long-distance wireless heads at 5G speeds at speeds of less than 60 km/h, it can provide 5G mobile network service over a long period of time over 5Gb/s. It is therefore assumed that IFoF technology can support 5G networks based on RAN radio, and can provide real multi-Gb mobile service [11].

Mekuria et al. (2019): The fifth wireless ICT ecosystem guarantees three pillars of complex innovation: (a) reliable and low-cost communications, (b) improved Gigabit Wireless broadband, and (c) large-scale machine-type communications. It is hoped that 5G will change the way people and devices communicate and service replacement, or restore the level of industrial automation. The purpose of this article is to discuss recent research on range sharing and unlicensed networks, as well as its extension to 5G investigation based on usage issues and related knowledge to meet the needs of affordable Wireless broadband. This article describes the search results of the Spectrum Toolbox, including smart range distribution for 5G and low-cost networks. The search results are presented along with an evidence-based web test platform urbanized in South Africa. And finally, this article challenges existing and future research on the technical standards of 5G to contribute to the implementation of a global and affordable network for the next billion users [12].

Dongfengfang et al. (2018): Enhanced efforts to secure 5G mobile network systems provide new challenges. This article investigates the safety of 5G wireless network systems in depth, comparing them to the security of older mobile networks. The article starts by discussing the specifics of 5G wireless networks, such as the latest necessities and motivations for bolstering wireless security in the 5G era. New services and use cases on 5G wireless networks are discussed, along with a summary of the possibility of attacks and security services. It is founded on state-of-the-art innovations and preexisting solutions to security issues, such as verification, access, data protection, fundamental management, and confidentiality. This article delves further into the new security features associated with a wide range of 5G-related technologies, including heterogeneous networks, D2D, numerous outputs, and more, as well as a network of defined software and the IoT. In response to these concerns, researchers and security experts have developed a novel approach to 5G wireless security that uses analysis of identity management data to verify user credentials. To demonstrate the advantages of the suggested safety architecture, they used the donation procedure and load-bearing technique as a case study [13].

Betzler et al. (2019): The large-scale deployment of small power station (SC) stations is critical to meeting the rations of future 5G networks. However, two barriers to SC deployment are: (a) the extremely limited and costly site having a cable backhaul resource and (b) the difficulty of managing the installation of the backhaul node software. In response to these challenges, the authors introduced integration of SODALITE into 3GPP 4G and 5G architectures, and demonstrated SODALITE's capabilities by testing it on the LTE network. They confirmed the reliability of SODALITE through a cross-sectional study of traffic from LTE networks and discussed the impact of cell or structural amplification. 5G systems in these studies. In addition, a secure backhaul solution for wireless links is integrated into SODALITE through an SDN-supported system. This system is able to reconfigure the data when it detects a link failure. Experiments in the field of LTE network testing have demonstrated its reliability, and network simulators have been studied in depth through rigorous simulations and evaluations [14].

Tahir et al. (2019): In the automotive world, weather forecasting and road condition predictions are elements that improve road safety. It can connect to remote drive networks based on IEEE 802.lip. Currently, mobile communications with future 4G and 5G are also considered intermediaries in in-car communications. In this work, it is looked at how the measurements from IEEE 802.11p and the measurements from 4G and 5G pilot networks are different. The company has set up a number of 4G and 5G networks based on IEEE 802.11p ITS-G5.

Using a 5G test system based on LTE 14, 4G (LTE), and IEEE 802.11p networks, they looked at the system-level latency as well as the latency and output from the point of view of the underlying infrastructure or the user's end device. In this comparison, the state-of-the-art 5G radio interface came out on top in many ways, including portability, durability, output, coverage, and data transfer rate, to name a few [15].

Koumaras et al. (2018): Visible light communication (VLC) has been a subject of a lot of research and development in recent years. It has recently come to the attention of experts in long-distance communication as a possible control tool for 5G networks. VLC is a great way to meet the performance, data rate, spectrum compatibility, energy efficiency, battery life, and latency needs of 5G systems. This is because VLC has some good qualities that help in these areas. bottom. Though, because the characteristics of the channel may not always be present in practice, this remarkable success is limited by negative reception. This paper presents and validates the VLC additive SDN system through experimentation. The system is paired with Wi-Fi access knowledge to recover the reliability of the VLC system and to ensure that during inconsistencies or roadblocks or when moving between two users, the quality of its reception has zero packet loss over time. VLC launcher continues and lives the "dead zone coverage" [16].

Zhang and Wang (2020): A number of cutting-edge technologies have been made so that 5G wireless media services can be used with reliable low-latency communication (URLLC). Low-latency QoS is one of these kinds of technologies. There is no longer any way to make the delay or the bit error rate better. Both of these measurements are as good as they can be. On the other hand, wireless (mmWave) CF multi-channel multi-input (m-MIMO) technology is one of the possible technologies that will restore QoS efficiency in 5G networks. This configuration uses many multi-input points (AP) so that millimeter processing services can be made available to all users who have access to the same source-time information. By taking advantage of how mmWave wireless pipelines scatter traffic, the mmWave wireless pipeline model built on the filter cluster may be able to spread traffic that comes in on the same stream to cut down on queue latency. Spreading out the traffic is one way to do this. But it is hard to get QoS efficiency data for the whole length of CF m-MIMO projects because it is hard to analyze the queue when there is mmWave cross-channel traffic. In this research, a new analysis model for evaluating the random QoS efficiency of CF m-MIMO initiatives through mmWave pipelines with low error rates is shown. This model helps solve some of the problems already talked about. The team has made a simulation of how a CF m-MIMO system works that covers all of its parts. This system works through a series of cable channels that are only a few millimeters wide. On top of that, the Mellin transform and the spatial multiplexing sequence model are used to figure out how much of a violation can be allowed. The results confirm and analyze the quality of the planned mmWave CF m-MIMO FBC-based project within the limits of QoS requirements and statistical/rate limits [17].

Keogh et al. (2018): Self-interference (SIC) is a major deliberation for 5G wireless transceiver architecture. Properly applied, SIC can increase spectrum efficiency through full-duplex operation. The major confrontation is to effectively eliminate the interference caused by the transmitted signal. This article presents a circular model of broadband analog monitoring using a near-field vector camera. When used in conjunction with instrumentation, spatial and large-scale application algorithms can provide effective optimization. The proposed technique provides 35 dB of RF repulsion over a 100 MHz channel based on 2.1 GHz [18].

Jin et al. (2020): Intelligent machine-based mobile and wireless applications need to have three things: low latency, high bandwidth, and a high connection rate. The above ratios are possible in 5G networks because of the improvements made to 5G wireless technology and the way network systems are built. This article suggests a cutting-edge, self-sufficient way to virtualize 4G mobile networks by using new developments in MEC technology and expertise in network slicing. In this argument, either an open 4G network or mature, low-cost mobile devices are used to build a private network that can meet the latency, bandwidth, and high-volume needs of modern connectivity while working in a mixed-network environment. This makes it possible for the private network to work in a setting with both open and closed networks. Also provided are wireless video, sensor data, and other health-related services in addition to HRLL [19].

17.2.1 WSN under different network

Anish Khan et al. (2021): Life and energy efficiency are of paramount importance in wireless sensor networks. WSN has a variety of sensors along with a stationary base. Sensor node power consumption is limited. The efficient use of this energy in this way allows the data to be transmitted efficiently from the sensor point to the current headache base station. LEACH is a standard road transport protocol used in WSNs. The entire LEACH transmission protocol has a few drawbacks, for example, it does not take into account the energy at the nodes found in the network during the host cluster selection process. In order to overcome these shortcomings, the conventional LEACH protocol has made great strides. The main purpose of this article is to briefly introduce an improved version of the LEACH protocol. Finally, comparisons were made between the updated versions of the LEACH protocol [20].

Sana and Noureddine (2019): The term "Wireless Sensor Network" (WSN) refers to a system of interconnected, battery-operated sensor nodes that can be deployed in unpopulated areas for scientific purposes. Many factors, including limited battery life, limit the usefulness of these sensor nodes. Additionally, sensors and communication operations are quite power-hungry, therefore, employing rapid strategies for power management might increase the node's range. More and more academics are examining wireless communication networks; the specifics of their investigations vary by application, network architecture, and, most importantly, the metrics employed in the communication network (distance, hop count, etc.) Then, every one of them is unique in some way. A well-designed routing protocol can boost transmission competence or extend the life of the entire network, thus it's crucial to pick the right one. The reliability of a wireless sensor network is crucial to any application that relies on it. In order to facilitate wireless communication in both isotropic links and anisotropic networks, a new method strategy is introduced in this research. When determining how to include a node into the flow table, this strategy prioritizes the insertion of nodes with the fewest possible hops. This algorithm is an advancement of the Prim algorithm, with its core concept originating in the MST (Minimum Spanning Tree) branch of graph theory. Simulation findings further demonstrate the efficacy of the suggested strategy in reducing the occurrence of single discharges in the substrate network comprising isotropic and anisotropic sensory regions [21].

Mohapatra et al. (2017): A wireless sensor network has randomly distributed, tiny sensor points with constrained processing speed, memory, and battery life. Due to the battery's limited capacity, the WSN should prolong the network's lifespan by lowering power usage. The total energy consumption in the data transmission path can be decreased because clusters have the ability to use less energy. They can balance energy consumption and have a high speed and size. In this article, examined are various LEACHEES (small energy adaptive cluster hierarchies) and dynamic protocols based on hierarchical energy-saving clusters, which contribute to the development of wireless communication networks. The advantages and disadvantages of LEACH and its descendants and compared them based on various metrics such as mobility, accuracy, accuracy, memory, and hop count are also talked about [22].

Chand et al. (2012): A newly developed initiative called OEERP (Optimal Energy-save Routing Protocol) is being proposed, which can extend the range of Wireless Wings (WSN) communications. It is a cluster-based protocol that each time a node changes, it changes to the head of the cluster. For two primary reasons, this strategy can help the WSN last longer. The first is that all the nodes use the same amount of battery power, and the second is that no node needed to use more than four transmissions to go to the old entry point. Data recording and data collection are performed in a way that reduces the number of messages sent at the input level. This protocol can be used for any time-to-time monitoring applications that use WSN. The network simulator (NS-2) supports the simulation of wireless networks and is used to build OEERP-based wireless sensor networks [23].

17.3 NETWORK MODEL

All of the N points that make up the structure are spread out over an infinite sphere. Each node's initial power, maximum packet size, message size, and threshold are broken down into their own specific parameters. The system is built on a network, which can be thought of as a path along which individual nodes are connected to each other in groups called clusters. To get a good idea of how valuable a sensor network is, you need to know how long it will last. Cluster modeling explains the functions of all network creation models. Each node is allocated an initial power value of 2.0J, and all nodes communicate in distributed mode. The maximum packet size is 4000 bits, or 200 bits per message, with 50 bits allotted to each node.

17.3.1 Key security challenges in 5G

5G needs robust safety architecture or explanation because it connects all aspects of the network. We therefore investigated and highlighted major security OR privacy challenges in the 5G complex and described potential solutions for a secure 5G system. The key 5G challenges generated by the Next Generation Mobile Network (NGMN) [6] are as follows:

- **Flash network traffic**: There will be a large number of end-user or innovative devices (IoT) devices.
- **Radio interface security**: The radio blocking key is sent to the unsecured channel.
- **User data aircraft**: User data aircraft do not have integrity security protection.
- **Compulsory security in the network**: Restricting the service driver to the security architecture allows you to choose to use security procedures.
- **Roaming security**: Users' security parameters are not efficient when they travel from one operator's complex to another operator's network, resulting in secure safety while roaming.
- **DoS attacks on infrastructure**: There are essentials that manage the perceptible network and an unlocked control channel.
- **Storm signals**: Circulated monitoring systems need to configure, for example, the Non-Access Stratum (NAS) layer in the 3rd Generation Partnership Project (3GPP) protocol.

- **DoS attacks on end-user devices:** There are no safety procedures against the operating system, software, or encryption data on user devices. NGMN's 5M proposes morality that goes beyond radio efficiency, including the creation of globally compiled concepts and the use of new information technologies or communication networks to simplify operations and management. Therefore, we focus on securing technologies (i.e. mobile cloud, SDN, and NFV) that comply with the design principles defined by NGMN.

17.4 CONCLUSION

The suggested work makes use of the completed work of finite automata by training the automata to learn in accordance with established guidelines and norms. To evaluate the viability of each possible route, the AFD-PSO suggestion is applied. The PSO is used to check and analyze calculations in a simplified manner. The number of people who took part in the study was cut because of what the AFD-PSO model and the networking model told us. Its effectiveness can be judged by a number of factors such as network life, throughput, power consumption, and data from operational nodes, to name a few. The experimental results indicate that the discharge of AFD-PSO is 16 percent higher than cluster-based IDS and 70 percent higher than others. As a result, it's safe to say that AFD-PSO is a great tool for keeping the WSN running smoothly and efficiently while also conserving energy.

DATA AVAILABILITY STATEMENTS

We have read lots of paper and taken my data from Google Scholar research papers.

FUNDING

No Funding

AUTHOR CONTRIBUTION

1. Jyoti Srivastava: Paper writing, Paper formatting, Literature survey
2. Jay Prakash: Supervision, Paper editing

CONFLICT OF INTEREST

No conflict of interest.

REFERENCES

1. M. Agiwal, A. Roy, N. Saxena, Next generation 5G wireless networks: A comprehensive survey, *IEEE Communications Surveys & Tutorials*, vol. 18, no. 3, pp. 1617–1655, 2016.
2. Jing He, Yanchun Zhang, Guangyan Huang, Yong Shi, Network lifetime of application-specific randomly deployed, *Wireless Sensor Networks in Arbitrary Sensor Density*. DOI:10.1109/ICIS.2007.139
3. Haiming Wang, Peize Zhang, Jing Li, Xiaohu You, Radio propagation and wireless coverage of LSAA-based 5G millimeter-wave mobile communication systems, *China Communications*, vol. 16, pp. 1–18, 2019.
4. Qiao, X. S. Shen, J. W. Mark, Q. Shen, Y. He, L. Lei, Enabling device-to-device communications in millimeter-wave 5G cellular networks, *IEEE Communications Magazine*, vol. 53, no. 1, pp. 209–215, 2015.
5. Shailesh Pramod Bendale, Jayashree Rajesh Prasad, Security Threats and Challenges in Future Mobile Wireless Networks, *2018 IEEE Global Conference on Wireless Computing and Networking (GCWCN)*, 2018 DOI:10.1109/IEEE Lonavala, India.
6. Mehmet Alp Ilgaz, Bostjan Batagelj, Application of an Opto-Electronic Oscillator in 5G Mobile and Wireless Networks with a Low Frequency Drift, a High Side-Modes-Suppression Ratio and without a Power Penalty Due to Chromatic Dispersion 2018, *European Conference on Networks and Communications (EuCNC)*, 2018.
7. Haiming Wang, Peize Zhang, Jing Li, Xiaohu You, Radio propagation and wireless coverage of LSAA-based 5G millimeter-wave mobile communication systems, *China Communications*, vol. 16, pp. 1–18, 2019.
8. J. Zhang, W. Xie, F. Yang, An Architecture for 5G Mobile Network based on SDN and NFV, *6th International Conference on Wireless, Mobile and Multi-Media (ICWMMN 2015)*, 2015, pp. 87–92.
9. Eungha Kim, Young-Il Choi, Traffic monitoring system for 5G core network 2019, *Eleventh International Conference on Ubiquitous and Future Networks (ICUFN)*, 2019.
10. Muhammad Farhan Khan, An Approach for Optimal Base Station Selection in 5G Het Nets for Smart Factories 2020, *IEEE 21st International Symposium on "A World of Wireless, Mobile and Multimedia Networks" (WoWMoM)*, 2020.
11. Minkyu Sung, Joonyoung Kim, Eon-Sang Kim, Seung-Hyun Cho, Young-Jun Won, Byoung-Chul Lim, Sung-Yeop Pyun, Hoon Lee, Joon Ki Lee, Jong Hyun Lee, RoF-Based Radio Access Network for 5G Mobile Communication Systems in 28 GHz Millimeter-Wave, *Journal of Lightwave Technology*, vol. 38, no. 2, pp. 409–420, 2020.
12. Fisseha Mekuria, Luzango Mfupe, Spectrum Sharing for Unlicensed 5G Networks 2019, *IEEE Wireless Communications and Networking Conference (WCNC)* 9, 2019.
13. Dongfeng Fang, Yi Qian, Rose Qingyang Hu, Security for 5G mobile wireless networks, *IEEE Access*, vol. 6, pp. 4850–4874, 2018.
14. August Betzler, Daniel Camps-Mur, Eduard Garcia-Villegas, Ilker Demirkol, Joan Josep Aleixendri, SODALITE: SDN Wireless Backhauling for Dense 4G/5G Small Cell Networks, *IEEE Transactions on Network and Service Management*, vol. 16, no. 4, pp. 1709–1723, 2019.

15. Muhammad Naeem Tahir, Kari Mäenpää, Timo Sukuvaara, Evolving Wireless Vehicular Communication System level comparison and analysis of 802, 11 p, 4G 5G 2019, *2nd International Conference on Communication, Computing and Digital systems (C-CODE)*, 2019.
16. Harilaos Koumaras, Dimitris Makris, Andreas Foteas, George Xilouris, Michael-Alexandros Kourtis, Vaios Koumaras, John Cosmas, A SDN-based WiFi-VLC Coupled System for Optimised Service Provision in 5G Networks, *2018 IEEE 19th International Symposium on "A World of Wireless, Mobile and Multimedia Networks" (WoWMoM)*, 2018.
17. Xi Zhang, Jingqing Wang, H. Vincent, Poor Statistical Delay/Error-Rate Bounded QoS Provisioning Across Clustered MmWave-Channels Over Cell-Free Massive MIMO Based 5G Mobile Wireless Networks in the Finite Blocklength Regime, *2020 54th Annual Conference on Information Sciences and Systems (CISS)*, 2020.
18. Brian Keogh, Anding Zhu, Wideband self-interference cancellation for 5G full-duplex radio using a near-field sensor array, *2018 IEEE MTT-S International Microwave Workshop Series on 5G Hardware and System Technologies (IMWS-5G)*, 2018.
19. Qibing Jin, Qing Guo, Mingshi Luo, Yu Ming Zhang, Wu Cai, Research on High Performance 4G Wireless VPN for Smart Factory Based on Key Technologies of 5G Network Architecture, *2020 International Wireless Communications and Mobile Computing (IWCMC)*, 2020.
20. Anish Khan, Nikhil Marriwala, A Literature Survey on LEACH Protocol and Its Descendants for Homogeneous and Heterogeneous Wireless Sensor Networks 20 February 2021.
21. Messous Sana, Liouane Noureddine, Multi-hop energy-efficient routing protocol based on Minimum Spanning Tree for anisotropic, *Wireless Sensor Networks 2019 International Conference on Advanced Systems and Emergent Technologies (IC_ASET)*, 2019.
22. Hitesh Mohapatra, A Survey on LEACH in WSN January 2017 Conference: Convergence of Technology-2017At: 1 Volume: 1.
23. Kishan Chand, Optimized Energy Efficient Routing Protocol for life-time improvement in Wireless Sensor Networks January 2012.

Chapter 18

Role of cloud computing and blockchain technology in paradigm shift to modern online teaching culture in the education sector

Nazreen Khanam
Jamia Millia Islamia, New Delhi, India

Md Safikul Islam
Ministry of Social Justice and Empowerment, New Delhi, India

18.1 INTRODUCTION

According to Mahatma Gandhi, (1942)

> a teacher, who establishes rapport with the taught, becomes one with them, learns more from them than he teaches them. He who learns nothing from his disciples is, in my opinion, worthless. Whenever I talk with someone, I learn from him. I take from him more than I give him. In this way, a true teacher regards himself as a student of his students. If you will teach your pupils with this attitude, you will benefit much from them.
> (A talk to Khadi Vidyalaya students, Sevagram, Sevak, 15 February 1942 CW 75, p. 269)

Teachers will adjust how they teach and learn to accommodate their teaching methods (Figure 18.1). Teaching is a labor of the heart as much as the head. If we imagine that the teacher is a magician who grants us the ability to speculate about achieving success in life, then we would enter a world of opportunities. The purpose of a competent teacher's story is to provide new views to his students. An emphasis on learning approaches and assisting the student in exploring new concepts is required.

Previous research has shown that an effective teacher and a democratic school have a positive impact on a student's academic success, thus the role of both a teacher and a school is completely proportionate. Teachers must use several strategies to design their lectures so that students enjoy the analysis (Marzano, 2007). Sergiovanni (1994) observed in his research that the instructor must share his efforts to ensure the success of teaching through modeling, happy mood, transparency, sensitive growth, honesty, and love of pupils (Singh & Mishra, 2017).

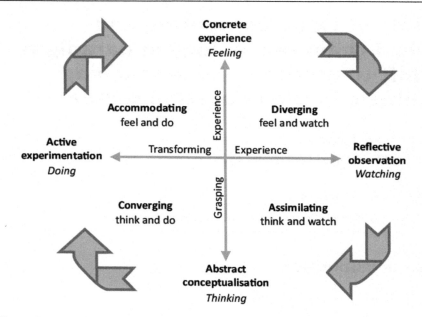

Figure 18.1 Model of learning process.

Source: Kolb's Experiential Learning Theory & Learning Styles (Kurt, 2020)

18.2 CLASSROOM TEACHING TO ONLINE TEACHING CULTURE: 360-DEGREE CHANGE IN TEACHING–LEARNING PERSPECTIVES

Learning or schooling is indeed an essential part of life because no human being can live effectively without it. In the coming days, a variety of paradigms for learning or studying are available. E-learning is one of the most promising educational models. It is the deliberate application of networked information and communication technology (ICT) in the teaching and learning of students (online teaching and learning with the help of cloud computing and blockchain technology). This method of teaching and learning is often referred to as "online learning," "interactive learning," "collaborative learning," "networking," and "web-based learning." Increased access to ICT and cheaper pricing are intrinsically linked to the growth of e-learning. A lack of or inadequate digital infrastructure can affect teachers, students, and the learning experience as a whole.

18.2.1 Teaching: From Ashram culture to online teaching culture

In ancient times, education was passed down through the Guru–Shishya system through the Ashram. It required the student to stay at the Ashram

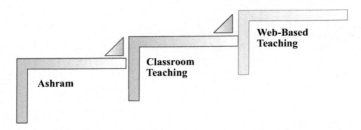

Figure 18.2 Paradigm shift in the teaching–learning community.
Source: **Prepared by authors**

during his or her education, which was a continuous process. The methods of disseminating knowledge have changed over time, but the principle has not. To move information, it requires a "teacher" and a "student." Physical infrastructure such as classrooms, labs, and other facilities are required by the current education system, which is governed by a syllabus and a graduation system. Therefore, the number of persons who can receive high-quality education is limited due to a lack of seats. Covid-19 has ushered in a new age in which there are no limits on the number of seats that may be purchased. The use of technology in the classroom referred to as "Digital Learning" has made it possible for students to learn more effectively (Figure 18.2).

Every technique has its method, which makes it useful and compassionate. The current great question is whether digital learning will become a permanent educational framework or only a stopgap measure. Many of the difficulties that have plagued our brick-and-mortar classroom structure for years will be addressed by digital learning.

- Affordability
- Extension
- Versatile learning environment
- Involved and engaging events
- The infinite capacity of the classroom

18.2.2 Cloud computing

The National Institute of Standards and Technology or NIST (Mell & Grance, 2011) defined that

> Cloud computing is a model that provides standardized, simple, on-demand network access to a common pool of configurable computing resources (e.g. networks, servers, storage, software, and services) that can be easily distributed and executed with minimal latency or service provider involvement.

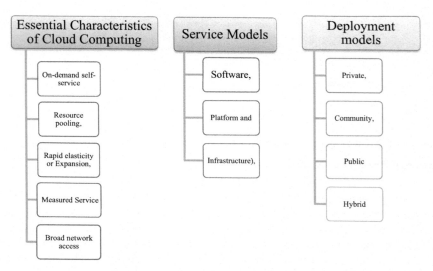

Figure 18.3 Essential characteristics, service models, and deployment models of cloud computing given by NIST.
Source: **Prepared by authors**

Moreover, NIST outlined the lists of key characteristics, service models, and deployment models for cloud computing to help people with categorizing different approaches to offer cloud services (Figure 18.3).

- Figure 18.4 demonstrates all the suitable platforms for cloud computing including Gmail, Google Drive, Hangouts, Google Docs, and Calendar which are some of Google's most well-known productivity apps. All of the G Suite apps are available in the cloud. Microsoft, like Google, offers a unique edition of their productivity software for educators and students called Office 365 Education. It is a simple-to-use cloud-based subscription edition of Microsoft Office.
- Through its cloud infrastructure, Coursera offers a diverse selection of online courses from recognized universities and teachers. It also offers courses on specialized job skills and awards accredited university degrees.
- For grades K-12, higher education, and government, Blackboard provides cloud-based learning technologies.
- Teachers may build interactive courses, quizzes, and events with Class Flow, which is built for linked whiteboards and displays, and then present or distribute them to students.
- Bright Space is a learning management system named D2L.
- A Cloud Guru is an online repository of cloud computing courses designed to educate and train people. The A Cloud Guru course, for example, is for people who want to learn how to use Amazon Web Services in their careers.

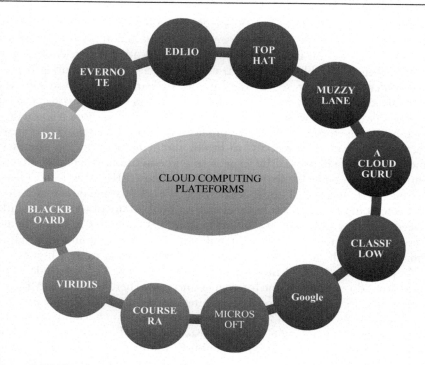

Figure 18.4 Cloud computing platforms.

Source: **Prepared by authors**

- Viridis has developed a cloud-based program that connects community college students with job databases in order to match them with appropriate postgraduate jobs.
- Muzzy Lane's goal is to make learning more like playing a game in order to keep children interested and engaged.
- Top Hat's educational software allows college students and lecturers to connect through quizzes, debates, and the distribution of material for reading.
- Edlio is a firm that creates school websites and social media platforms. Evernote allows users to take notes on their phones, tablets, or laptops, save them in the cloud, and sync them across all devices, removing the need for multiple notebooks. Audio files, photos, and URLs may also be saved.

18.3 IS CLOUD COMPUTING WIDENING THE DIGITAL DIVIDE?

People are classified into two parts when it comes to learning about this technology: those who are uninterested in learning how it works and those who are keen to learn but have few opportunities to do so. The latter is usually made up of younger people who are eager to learn and apply their previous knowledge to the business. However, due to educational restrictions, only a few colleges offer blockchain-focused degrees. As a result, many prospective students have resorted to the Internet to learn about blockchain via manuals and video courses. The digital divide refers to the growing societal divide between those who have connection to the web and those who do not. Members of the former community, predictably, perform better in other areas because they have better access to health care, employment opportunities, and, of course, education. The same is true in education, where cloud computing is reliant on technology availability. In August 2018, American College Testing (ACT) Inc. polled students all around the country and determined that access to computers and the Internet was unequal. Many of the 15% of kids polled said they only had one Internet-connected computer at home. Some were compelled to compete with family members for computer and online time. Others in the neighborhood have very limited access to their homes.

As per the National Center for Education Statistics, 18% of students in rural areas do not have home Internet connection. There were 41% African-Americans, 26% Latinos, and 13% Whites among the 18%. A new post on the Digital Divide Council's website promotes "recommended" measures to close the digital divide, which include teaching digital literacy and collaborating with big tech companies to establish resource networks, in addition to providing kids with tablets. "Bridging the digital divide in education is a lengthy and expensive process that is necessary for progress," the article says (Robb, 2016).

18.4 BLOCKCHAIN TECHNOLOGY

Blockchain is essentially a means of resolving issues with self-esteem (Figure 18.5). Blocks of transactions are incorporated in a collective, online ledger by a peer-to-peer network of computer stores and chains, making it difficult to modify unrecognized entries. Technology has non-monetary benefits, with the primary draws being cost reductions, enhanced privacy, and

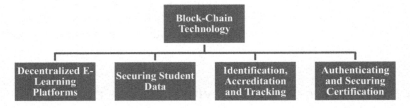

Figure 18.5 Benefits of blockchain technology.

Source: **Prepared by authors**

protection. This is true in any situation where records are held by a third party. Satoshi Nakamoto came up with the concept of blockchain in his thesis "Bitcoin: A Peer-to-Peer Electronic Cash System," which was published in 2008. The online education and professional development market—think Udemy, Lynda, and Coursera—is one area where it might solve genuine problems. With the Internet, professional growth has grown to more than US$ 300 billion. It has aided millions of people in learning new skills and knowledge from the convenience of their own homes.

Decentralized education networks can be managed by consensus rather than by a central authority thanks to blockchain. This allows for a true peer-to-peer learning experience. This is the next phase of online learning, in which centralized systems such as Udemy and Lynda are replaced by a new generation of instructors who work on a single, open platform and share a common currency, regardless of their geographic, industrial, or technical demographics. Using blockchain to verify authenticity is a quick process. When the teacher affixes the individual's certificate(s) to the ledger, the records become not only irreversible but also open to the public and apart from the issuing body. Employers can simply check the online record to confirm a new employee's learning history. The same measure is frequently used by students to gain complete control over their data. Because all of their information is available online, students no longer need to carry certificates to demonstrate their abilities.

18.5 CLASSROOM TEACHING TO ONLINE TEACHING CULTURE: A PARADIGM SHIFT IN TEACHING–LEARNING PERSPECTIVES WITH THE HELP OF CLOUD COMPUTING AND BLOCKCHAIN TECHNOLOGY

Online education has increased considerably since the turn of the century, due to the fast expansion of Internet technology. The term "online education," sometimes known as "remote learning" or "online learning," refers to a web-based educational tool that employs computer and Internet technology to spread information and speed up learning. Through the use of the Internet as a medium, online teaching transcends the boundaries of time, space, place, and teachers by offering students access to engaging learning opportunities whenever they need it. Occupational training and online education are two forms of online education.

MOOCs (massive open online courses) have become increasingly popular in recent years. MOOCs were established in the United States by top content providers such as Udacity, Coursera, and edX. Top American university affiliations have had online learning platforms and free online courses since 2012. MOOCs are self-contained management systems that provide high-quality instructional programs for higher education.

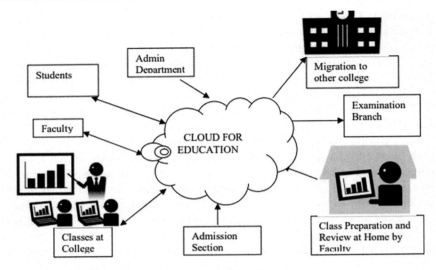

Figure 18.6 Sevices attached to education cloud.

Source: **Margianti & Mutiara (2015)**

18.5.1 Cloud computing in education industry

In education, cloud computing benefits students, teachers, and administrators equally (Figure 18.6). Students may access assignments from anywhere they have an Internet connection, teachers can post learning materials right away, and administrators can collaborate swiftly while saving money on data storage (Pund et al., 2012). The necessity to adapt traditional educational practices to suit the demands of the twenty-first century drives modern education. This is largely focused on bringing education out of the classroom and into the comfort of our own homes or workplaces.

Although cloud computing has existed for some time (Garcés et al., 1997; Sharma et al., 2014), the notion became "common" in October 2007 when IBM and Google announced cooperation in the industry (Bokhari & Makki, 2018). Following that, IBM announced the "Blue Cloud" initiative (Ise, 2015). Since then, "Cloud Computing" has been discussed often. Obviously, the necessary Wikipedia entry is present. In the education sector, cloud computing solutions are employed in the following areas.

18.5.1.1 Software as a service (SaaS)

SaaS refers to the apps which may be utilized from anywhere and at any time. This is one of the most serious educational issues right now. Both the program and the data are stored on the cloud, and all that is required for access is a web browser. One of the most well-known examples is Microsoft Live@edu, while another is Google Apps for Education.

18.5.1.2 Platform as a service (PaaS)

Such software programs run on the operating system. PaaS gives programmers the ability to construct new cloud services or applications that don't depend on a particular platform and make them widely accessible to users over the Internet. Some of the PaaS options available are Microsoft's Azure Services Framework, Amazon's Relational Database Services, Rackspace Cloud, Google Apps Engine, and Salesforce's Force.com Development Platform.

18.5.1.3 Infrastructure as a service (IaaS)

There is always access to these data centers. For the purpose of running their own operating systems and applications, customers can rent basic computing resources on this server, such as CPUs and storage. The on-demand platform must be monitored, managed, and patched; however, you only pay for the services that you actually use. The service provides all of the capabilities you require. Through the use of cloud computing, users can view and access data online. The important users of a traditional higher education cloud include students, instructors, administrators, the Test Branch, and the Admission Branch (Figure 18.3). The educational system would make it simpler for teachers to identify problem areas where students appear to make mistakes by examining student data. In addition to using it in the classroom, students will be able to access educational material online at home to study and then be ready for lectures.

18.5.2 Blockchain technology in education

Education is a crucial sector, just like health care and banking, and it compasses numerous areas that could benefit from technological advancements. The global market for educational technology is anticipated to reach $93.76 billion by 2020. Virtual reality and customized learning with artificial intelligence are already assisting students at all levels in improving their learning experiences. We may begin to make assumptions about how these networks might be used in an educational setting by looking at other industries that are actively adopting blockchain technologies. Although there are many potential applications in education, there is a particularly intriguing couple.

- Test Prep and Learning
- Data and Library Facilities
- Transportation of products

The inadequacy of accreditation, lack of acceptance, and data security are among the issues with online education that blockchain technology can address. Currently, the Internet of Things (IoT), finance, and other sectors

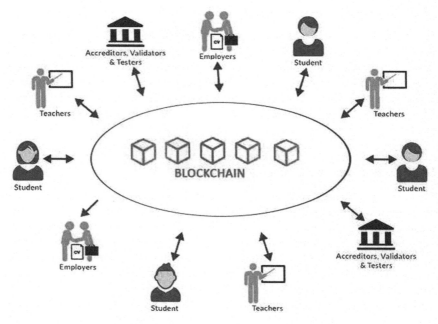

Figure 18.7 Blockchain technology in education.
Source: **Atienza-Mendez & Bayyou (2019)**

primarily use this technology. Digital money, currency conversion, trade, and payment systems are all common financial applications. Smart contracts might range from equities to bank loans and can be implemented automatically without the need for human interaction (Singh & Mishra, 2017).

This technology has also been put to the test in the field of education (Figure 18.7). Credit card authentication, safe data encryption, and distributed data storage have all been proposed as applications for blockchain (Sharples & Domingue, 2016). Based on Mozilla's open badge system, the MIT Media Lab has built a blockchain-based digital learning certificate system. The Sony Corporation of Japan's Blockchain Technology Infrastructure Platform would enable students to freely and securely exchange learning courses and data without having to divulge it to the Education Management Authority, assuring educational equity and digitalization. University College London is utilizing blockchain technology to assist postgraduate students in reducing their financial risks by verifying their academic credentials.

Blockchain is a rapidly expanding technology with several applications that affect every area of our life, including how we conduct transactions and maintain data. Analysts think blockchain technology has the ability to disrupt standard transactional procedures and data collecting facilities since blockchains are scattered, decentralized, and permanent bits of information.

Student records include assets such as attendance, classes, school payments if they attend a private school, grades, coursework, and even graduation. Because these papers are permanent, they can't be destroyed, which helps to safeguard data. The student, not the institution, is equally responsible for this. After a record is stored using blockchain, no one can tamper with it. If there is a mistake in the record, a new record must be added to fix the problem, and both the defective and correct records will be available.

18.6 ADVANTAGES AND DISADVANTAGES OF ONLINE TEACHING USING CLOUD COMPUTING AND BLOCKCHAIN TECHNOLOGY

18.6.1 Advantages of blockchain technology

Open-source software is a form of computer program whose source code is made publicly accessible under a license that permits anyone to study, modify, and distribute it for any purpose (Laurent, 2004). According to a study conducted by the European Union on Blockchain in Education, "further technological advancement in education should be viewed as a combined capability of the market and public authorities to achieve an acceptable balance between private sector innovation and public interest security."

18.6.2 Advantages of cloud computing

Personalized learning: Cloud computing makes materials available to a greater range of students.

Option in learning: A computer with an Internet connection can provide students with access to a variety of resources and software applications tailored to their learning preferences and areas of interest.

Storage space: Massive files may be stored and exchanged without the use of storage devices via the Cloud. Many organizations, including Google, Apple, Dropbox, and others, provide cloud services.

Reduced costs: Cloud-based technology may enable businesses to cut costs while increasing their use of cutting-edge tools to meet changing educational requirements. Office software can be used by students for free by purchasing, installing, and maintaining it on their computers.

Accessibility: The most essential and helpful element of adopting a cloud-based education system for the user is the consistency of materials. This system must be adaptable at all times, 24 hours a day, seven days a week. By logging in, you may access the information from anywhere.

No extra infrastructure: The colleges and the governments may now concentrate on their aims of growing research opportunities for students and improving the global environment without having to worry about dorms, laboratories, instructors, or other concerns.

Go green: The cloud will undoubtedly minimize the carbon footprint in the education sector.

User friendly: The new facility is simple to use, so there is no need to worry about its shortcomings. It is simple to use and understand (Ercan, 2010).

18.6.3 Disadvantages

The current frameworks and models for online education have a number of issues in light of an increasingly digitized and accessible Internet. The intellectual property of students cannot be sufficiently safeguarded due to the openness and hackability of the Internet, and students' privacy is at risk because courses and data protection are exclusively dependent on a centralized online education platform. A distributed and reliable data storage system must be created to track students' learning processes, make all learning data accessible to the public, provide data confidentiality and non-tamperability, and ensure that the learning process and outcomes are correct (Sun et al., 2018). Students, teachers, and their families are experiencing difficulties as a result of the new online teaching–learning culture.

- **Emotionally**: Students do not have access to personal space or time
- **Economically**: The digital divide is widening all around the world.
- **Intellectually**: Lack of expertise and understanding of digital equipment.

18.7 CONCLUSION

As the online learning culture grew in popularity, data protection became a critical security issue. If data are hosted within the institution, educational institutions will feel safer about their data. A risk arises when data are sent to a mediator for storage at an unidentified, faraway data center that is not under the company's control. Arguably, hiring numerous cloud providers is the best approach to reduce risk because using a single cloud provider creates a single point of failure. Another security risk is unwanted ads, which occur when cloud providers threaten consumers with unsolicited emails or adverts (Ercan, 2010).

Applications are now a part of cloud computing, which has gone beyond basic infrastructure services like storage and processing tools. In educational settings, the use of cloud computing is growing. Students and teachers commonly use free or inexpensive cloud services to help with learning, social networking, content development, publication, and collaboration. However, many global organizations are using it to help teach the next generation about technology[1]. The application of blockchain and cloud computing in education is still new and uncharted territory, and there are several unanswered questions. What is obvious, though, is that the possibilities are infinite. When school leaders look to the future, they would do well to think about how to use this innovative technology.

NOTE

1 http://www.ariadne.ac.uk/issue42/fraser-rvw/

REFERENCES

Atienza-Mendez, C., & Bayyou, D. G. (2019). Blockchain technology applications in education. *International Journal of Computing and Technology*, 6(11), 68–74. https://www.researchgate.net/publication/337670514_Blockchain_Technology_Applications_in_Education

Bokhari, M. U., & Makki, Q. (2018). A survey on cloud computing security. In *Springer, Singapore* (Vol. 654, pp. 149–164).

Ercan, T. (2010). Effective use of cloud computing in educational institutions. *Procedia – Social and Behavioral Sciences*, 2(2), 938–942. https://doi.org/10.1016/j.sbspro.2010.03.130

Garcés, M., Krijgsman, W., Van Dam, J., Calvo, J. P., Alcalá, L., & Alonso-Zarza, A. M. (1997). Late Miocene alluvial sediments from the Teruel area: Magnetostratigraphy, magnetic susceptibility, and facies organization. *Acta Geologica Hispanica*, 32 (3–4), 171–184.

Ise, O. A. (2015). A novel framework for student result computation as a cloud computing service. *American Journal of Systems and Software*, 3(1), 13–19. https://doi.org/10.12691/ajss-3-1-2

Kurt, S. (2020, December 28). *Kolb's Experiential Learning Theory & Learning Styles*. Educational Technology. https://educationaltechnology.net/kolbs-experiential-learning-theory-learning-styles/

Laurent, A. M. S. (2004). Understanding open source and free software licensing. In *Ariadne*. O'Reilly Media, Inc. http://www.ariadne.ac.uk/issue42/fraser-rvw/

Margianti, E. S., & Mutiara, A. B. (2015). Applications of cloud computing in education. *The 1st International Joint Conference Indonesia-Malaysia-Bangladesh-Ireland (IJCIMBI)*. https://doi.org/10.13140/RG.2.1.3506.0247

Marzano, R. J. (2007). *The Art and Science of Teaching: A Comprehensive Framework for Effective Instruction*. Association for Supervision and Curriculum Development. https://books.google.co.in/books?id=JBeFD6sTx_IC&printsec=frontcover&source=gbs_ge_summary_r&cad=0#v=onepage&q&f=false

Mell, P., & Grance, T. (2011). The NIST Definition of Cloud Computing. In *Special Publications* (Vols. 800–145).

Pund, B. G., Nair, S. S., & Deshmukh, P. P. (2012). Using cloud computing on E-learning. *International Journal of Emerging Trends & Technology in Computer Science (IJETTCS)*, 1(2), 202–209.

Robb, B. E. (2016). *A Paradigm Shift in Classroom Learning Practices to Propose Methods Aligned with a Neuroeducation Conceptual Framework*, 228.

Sergiovanni, T. J. (1994). *Building Community in Schools*. Jossey-Bass, San Francisco.

Sharma, S., Kaur, K., & Singh, A. (2014). Role of Cloud Computing in Bioinformatics. *International Journal of Computer Techniques*, *3*(3), 1–4. www.ijctjournal.org/Volume3/Issue3/IJCT-V3I3P1.pdf

Sharples, M., & Domingue, J. (2016). The blockchain and kudos: A distributed system for educational record, reputation and reward. In *Lecture Notes in Computer Science (including subseries Lecture Notes in Artificial Intelligence and Lecture Notes in Bioinformatics): Vol. 9891 LNCS* (pp. 490–496). https://doi.org/10.1007/978-3-319-45153-4_48

Singh, B., & Mishra, P. (2017). Process of teaching and learning: a paradigm shift. *International Journal of Education*, *7*(63022), 31–38.

Sun, H., Wang, X., & Wang, X. (2018). Application of blockchain technology in online education. *International Journal of Emerging Technologies in Learning*, *13*(10), 252–259. https://doi.org/10.3991/ijet.v13i10.9455

Chapter 19

Artificial intelligence-based communication systems used in industry 4.0

For multiple input and multiple output antenna 5G wireless devices

Suverna Sengar and Praveen Kumar Malik
Lovely Professional University, Jalandhar, India

Monika Agarwal
IIMT University, "O" Pocket, Meerut, India

19.1 INTRODUCTION

As the Internet of Things (IoT) develops, it becomes clear that it is a key component of many applications, including smart cities, smart homes, smart healthcare, and smart agriculture. The impressive advancement of IoT technology encounters adopting different benchmarks and advancements. Interoperability among specific devices and businesses is a big challenge with IoT devices [1]. A fully functional and effective 5G network for the IoT cannot be completed without taking AI (artificial intelligence) scheduling into account. Existing 4G networks' all-IP (Internet Protocol) broadband availability relies on responsive origination, which results in insufficient range production for IoT applications [2].

A few crucial areas of 5G network management, such as increased administration quality, working on transmitting, improved network security, and higher network productivity, will be significantly impacted by computer-based intelligence. Simulated intelligence will therefore enhance the 5G architecture from beginning to end for IoT frameworks. The key to how artificial intelligence can improve 5G is radio mindfulness, rather than a hand-made calculation, which is the best tool to make sense of the complex RF signals around the device. Enhanced radio awareness enables a number of improvements, including enhanced device usability, enhanced system performance, and enhanced radio security [3, 4].

MIMO is a requirement for networks like 5G as it is the primary air interface for 5G and 4G broadband distant communications. With the growth in the number of distant devices and the number of new adaptable clients over the past few years, interest in remote throughput has increased significantly.

19.1.1 Introduction to MIMO Technology

In single antenna systems, multipath fading and co-channel interference are common issues. These can be reduced by using MIMO technology, by placing multiple numbers of antenna elements on the receiver and transmitter sides as depicted in Figures 19.1 and 19.2. The data rates, channel capacity, and coverage area are very well enhanced, just by adding the MIMO system. The good thing about MIMO is that it can provide all said points without any additional increment in power and frequency spectrum. Depending upon the arrangement of antennae on the transmitter and receiver sides, usually, there are four types of configurations possible. In a conventional radio communication system, a single antenna is present on both the transmitter side and the receiver side, which is commonly known as the single input single output (SISO) configuration. This configuration is very simple and does not require any additional processing. But the SISO system has more chances of receiving interference and fading with the limited amount of bandwidth. In multiple input and single output (MISO) configuration,

Figure 19.1 Applications of 5G in different areas [5].

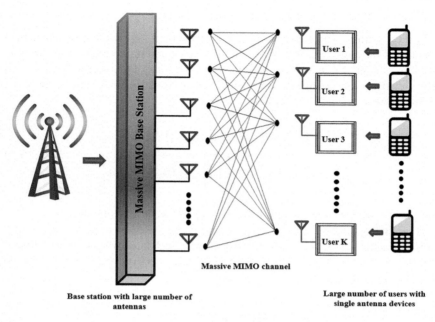

Figure 19.2 General outline of multiple input multiple output system.

multiple antennas are placed on the transmitter side and a single antenna is placed on the receiver side. In this configuration, the same data are transmitted from multiple antennas and the receiver will receive the optimum signal; thereafter it would extract the required data [6].

According to the Shannon–Hartley capacity theorem, the SISO system's capacity of the error-free channel is given by

$$C = B \log_2 \left(1 + \mathrm{SNR}\right) \tag{19.1}$$

19.1.2 MIMO diversity parameter

In this section, various diversity parameters which decide the effective performance of the MIMO antenna are described.

19.1.2.1 Envelope correlation coefficient (ECC)

In the MIMO system, the capacity in a multipath environment depends upon the number of antenna elements. If more antennas are placed in closed proximity, then very high are the chances of coupling occurring between antenna elements, which will decrease the capacity of the MIMO antenna system and antenna efficiency. There are three ways to calculate the ECC parameter [7].

19.1.2.1.1 Method 1: From far-field radiation pattern

The ECC value can be evaluated using a far-field [8]. During the measurement of ECC, all the ports of the antenna are terminated except on the port that is excited by a load impedance of 50 ohms. This method uses both the elevation and azimuthal radiation field pattern for calculating the ECC value between i^{th} and j^{th} elements.

$$\rho_e(i,j) = \frac{\left|\iint_{4\pi}^{0} \left(\vec{\beta}_i(\theta,\varnothing)\right) \times \left(\vec{\beta}_j(\theta,\varnothing)\right) d\Omega\right|^2}{\left[\iint_{4\pi}^{0} \left|\left(\vec{\beta}_i(\theta,\varnothing)\right)\right|^2 d\Omega\right]\left[\iint_{4\pi}^{0} \left|\left(\vec{\beta}_j(\theta,\varnothing)\right)\right|^2 d\Omega\right]} \quad (19.2)$$

where β_i and β_j are field radiation patterns of two radiating elements.

The ECC calculation from this method is very tedious in terms of calculation, measurement, and time consumption and is more expensive but is more accurate and exact. In this method, an anechoic chamber is used to extract the parameters.

19.1.2.1.2 Method 2: From Scattering parameter

There is another simpler, easy, and fast way to evaluate the ECC value directly from the S-parameter value [9]. The scattering parameters are extracted from each port with the help of a vector network analyzer (VNA). The ECC between the ith and jth element for N number of antenna elements is evaluated as follows [5]:

$$\rho_e(i,j,N) = \frac{\left|\sum_{n=1}^{N} S_{i,j}^* S_{n,j}\right|^2}{\pi_{k=i,j}\left[1 - \sum_{n=1}^{N} S_{k,n}^* S_{n,k}\right]} \quad (19.3)$$

19.1.2.1.3 Method 3: S-parameter and radiation efficiency

This method is suitable for the correct measurement of lossy antenna systems. It is a very complex method and not suitable for pattern shape and tilt beam for ECC calculation. This method for ECC calculation requires the radiation efficiency of each radiating element and is reliable for high efficiency. The ECC value for the N-port MIMO antenna is calculated as follows [10]:

$$\rho_e(i,j,N) = \frac{\sum_{n=1}^{N} S_{i,j}^* S_{n,j}}{\sqrt{\left(1-\sum_{n=1}^{N}|S_{n,i}|^2\right)\left(1-\sum_{n=1}^{N}|S_{n,j}|^2\right)\eta_{\text{rad},i}\,\eta_{\text{rad},j}} + \sqrt{\left(\frac{1}{\eta_{\text{rad},i}}-1\right)\left(\frac{1}{\eta_{\text{rad},j}}-1\right)}} \quad (19.4)$$

where $\eta_{\text{rad}, i}$ and $\eta_{\text{rad}, j}$ are radiation efficiency for *i*th and *j*th elements, respectively.

19.1.2.2 Diversity gain (DG)

Diversity is a very important factor for the MIMO antenna. It is achieved when the transmitter and the receiver transmit or receive multiple signals simultaneously in a multiple-channel path. The diversity of the whole communication system is decided by diversity gain. The ideal theoretical value for the uncorrelated antenna is −9.95 dB. The higher the value of diversity gain, the better the performance of the MIMO system. It is used to describe how a MIMO system increases gain in a combined signal with respect to time average SNR. The diversity gain is evaluated as follows [11]:

$$G_{\text{DG}} = 10 \times \sqrt{1 - |\rho_e|^2} \tag{19.5}$$

19.1.2.3 Mean effective gain (MEG)

The ratio of P_r (the average power received by the antenna being tested) to P_{ref} (the average power received by an isotropic antenna) in the same environment can be used to express the MEG [12].

$$G_e = \frac{P_r}{P_{\text{ref}}} \tag{19.6}$$

For the MIMO system, MEG can be evaluated as follows [9]:

$$\text{MEG} = \int_0^{2\pi}\int_0^{\pi} \left\{ \frac{\text{XPR}}{1+\text{XPR}} G_\theta(\theta,\varnothing) \times P_\theta(\theta,\varnothing) \right. \\ \left. + \frac{\text{XPR}}{1+\text{XPR}} G_\varnothing(\theta,\varnothing) \times P_\varnothing(\theta,\varnothing) \right\} \sin\theta \, d\theta \, d\varnothing \tag{19.7}$$

where $G_\theta(\theta, \varnothing)$ represents the θ part of the antenna power gain, $P_\theta(\theta, \varnothing)$ represents the θ and $P_\varnothing(\theta, \varnothing)$ represents the ∅ component of the angular density function, and XPR denotes the cross-polarization coupling.

$$\text{XPR}\,(\text{cross polarization coupling}) = \frac{P_1}{P_2}$$

in which P_1 and P_2 are the mean received powers of vertically polarized and horizontally polarized isotropic antennae, respectively.

The MEG can also be calculated from the knowledge of scattering parameters [13].

$$\text{MEG}_i = 0.5\left[1 - \sum_{j=1}^{N}\left|S_{i,j}\right|^2\right] \tag{19.8}$$

The MEG for any antenna element must not be more than −3 dB, and the difference in MEG between any two components must be less than 3 dB for a MIMO antenna to function effectively.

19.1.2.4 Channel capacity loss (CCL)

The CCL is defined as the maximum reliable data transmission over a transmission channel. In the MIMO system, the practical accepted limit is less than 0.4 bits/s/Hz through the entire operating frequency band. The CCL can be evaluated as follows [14]:

$$C_{\text{Loss}} = -\log_2 \det(\psi^R) \tag{19.9}$$

where ψ^R is the receiving antenna's correlation matrix. For the N-port MIMO system, the process of computing is mentioned below.

$$\psi^R = \begin{bmatrix} \psi_{11} & \psi_{12} & \cdots & \psi_{1N} \\ \psi_{21} & \psi_{22} & \cdots & \psi_{2N} \\ \psi_{31} & \psi_{12} & \cdots & \\ & & \cdots \cdots & \\ \psi_{N1} & \psi_{N2} & \cdots & \psi_{NN} \end{bmatrix} \tag{19.10}$$

$$\psi_{nn} = 1 - \sum_{m=1}^{4}\left|S_{nm}\right|^2 \tag{19.11}$$

$$\psi_{nm} = -\left(S_{nn}^* S_{nm} + S_{mn}^* S_{nm}\right) \tag{19.12}$$

19.1.2.5 Total active reflection coefficient (TARC)

For a single antenna element, the reflection coefficient ($|S_{11}|$) is sufficient to analyze the performance of the antenna. But for the MIMO system, TARC is a very essential parameter to measure radiation characteristics and bandwidth for different polarization operations. It is the ratio of the square root of the difference between available powers generated by all excitation minus radiated power divided by available power [15].

$$\Gamma_a^t = \sqrt{\frac{\text{available power} - \text{radiated power}}{\text{available power}}} = \sqrt{\frac{P_a - P_r}{P_a}} \qquad (19.13)$$

TARC is a real number and its value normally ranges between 0 and 1. If all the delivered power is radiated by the antenna element, then it will be 0, and it is 1 if completely reflected back. TARC is generally expressed on a decibel scale. The practical accepted value for effective performance is less than −10 dB through the entire operating band. For the N-port MIMO system, if the ith port of the antenna is excited and the remaining ports are at matched loads, then TARC can be directly calculated from the scattering matrix by equation (19.13).

$$\Gamma_{ai}^t = \sqrt{1 - P_{ri}} = \sqrt{\sum_{j=1}^N |S_{i,j}|^2} \quad i = 1, 2, \ldots N \qquad (19.14)$$

It can alternatively be written as the ratio of the square root of the total power that is reflected to the square root of the total power that is incident [16].

$$\Gamma_a^t = \frac{\sqrt{\sum_{i=1}^N |b_i|^2}}{\sqrt{\sum_{i=1}^N |a_i|^2}} \qquad (19.15)$$

where b_i indicates the reflected power and a_i denotes the incident power. For the $N \times N$ MIMO antenna, the scattering matrix can be described as follows:

$$\begin{bmatrix} b_1 \\ b_2 \\ \vdots \\ b_N \end{bmatrix} = \begin{bmatrix} S_{11} & S_{12} & \ldots & S_{1N} \\ S_{21} & S_{22} & \ldots & S_{2N} \\ \vdots & \vdots & & \vdots \\ S_{N1} & S_{N2} & \ldots & S_{NN} \end{bmatrix} \begin{bmatrix} a_1 \\ a_2 \\ \vdots \\ a_N \end{bmatrix} \qquad (19.16)$$

If the magnitude of the reflected signal is assumed to be unity but the phase is varying at different angles, the TARC for the N-port MIMO antenna can be evaluated at different phase angles by [17]:

$$\Gamma_{ai}^t = \frac{\sqrt{\left|S_{11} + S_{12}e^{j\theta_1} + \ldots + S_{1N}e^{j\theta_{N-1}}\right|^2 + \left|S_{21} + S_{22}e^{j\theta_1} + \ldots + S_{2N}e^{j\theta_{N-1}}\right|^2 + \ldots + \left|S_{N1} + S_{N2}e^{j\theta_1} + \ldots + S_{NN}e^{j\theta_{N-1}}\right|^2}}{\sqrt{N}} \qquad (19.17)$$

19.1.3 Antenna basic parameters

Following are the parameters on which basis antenna performance is analyzed [18–26].

19.1.3.1 Antenna gain

An antenna's ability to send and receive radio waves in a specific direction is referred to as its gain. This is more advantageous than an isotropic antenna. When it comes to transmitting antennas, the gain is the ratio of the input energy that is turned into radio waves in a single direction. In the case of receiving antennas, the gain is the proportion of radio frequency waves that are converted into electrical signals. The gain is a consequence of both the antenna's efficiency and directivity. The connection between directivity and gain is represented graphically, called the radiation characteristics. Mathematically, the gain is expressed as a product of directivity and efficiency:

$$G = \eta D \qquad (19.18)$$

The latter of which is dimensionless and lies between 0 and 1, with 1 representing a perfectly lossless antenna. In reality, antenna gain is always lower than directivity [27].

19.1.3.2 Input impedance

The input impedance of an antenna is described as "the impedance presented by an antenna at its terminals or the ratio of the voltage to the current at the pair of terminals or the ratio of the appropriate components of the electric to magnetic fields at a point". It is represented by the mathematical expression in equation (19.19):

$$Z_{im} = R_{im} + jX_{im} \qquad (19.19)$$

The amount of energy stored in the vicinity of an antenna is determined by the imaginary part of its input impedance, Z_{im}, which is composed of R_{im} (the antenna's resistance) and X_{im} (the antenna's reactance). Further, R_{im} is divided into two components: resistance to radiation (R_r) and resistance to loss (R_L). The actual power transmitted by the antenna is related to the radiation resistance, while the power lost as heat is due to either dielectric or antenna conductive losses [28].

19.1.3.3 Return loss

The ratio of reflected power (P_r) to the source transmitted power (P_t) is known as return loss; this is normally measured at the coaxial cable's input end where the antenna is connected. The return loss should be measured in dB and should be a negative number that is as huge as possible for maximum power transfer, with the desired value being as close to zero as possible [29, 30]. A large negative value indicates a good return loss (19.19):

$$R_L(\text{dB}) = -20\log 10\,\Gamma \qquad (19.20)$$

where |Γ| represents the reflection coefficient.

19.1.3.4 VSWR

How well the antenna impedance is matched to the transmission line to obtain the most power is determined by the voltage standing wave ratio (VSWR) parameter. The impedances at both ends must be equal for the source to transmit the most power to the load. The ending point experiences full or partial reflection of a wave if the transmission line is not concluded properly. The VSWR is produced in the line as a result of the interaction of reflected and incident waves. The maximum to minimum voltage amplitude ratio is known as the VSWR [31, 32], and it can be calculated using the equation in (19.20):

$$\text{VSWR} = V_{\max}/V_{\min} \qquad (19.21)$$

The VSWR is a measure derived from the reflection coefficient which describes the amount of power that is reflected from the antenna. This ratio can be calculated using a specific equation (19.21):

$$\text{VSWR} = \frac{1+|\Gamma|}{1-|\Gamma|} \qquad (19.22)$$

The VSWR value lies between 2 and 1. When the VSWR is low, the antenna is optimized for the transmission line, allowing for more power delivery with no reflected signals (Figure 19.3).

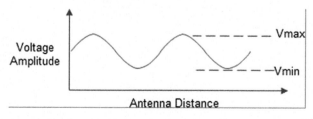

Figure 19.3 Generalized VSWR along the transmission line.

19.1.3.5 Bandwidth

It is written as "the range of frequencies within which the performance of the antenna, with respect to some characteristic, conforms to a specified standard". The frequencies on either side of the cut-off frequency are considered as needing antenna performance criteria such as radiation pattern, input impedance, beamwidth, and polarization that are within the acceptable range when it comes to the center frequency.

19.1.3.6 Radiation pattern

The radiation pattern of an antenna describes how the radio waves emitted from the antenna or received by the antenna depend on the angular direction. This is shown graphically and can be a 2D or 3D plot. It is comprised of three parts: the main lobe, the side lobe, and the back lobe. The main lobe is the major part of the power emitted from the antenna and is the most directional, while the side and back lobes represent the energy that is wasted in other directions. The side lobes are minor and spread the power outward, while the back lobe is opposite to the main lobe and reflects energy in the opposite direction (Figure 19.4).

The most commonly used radiation patterns for antennas are the omnidirectional/non-directional pattern, pencil-beam pattern, and fan-beam pattern. The omnidirectional pattern appears as a figure of eight when viewed in 2D, and gives a doughnut-like geometry in 3D. The pencil-beam pattern has a sharp, directional beam with the shape of a pencil, while the fan-beam pattern has a fan-shaped beam.

Some antenna designs are shown in Table 19.1.

The identified research gaps offer great potential for a future roadmap for the 5G and IoT businesses. The conventional approaches of managing radio resources and interference in single-level networks will probably not perform

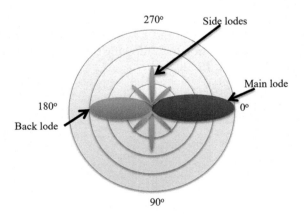

Figure 19.4 Generalized radiation pattern of an antenna.

Table 19.1 Comparison of various MIMO antenna designs

Ref. No.	Authors	Year	Work done	Findings	Application
[33]	Yasar Amin, Seyed Sajad Mirjavadi, Muhammad Jamil Khan, Niamat Hussain, Mahnoor Khalid, Syeda Iffat Naqvi, and MuhibUr Rahman	2020	Here, the authors designed a 4-port MIMO antenna array that uses the mm-wave frequency spectrum. Size of antenna: 30 × 35 × 0.76 mm^3. Substrate used: FR4	Frequency range: 24.25–27.5 GHz & 27.5–28.35 GHz. Gain of antenna: 8.3 dBi. ECC is less than 0.01. DG is greater than 9.96 dB.	Used mm-wave spectrum for 5G applications
[34]	Philip Ayiku Dzagbletey and Young-Bae Jung	2018	For the millimeter wave (mm-wave) spectrum, a parasitic patch antenna with 42 elements was created.	Frequency range: 27.5–28.5 GHz. Gain of antenna: 21.4 dBi. Cross-polarization is greater than −12 dB. Side-lobe level (SLL) is −19.1 dB.	5G-mobile communication system for base station applications.
[35]	Mian Muhammad Kamal, Shouyi Yang, Xin-cheng Ren, Saad Hassan Kiani, Ahsan Altaf, Muhammad Rizwan Anjum, Amjad Iqbal, Muhammad Asif, and Sohail Imran Saeed	2021	Here, a single-layer MIMO antenna array was suggested as a novel design. Size of antenna: 30 × 30 mm^2. Substrate used: Rogers RT/Duroid 5880, ε_r = 2.2.	Frequency range: 28 GHz. Gain of the antenna is 5.5 dBi. ECC is less than 0.16 dB. Efficiency (%η) improved by 92%. Isolation is −29 dB.	5G mm-wave communication applications.
[36]	Ajay Kumar Dwivedi, Anand Sharma, Akhilesh Kumar Pandey, Vivek Singh	2021	A 2-port wide-band circularly polarization (CP) small-size MIMO antenna array was created in this case. Substrate used: FR4, ε_r = 4.4.	Frequency range: 3.3–4.2 GHz. ECC is less than 0.10 dB. Isolation is greater than 15 dB. TARC is −15 dB. DG is 9.94 dB. Efficiency (%η) improved by 95%.	Used in sub-6 GHz applications.
[37]	Evizal Abdul Kadir	2017	The author of this article suggested a reconfigurable antenna system to overcome the MIMO's data throughput limits. Substrate used: FR4 epoxy, ε_r = 4.7.	Frequency range: 2.4–2.6 GHz. Isolation is greater than 15 dB. Reflection coefficient is −24.3 dB.	WLAN and LTE.

well in this situation, forcing the creation of alternative strategies. MIMO and 5G technologies are not yet capable of addressing the heterogeneity and scalability issues of IoT. Given the resource-constrained nature of IoT devices, multi-traffic data transfer via 5G is a difficult task. In this chapter, the 5G technologies were integrated with the IoT and artificial intelligence to develop RF devices.

REFERENCES

[1] Alhayani, B. and Abdallah, A.A. "Manufacturing intelligent Corvus corone module for a secured two way image transmission under WSN", *Engineering Computations*, 38.4, 1751–1788, 2021. https://doi.org/10.1108/EC-02-2020-0107

[2] A. A. R. Alsaeedy and E. K. P. Chong, "Mobility management for 5G IoT devices: Improving power consumption with lightweight signaling overhead", *IEEE Internet of Things Journal*, 6.5), 8237–8247, 2019. https://doi.org/10.1109/JIOT.2019.2920628

[3] Wang, J., Weitzen, J., Bayat, O. et al. Interference coordination for millimeter wave communications in 5G networks for performance optimization. *Journal on Wireless Communications and Networking*, 46, 2019. https://doi.org/10.1186/s13638-019-1368-6

[4] Gao, X., Dai, L., Chen, Z., Wang, Z., Zhang, Z. Near-Optimal Beam Selection for Beamspace MmWave Massive MIMO Systems, 2015 IEEE, 2015.

[5] Alhayani, B., Kwekha-Rashid, A.S., Mahajan, H.B. et al. "5G standards for the Industry 4.0 enabled communication systems using artificial intelligence: perspective of smart healthcare system", *Applied Nanoscience*, 2022. https://doi.org/10.1007/s13204-021-02152-4

[6] Balevi, Eren, Andrews, Jeffrey G., "Wideband channel estimation with A generative adversarial network", *IEEE Transactions on Wireless Communications*, 20.5, 3049–3060, 2021.

[7] Balevi, Eren, Doshi, Akash, Andrews, Jeffrey G., "Massive MIMO channel estimation with an untrained deep neural network", *IEEE Transactions on Wireless Communications*, 19.3, 2079–2090, 2020.

[8] Zheng, Kan, Zhao, Long, Mei, Jie, Shao, Bin, Xiang, Wei, Hanzo, Lajos, "Survey of Large-Scale MIMO Systems", *IEEE Communications Surveys and Tutorials*, 17.3, 1738–1760, 2015.

[9] Gupta, Akhil, Jha, Rakesh Kumar, "Power optimization using massive MIMO and small cells approach in different deployment scenarios", *Wireless Networks*, 23.3, 959–973, 2017.

[10] Abd El-Hameed, Anwer S., Wahab, Mohamed G., Elshafey, Nashwa A., Elpeltagy, Marwa S., "Quad-Port UWB MIMO antenna based on LPF with vast rejection band", *AEU – International Journal of Electronics and Communications*, 134, 153712, 2021.

[11] Ahmad, Ashfaq, Choi, Dong-You, Ullah, Sadiq, "A compact two elements MIMO antenna for 5G communication", *Scientific Reports*, 12.1, 3608, 2022.

[12] Khan, Aqeel Ahmed, Naqvi, Syed Aftab, Khan, Muhammad Saeed Ijaz, Bilal, "Quad port miniaturized MIMO antenna for UWB 11 GHz and 13 GHz

frequency bands", *AEU – International Journal of Electronics and Communications*, 131, 153618, 2021.
[13] Mahmood, Sarmad Nozad, Ishak, Asnor Juraiza, Ismail, Alyani, Soh, Azura Che, Zakaria, Zahriladha, Alani, Sameer, "ON-OFF body ultra-wideband (UWB) antenna for wireless body area networks (WBAN): A review", *IEEE Access*, 8, 150844–150863, 2020.
[14] Kommuri, Ramesh R., Kiran, Usha, "Isolation enhancement for dual-band MIMO antenna system using multiple slots loading technique", *International Journal of Communication Systems*, 33.12, 2020. https://doi.org/10.1002/dac.4470
[15] Saxena, Gaurav, Jain, Priyanka, Awasthi, Yogendra Kumar, "High diversity gain MIMO-antenna for UWB application with WLAN notch band characteristic including human interface devices", *Wireless Personal Communications*, 112, 105–121, 2020.
[16] Naktong, Watcharaphon, Ruengwaree, Amnoiy, "Four-port rectangular monopole antenna for UWB-MIMO applications", *Progress in Electromagnetics Research B*, 87, 19–38, 2020.
[17] Tirado-Méndez, José Alfredo, Jardón-Aguilar, Hildeberto, Rangel-Merino, Arturo, Vasquez-Toledo, Luis Alberto, Gómez-Villanueva, Ricardo, "Four ports wideband drop-shaped slot antenna for MIMO applications", *Journal of Electromagnetic Waves and Applications*, 34.9, 1159–1179, 2020.
[18] Kiem, Nguyen Khac, Phuong, Huynh Nguyen Bao, Chien, Dao Ngoc, "Design of compact 4 × 4 UWB-MIMO antenna with WLAN band rejection", *International Journal of Antennas and Propagation*, 2004, 539094, 2014.
[19] Zhao, Xiongwen, Riaz, Sharjeel, "A dual-band frequency reconfigurable MIMO patch-slot antenna based on reconfigurable microstrip feedline", *IEEE Access*, 6, 41450–41457, 2018.
[20] Kumar, Pawan Urooj, Shabana Alrowais, Fadwa, "Design and implementation of quad-port MIMO antenna with dual-band elimination characteristics for ultra-wideband applications", *Applied Sciences* (Switzerland), 10.5, 1715, 2020.
[21] Stutzman, W. L., *Polarization in Electromagnetic Systems*, MA: Artech House, 1993.
[22] James, J. R., Hall, P. S., *Handbook of Microstrip Antennas*, Vol. 1, London: Peter Peregrinus Ltd., 1989.
[23] Prasad, K. D., *Antenna wave and propagation*, Satya Parkashan, 1983.
[24] Bhatacharya, A. K., Garg, R., "Generalized transmission line model for microstrip patches", *IEE Proceedings Microwaves, Antennas and Propagation*, Pt. H, Vol. 132, No. 2, pp(s): 93–98, 1985.
[25] Khalid, Mahnoor, Naqvi, Syeda Iffat, Hussain, Niamat, Rahman, MuhibUr, Fawad, Seyed Sajad Mirjavadi, Khan, Muhammad Jamil, Amin, Yasar, "4-Port MIMO antenna with defected ground structure for 5G millimeter wave applications", Nov 15 2022.
[26] Dzagbletey, Philip Ayiku, Jung, Young-Bae, "Stacked microstrip linear array for millimeter-wave 5G baseband communication", *IEEE Antennas and Wireless Propagation Letters*, 17, 780–783, 2018 IEEE.
[27] Kamal, Mian Muhammad, Yang, Shouyi, Ren, Xin-Cheng, Altaf, Ahsan, Kiani, Saad Hassan, Anjum, Muhammad Rizwan, Iqbal, Amjad, "Infinity Shell Shaped MIMO Antenna Array for mm-Wave 5G Applications", *Electronics*, 10.2), 165, 2021.

[28] Dwivedi, Ajay Kumar, Sharma, Anand, Pandey, Akhilesh Kumar, Singh, Vivek, "Circularly polarized two port MIMO cylindrical DRA for 5G applications", October 3, 2020 IEEE Xplore.

[29] Kadir, Evizal Abdul, "A reconfigurable MIMO antenna system for wireless communications", 19–21 September 2017.

[30] Malik, P., Lu, J., Madhav, B.T.P., Kalkhambkar, G., Amit, S. (Eds.) *Smart antennas: Latest trends in design and application*, Springer, ISBN 978-3-030-76636-8, https://doi.org/10.1007/978-3-030-76636-8

[31] Roges, R., Malik, P.K. "Planar and printed antennas for Internet of Things-enabled environment: Opportunities and challenges", *The International Journal of Communication Systems*, 34.15, e4940, 2021. https://doi.org/10.1002/dac.4940(IF: 2.047)

[32] Rahim, Abdul, Malik, Praveen Kumar, "Analysis and design of fractal antenna for efficient communication network in vehicular model", *Sustainable Computing: Informatics and Systems*, 31, 100586, 2021. https://doi.org/10.1016/j.suscom.2021.100586

[33] Shaik, N., Malik, P.K. "A comprehensive survey 5G wireless communication systems: Open issues, research challenges, channel estimation, multi carrier modulation and 5G applications", *Multimedia Tools and Applications*, 2021. https://doi.org/10.1007/s11042-021-11128-z

[34] Malik, P.K., Wadhwa, D.S., Khinda, J.S. "A survey of device to device and cooperative communication for the future cellular networks", *International Journal of Wireless Information Networks*, 27, 411–432, 2020. https://doi.org/10.1007/s10776-020-00482-8

[35] Tiwari, P., Malik, P. K. "Wide band micro-strip antenna design for higher 'X' band", *International Journal of e-Collaboration (IJeC)*, 17.4, 60–74, 2021. http://doi.org/10.4018/IJeC.2021100105

[36] Wadhwa, D.S., Malik, P.K., Khinda, J.S. "High gain antenna for n260- & n261-bands and augmentation in bandwidth for mm-wave range by patch current diversions", *World Journal of Engineering*, 2021, ahead-of-print. https://doi.org/10.1108/WJE-03-2021-0133

[37] Kaur, Amandeep, Malik, Praveen Kumar, "Multiband elliptical patch fractal and defected ground structures microstrip patch antenna for wireless applications", *Progress in Electromagnetics Research B*, 91, 157–173, 2021. https://doi.org/10.2528/PIERB20102704

Chapter 20

Wireless technologies and the Internet of Things

Vikrant Pachouri, Samta Kathuria, Shweta Pandey, Rajesh Singh, Anita Gehlot and Shaik Vaseem Akram
Uttaranchal University, Dehradun, India

Praveen Kumar Malik
Lovely Professional University, Phagwara, Punjab, India

Safia Yasmeen
Alfaisal University, Saudi Arabia

20.1 INTRODUCTION

The Internet of Things (IoT) would link 500 billion gadgets by the year 2050. Today, there is an increasing number of real-world applications in various fields, like security, agriculture, intelligent cities, and houses, as an outcome of the exponential development of IoT technology. IoT uses typically have certain criteria, such as a low enough bitrate for data sharing, low energy consumption, significant ranges, and economic considerations. Many times, short-range techniques like Bluetooth and ZigBee are insufficient for long-range transmission applications, particularly in surveillance uses where the transfer of linked things is crucial. Cellular communication-based solutions like 3G, 4G, and 5G provide a wide coverage area. However, the consumption of power from sources is very high while causing additional costs of connection imposed by the operators [1].

Some developing technologies, including IoT and the fifth-generation (5G) cellular network, have changed the globe during the past ten years and enable anything, anyone, anytime, and anywhere (4A) [2].IoT is now widely used in various fields, including connectivity & communication, the living & homely environment, medical health care, agriculture, logistics & transport, infrastructure, intelligent cities, industries, education, and government, among others. In addition to these sectors, the fourth industrial revolution's core idea—the IoT—points to a new level in the management of the entire chain of value creation. But rather than introducing a novel theoretical idea, IoT has started a revolution in the communication & information sector [3]. The capacity to connect and interact with smart, autonomous, and heterogeneous devices is a crucial component of the IoT idea. Such linked gadgets connect to the Internet, a global network with many diverse topologies.

The capacity of wireless data transfer from one point to another is the definition of wireless communication technology. Data are communicated from one source to another at various locations, which can vary from a few millimeters to several kilometers, depending on the realm of the use of the IoT system, over explicitly defined transmission channels. There isn't a single technique for wireless transmission because of the diversity of the market and the technologies now in use; instead, they may typically be classified according to the size of the electromagnetic waves they emit and the frequencies at which function. Radio frequency, infrared, microwave, and light wave are the four fundamental signal types utilized for wireless communication [4].

The approach used in the current study was used to assess earlier studies that looked at the importance of wireless connection techniques. In this study, the evolution of intelligent technology is evaluated. The key contribution of the study is as follows:

- The significance of various wireless technologies enabled in IoT systems to overcome the challenges faced by the technology.
- The characteristics and criteria analysis has been discussed which gives a brief about the selection of various wireless communication technologies.
- Finally, based on the study, the chapter presents the discussion and recommendations that are vital for future enhancement.

20.2 BACKGROUND

The IoT system has found usage in a variety of routine activities as well as in niche applications. The fundamental requirements that IoT imposed in terms of extended, low data range, low energy use, and cost-effectiveness set the framework for system designers but issues need to be addressed [5]. Due to their diversity, the heterogeneous communication technologies that are now available span the whole sector, but their applications are inadequate. In order to satisfy the demands of a certain user and application, it is required to periodically examine the technologies that are now accessible. The many types of wireless networks utilized in the IoT across regions of short range, long range, and cellular are classified based on the details of IoT applications [6]. Bluetooth, ZigBee, NFC, and Wi-Fi are examples of short-range communication technologies that have been thoroughly studied as well as technologies found. The overall inference is that long-distance transmission scenarios are not compatible with short-range radio systems. The most widely used long-range communication technology in the IoT is represented by the main technologies currently available [7].

In order to facilitate a number of applications, including smart cities, agriculture, and energy, the 5G-IoT is designed to achieve a scene where objects are seamlessly linked into the real world and information networks. A new technology trend is emerging with 5G [8].

20.3 COMPARATIVE ANALYSIS CRITERIA

The selection of the best way in terms of an acceptable software & hardware architecture is significantly influenced by the range of different IoT network applications. The broad range of IoT applications, where each one explicitly defines the hardware, software, and communication requirements, establishes the elements that must be taken into account throughout the analysis process. The process of design and implementation of an IoT network depends on various features those are:

A. *Reach*: The use of IoT in smart city environments goes against the specifications for the range of a communications network and can range from a few kilometers to a few millimeters away. Communication networks may be categorized into three broad groups based on the distance between IoT nodes. Short-, medium-, and long-range networks are all available [9].

B. *Bandwidth*: Nodes may provide information regularly depending on the IoT network application; for instance, in smart home systems, fire detection sensors connect to the network and send information if smoke or fire is detected. Security applications that leverage IoT nodes, such as security cameras, require ongoing connectivity as well as the delivery of large volumes of data. When choosing an IoT network implementation, bandwidth is a crucial component to consider since it influences power consumption and reach, which impacts other aspects like the cost and cost-effectiveness of the overall system [10].

C. *Coverage*: Wireless coverage is defined as the extent of the area to which the wireless signals are transmitted and generally can be divided into five categories that differ in distance and speed of access:
   - Wireless Body Area Network (WBAN)
   - Wireless Personal Area Network (WPAN)
   - Wireless Local Area Network (WLAN)
   - Wireless Metropolitan Area Network (WMAN)
   - Wireless Wide Area Network (WWAN)

   Coverage ranges from a few centimeters up to several kilometers. Coverage is a requirement but the conditions are not sufficient for transmission of data and sharing to be possible [11].

D. *Energy efficiency*: When developing an IoT network, one of the crucial characteristics is energy efficacy, which is the consumption of energy by nodes of IoT. The IoT idea includes autonomous sensor nodes with a longer lifespan, which presents significant development challenges. Energy efficiency continues to be a crucial component in the design process even as the efficiency and autonomous power source's quality are improved. The consumption of sensor components must be monitored since it can drastically shorten an IoT node's lifespan. The atmosphere of the area where the IoT nodes are placed may also have an impact on energy usage [12].

E. *Reliability*: The most crucial element is that the information can be delivered successfully to the final location as quickly as feasible if the created system of IoT is utilized in medical systems to the patient's critical parameters monitoring. The transmission or deletion of information about the humidity of soil at a certain period will not significantly harm an IoT system used in agriculture. Reliability is at the forefront of an important consideration for the selection of implementation of IoT networks since it has a substantial influence on both growing bandwidth and cost [5].

F. *Latency*: The time taken for a request to go from a transmitting object to a receiving object, including the time taken by the destination source to process it, is known as network latency. It is crucial that the information be sent to the intended location when IoT nodes are used in industrial settings to deliver signals and control information. Reduced latency has the impact of shortening the lifespan of autonomous IoT nodes, making it a crucial consideration in the construction of an IoT network.

G. *Cost effectiveness*: The budget, requirements, and capabilities that the developed system of IoT which must have proportionately governed by the cost-effectiveness criteria, which is an overall criterion that is strongly tied to the business model. According to all other criteria that depend on the topology of network nodes, cost-effectiveness is evaluated. To choose the network that will be most cost-effective for a certain IoT system, it is important to first comply with the system's fundamental needs [13] (Figure 20.1).

20.4 WIRELESS TECHNOLOGIES USED FOR IoT

The recently developed technologies used for wireless communication technologies in IoT systems are given below:

A. *LoRa*: The French start-up Cycleo in Grenoble debuted LoRa in 2009. It was brought after three years by Semtech (USA). The LoRa Alliance, which utilized it in 42 countries, carried out its standardization later, especially in 2015. Due to the growing investments made by several mobile businesses in France, the Netherlands, South Africa, etc., this technology has been deployed in more nations. Using a unique spread spectrum technique, LoRa is a physical layer technology that transforms communications into the sub-GHZ ISM band [14].

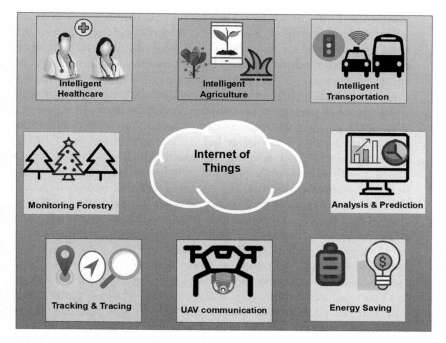

Figure 20.1 Various IoT applications.

To strike a balance between data throughput and geographic interval, LoRa uses six different spreading parameters. A longer range comes at the expense of a lower data rate when the spreading factor is raised, and vice versa. The LoRa physical layer data rate varies from 300 bps to 50 kbps depending on the channel's bandwidth and spreading factor. Additionally, each message's largest payload size is 243 bytes. The LoRa-Alliance standardized the first iteration of LoRaWAN, a LoRa radio layer-based network communication protocol, in 2015. Messages sent by a terminal using LoRaWAN may be received by several base stations nearby. This technique improves the message reception quality by analyzing such redundant reception [15]. One of the most well-known LPWAN (Low Power Wide Area Network) technologies is LoRaWAN, a form of wireless network. The LoRaWAN idea is the foundation of many commercial and industrial applications today. It employs a network and application server, also known as the LoRa Network and LoRa Application Server as of November 2019, to connect with the backend system. The server was renamed ChirStack since SemTech had trademark protection on the term LoRa[16].

B. 5G: IoT devices are utilized in both public and private locations, including farms, industries, houses, and streets. By using a 4G network, certain locations cannot be connected to smartphones. In the meantime, the development of 5G networks is increasingly serving as a primary engine for the expansion of IoT. With the increased coverage regions, fast throughput, reduced latency, and tremendous bandwidth connection density of 5G, the Internet can now link billions of sensors. The public's rising interest in 5G-IoT compels people to carefully comprehend its principles, possibilities, and difficulties [17].

C. *Bluetooth*: A wireless communication technology called Bluetooth tries to link mobile devices together. An essential criterion for BLE is its energy usage. The low energy standard, which was initially introduced in Bluetooth v4.0, provides wireless communications with lower consumption and cost. The recently created Bluetooth standard, version 5.3, keeps the same low Energy architecture as Bluetooth versions 4.0, 5.0, 5.1, and 5.2 with a little modification that increases efficiency, security, and stability. With a central device, peripherals can list their preferred channels. BLE is specific to low-bandwidth devices and those fuelled by autonomous power sources [18].

D. *Near-field communication (NFC)*: Sony and Philips created the wireless communication technology known as NFC. It allows for two-way data transfer when two devices are put side by side at a distance of 10 cm, and operates on the idea of magnetic coupling between the devices. NFC uses RFID technology and runs at up to 13.56 MHz. includes a tag that can be read-only or read-write depending on how the device rewrites the data. The most widespread use of personal data storage and access is in mobile devices for contactless payments [19].

E. *SigFox*: In order to link low-power devices, such as energy meters and smartwatches, which must be continually powered on and have very slow communication rates, SigFox technology was launched in the 2010s. SigFox operates in the ISM radio band, which has a channel capacity of 100 MHz and operates at 868 MHz in Europe and 903 MHz in the USA. SigFox is utilized to traverse wide areas and access items buried underneath. The coverage range of SigFox cells is less than 10 km in densely populated regions and between 30 and 50 km in rural areas. Overall, SigFox makes it possible to offer a wide area network with little power usage. Over 1.3 billion people live in the 72 nations that the SigFox IoT system now covers. Based on its own proprietary technology, SigFox offers an end-to-end IoT connectivity solution [20].

F. *NB-IoT*: In order to offer huge connections, wide-area coverage, extremely low power consumption, and cheap cost for IoT in 5G, the 3GPP created a new LPWAN radio technology called narrowband IoT. A promising new IoT connectivity technique for 5G is NB-IoT. Specific indoor coverage, low cost, extended battery life, and high connection density are the main goals of NB-IoT. It employs SC-FDMA for uplink communication and a 200 kHz bandwidth to narrow band

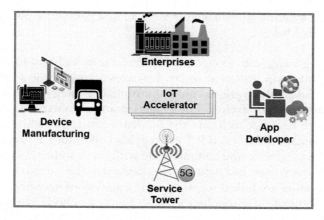

Figure 20.2 IoT accelerator and developer.

with OFDM modulation for downlink communication. An open source for IoT in 5G applications is NB-IoT. Based on the open source ENB of LTE technology, the open source NB-IoT is the outcome of collaboration between three providers, namely EURECOM, B-COM, and NTUST [21].

G. *ZigBee*: The radio frequency ranges used for industry, science, and medicine are where ZigBee communicates. It offers an IoT network solution with low power consumption and large-area connectivity. Compared to other communication technologies, ZigBee technology in IoT networks offers benefits due to its ease of use, adaptability, and low cost. ZigBee has a data rate of roughly 250 kbps and a transmission distance of about 100 m, depending on the environment and power supply. ZigBee is utilized in applications including home automation, data gathering, and industrial equipment control that need extremely low data rates, short range, and long-lasting batteries [22].

H. *Wi-Fi*: Wi-Fi is a well-known family of standard-based wireless communication systems. It is frequently used for Internet access and local area networks of devices within 100 m. It works in the 2.4–5 GHz range of frequencies. Wi-Fi is a viable connecting option for IoT networks and is ideal for short-range communication. The non-profit Wi-Fi Alliance, which accredits Wi-Fi interoperability, promotes the technology [23] (Figure 20.2).

20.5 CHARACTERISTICS OF WIRELESS TECHNOLOGIES WITHIN IoT

The characteristics of various technologies were taken into consideration to make an effective IoT environment and increase the system's performance.

The efficient solution to manage the demand and gain more popularity has been addressed below:

A. *Heterogeneity and cooperation of devices*: In addition to measuring different items, different smart devices working in the sensing layer may also employ various communication and energy-saving methods.
B. *Collaboration of devices*: The gadgets in an IoT system should be able to work together to locate the needed information and meet a specific demand. The Social IoT concept also takes into account how the devices cooperate and communicate with one another [24].
C. *Diverse network and networking standards*: The various sensors and aggregators are linked together using a variety of networking topologies. Some of these topologies use an aggregator node as a gateway to a deeper layer of the architecture, which relays the data sent to them via the network.
D. *Limitation of devices*: The smart devices that coexist in the IoT system of today meet high capacity requirements, including a variety of communication technologies, operate at a low cost, and consume little power, all with the goal of extending the life cycle of the device before it needs to be recharged or stops working and needs to be replaced. Utilizing renewable energy sources to harvest energy would considerably help to overcome these constraints and improve an IoT system's functionality [25].
E. *Self-organization, self-configuration, and autonomy*: An IoT system's nodes must be able to adjust to any environmental risks as well as any network changes that may have an impact on the system's functioning and operation. In order to identify prospective intruders and assist the system in staying up to date and defending against hacking threats, the self-configuration feature is crucial against any potential privacy problems.
F. *Unpredictable mobility pattern*: The performance of the system shouldn't be impacted by the fact that the numerous sensors might be static or mobile. Data obtained by any IoT application can advise the user about parking spots nearby that don't travel along a set route but instead follow the user's search for an urban environment [26].
G. *Multi-hop communication*: The sensors should use the Internet to interact with other sensors if necessary because, depending on the network topology and the IoT architecture they have chosen, they may need to relay data to an aggregator node before sending it to the gateway [27].

20.6 DISCUSSION AND RECOMMENDATIONS

The objective of wireless technologies in IoT has been discussed. This section gives a throughput of the challenges faced by the technology and future recommendation for further research in wireless technologies of IoT.

- It has been argued that the IoT would revolutionize many different areas to benefit humanity with 5G. In order to support and advance IoT applications and make them more widely used, low-power communication technologies will be crucial. Communication technologies like ZigBee, LoRa, SigFox, and NB-IoT have certain benefits including lower consumption of energy, wide coverage, usage of frequency bands which is unlicensed & compatibility with IoT networks, which are becoming more and more popular. While implementing these technologies, there are numerous challenges, but the two biggest ones are energy efficiency and security awareness. This offers potential research directions for the adoption of 5G in IoT applications in the future.
- The development of security mechanisms for IoT devices, gateways, edge devices, and cloud servers can address security and privacy concerns. The most pressing issue with communication solutions in the IoT 5G Network is security. And technology-based communication and cloud-based IoTnetworks may be able to tackle the problem of efficient energy.
- These include (1) scalability, (2) heterogeneity, (3) interoperability, (4) reliability & availability, (5) QoS, and (6) improvement in linked capabilities. The different open issues of IoT are also in front of us.

20.7 CONCLUSION

The current chapter has covered wireless technologies in IoT systems and research timetables. The properties of wireless technologies inside the IoT have been outlined in general and a brief review of several wireless technologies has been conducted. According to the survey's findings, the IoT will shape humanity's future since everything is connected via the Internet. Various applications benefiting mankind, including smart homes, smart agriculture, smart cities, smart factories, and IoT systems, have emerged as a result of the emergence of IoT in 5G networks. The choice of a certain technology depends on factors like range, capacity, and linked devices. IoT has an inherent trait called heterogeneity. It includes a number of difficulties including re-configuration and interoperability that raise the problem of data validity. IoT is about making effective decisions to make the best use of resources connected through the Internet. In the end, using 5G enables wireless mobile to completely connect with the human and gadgets that will be connected via the Internet and a wide area network solution will be presented on a worldwide scale. Additionally, the difficulties posed by the technologies enabled have been analyzed and described, and recommendations have been made in accordance with those findings.

REFERENCES

1. Rahimi, H., Zibaeenejad, A., & Safavi, A. A.A novel IoT architecture based on 5G-IoT and next generation technologies. In *2018 IEEE 9th annual information technology, electronics and mobile communication conference (IEMCON)* (pp. 81–88). IEEE, 2018.
2. Agiwal, M., Roy, A., & Saxena, N.Next generation 5G wireless networks: A comprehensive survey. *IEEE Communications Surveys & Tutorials*, 18(3), 1617–1655, 2016.
3. Alzubi, J. A., Manikandan, R., Alzubi, O. A., Gayathri, N., & Patan, R.A survey of specific IoT applications. *International Journal on Emerging Technologies*, 10(1), 47–53, 2019.
4. Sheng, Z., Mahapatra, C., Zhu, C., & Leung, V. C. Recent advances in industrial wireless sensor networks toward efficient management in IoT. *IEEE Access*, 3, 622–637, 2015.
5. Sinha, R. S., Wei, Y., & Hwang, S. H.A survey on LPWA technology: LoRa and NB-IoT. *ICTExpress*, 3(1), 14–21, 2017.
6. Sjöström, D. *Unlicensed and licensed low-power wide area networks: Exploring the candidates for massive IoT*, 2017.
7. Mekki, K., Bajic, E., Chaxel, F., & Meyer, F.A comparative study of LPWAN technologies for large-scale IoT deployment. *ICT Express*, 5(1), 1–7, 2019.
8. Shah, S. H., & Yaqoob, I.A survey: Internet of Things (IOT) technologies, applications and challenges. *2016 IEEE Smart Energy Grid Engineering (SEGE)*, 381–385, 2016.
9. Mohamed, K. S., & Mohamed, K. S. *The era of internet of things: Towards a smart world* (pp. 1–19). Springer International Publishing, 2019.
10. Singh, Devendra, Singh, Rajesh, Gehlot, Anita, Akram, Shaik Vaseem, Priyadarshi, Neeraj, and Twala, Bhekisipho. An imperative role of digitalization in monitoring cattle health for sustainability. *Electronics* 11, no. 17 (2022): 2702.
11. Mahmoud, M. S., & Mohamad, A. A. A study of efficient power consumption wireless communication techniques/modules for internet of things (IoT) applications, 2016.
12. Kimothi, Sanjeev, Singh, Rajesh, Gehlot, Anita, Akram, Shaik Vaseem, Malik, Praveen Kumar, Gupta, Anish, and Bilandi, Naveen. "Intelligent energy and ecosystem for real-time monitoring of glaciers." *Computers and Electrical Engineering* 102 (2022): 108163.
13. Vujović, V. Modelom upravljani razvoj arhitekture Senzor Veb mreža (Doctoral dissertation, University of Novi Sad (Serbia)), 2016.
14. Abdallah, W., Mnasri, S., & Nasri, N.Emergent IoT wireless technologies beyond the year 2020: A comprehensive comparative analysis. In *2020 International Conference on Computing and Information Technology (ICCIT-1441)* (pp. 1–5). IEEE, 2020.
15. Priyadarshini, I., Mohanty, P., Kumar, R. et al. A study on the sentiments and psychology of twitter users during COVID-19 lockdown period. *Multimedia Tools and Applications*, 2021. https://doi.org/10.1007/s11042-021-11004-w
16. Azad, C., Bhushan, B., Sharma, R. et al. Prediction model using SMOTE, genetic algorithm and decision tree (PMSGD) for classification of diabetes mellitus. *Multimedia Systems*, 2021. https://doi.org/10.1007/s00530-021-00817-2

17. Priyadarshini, I., Kumar, R., Tuan, L.M. et al. A new enhanced cyber security framework for medical cyber physical systems. *SICS Software-Intensive Cyber-Physical Systems*, 2021. https://doi.org/10.1007/s00450-021-00427-3
18. Priyadarshini, I., Kumar, R., Sharma, R. Singh, Pradeep Kumar, & Satapathy, Suresh Chandra, Identifying cyber insecurities in trustworthy space and energy sector for smart grids, *Computers & Electrical Engineering*, 93, 2021, 107204.
19. Singh, R., Sharma,R., Akram, S.V., Gehlot, A., Buddhi, D., Malik, P.K., & Arya, R., Highway 4.0: Digitalization of highways for vulnerable road safety development with intelligent IoT sensors and machine learning, Safety Science, 143, 105407, 2021. ISSN 0925-7535
20. Singh, Digvijay, Akram, Shaik Vaseem, Singh, Rajesh, Gehlot, Anita, Buddhi, Dharam, Priyadarshi, Neeraj, Sharma, Gulshan, & Bokoro, Pitshou N. Building integrated photovoltaics 4.0: Digitization of the photovoltaic integration in buildings for a resilient infra at large scale. *Electronics*, 11, no. 17, 2022, 2700.
21. Sharma, R., Kumar, R., Sharma, D.K. et al. Water pollution examination through quality analysis of different rivers: a case study in India. *Environment, Development and Sustainability*, 2021. https://doi.org/10.1007/s10668-021-01777-3
22. Ha, D.H., Nguyen, P.T., Costache, R. et al. Quadratic discriminant analysis based ensemble machine learning models for groundwater potential modeling and mapping. *Water Resources Management*, 2021. https://doi.org/10.1007/s11269-021-02957-6
23. Sharma, Sameer Dev, Sharma, Sonal, Singh, Rajesh, Gehlot, Anita, Priyadarshi, Neeraj, & Twala, Bhekisipho. Stress detection system for working pregnant women using an improved deep recurrent neural network. *Electronics* 11, no. 18, 2022, 2862.
24. Sharma, R., Gupta, D., Polkowski, Z., & Peng, S.-L. Introduction to the special section on big data analytics and deep learning approaches for 5G and 6G communication networks (VSI-5g6g). *Computers & Electrical Engineering*, 95, 107507, 2021.https://doi.org/10.1016/j.compeleceng.2021.107507
25. Kogias, D. G., Michailidis, E. T., Tuna, G., & Gungor, V. C. Realizing the wireless technology in internet of things (IOT). In *Emerging Wireless Communication and Network Technologies: Principle, Paradigm and Performance* (pp. 173–192), Springer International Publishing. 2018.
26. Mieczkowski, G., Borawski, A., & Szpica, D. Static electromechanical characteristic of a three-layer circular piezoelectric transducer. *Sensors*, 20, no. 1, 222, 2019.
27. Kim, S. W., Jeong, M. S., Lee, I., & Kwon, I. B. Static mechanical characteristics of tin-coated fiber Bragg grating sensors. *Sensors and Actuators A: Physical*, 214, 156–162, 2014.

Chapter 21

Artificial intelligence and blockchain-based intervention in building infrastructure

Vikrant Pachouri, Shweta Pandey, Samta Kathuria, Rajesh Singh, Anita Gehlot and Shaik Vaseem Akram
Uttaranchal University, Dehradun, India

Praveen Kumar Malik
Lovely Professional University, Phagwara, Punjab, India

Ahmed Alkhayyat
The Islamic University, Najaf, Iraq

21.1 INTRODUCTION

The UN Agenda states that by 2030, everyone would have access to essential services and decent, safe, and affordable housing. To achieve this, there is an increasing daily need for smart buildings. People who are wealthy, have a higher chance of developing their human capital, and make use of the possibilities offered by metropolitan areas are more likely to have suitable accommodation. In addition to assisting linked industries and fostering economic growth, job creation, service provision, and general poverty reduction, a prosperous housing sector also serves as a development multiplier [1]. The world has now witnessed Industry 4.0, the most recent version of the industrial ecosystem. In order to fulfil the escalating system protection needs, this fourth revolution necessitates the introduction of linked technologies. The ultimate objectives of creating such automatic relationships range from generating income to improving productivity. The most recent industrial infrastructure has established the groundwork for the desired smart manufacturing system, where the convergence between humans and robots relies on data, after fusing modern technologies such as artificial intelligence (AI), etc. The edge sensor's data is crucial for forecasting maintenance, tracking equipment irregularities, and monitoring the production process. Data is a crucial component, and if it doesn't meet security requirements, all of its activities will surely have an impact on or even shut down the entire industrial ecosystem. The Guardian and ABC's new security articles reinforce the US and Australian assertions that other nations are stealing their copyright data for their industries [2].

AI is a branch of science that conducts extensive study in fields which include robotics, ML, processing of natural language, and processing of image. Industrial AI, on the other hand, is a methodical field that focuses on creating, testing, and implementing diverse ML algorithms for industrial uses with long-term efficiency. It serves as a bridge between academia findings in AI to industry users and serves as a technique and discipline to offer answers for industrial uses [3]. Pillars of the digital physical systems supporting Industry 4.0 have been growing as AI and ML technologies combined with large data obtained through contemporary cyber technologies. Almost every industrial sector, including manufacturing, finance, transportation, healthcare, and research, is currently exploring AI-enabled processes. AI technology has significantly changed how some industries' function [4].

The industrial world gains new, cutting-edge features thanks to blockchains. Many current industrial processes are improved, optimised, secured, and streamlined with the use of these qualities. A blockchain is a decentralised, continually expanding list of data points, or "blocks," spread throughout a peer-to-peer network that are connected and safeguarded using encryption. Being immune to data tampering or modifications is a key characteristic of a blockchain. The recorded data in any one block "cannot be altered retrospectively without the changing of all following blocks, which requires collaboration of the network majority," according to the cryptographic methods used in blockchain transactions [5]. There are various attempts in investigations and use of blockchain beyond payments and to other domains and contexts. The fundamental application of blockchain is to save data as a secure exchange. Blockchain 2.0 is primarily connected with automatic cyber banking utilising contracts smartly as opposed to Blockchain 1.0, which is generally associated with crypto-currency and payments. Blockchain 3.0, a current era, is centred on the requirements of the cyber society, such as industry 4.0 and smart cities[6].

The key contribution in the current study is as follows:

1. An overview of SDGs as well as their integration into Industry 4.0 and the importance of AI and blockchain.
2. Recent scopes and trends for sustainability of manufacturing in Industry 4.0 have been described.
3. Various issue discussion and future recommendation have been provided accordingly.

The chapter is divided as follows: Section 21.2 deals with background regarding Industry 4.0, Section 21.3 discusses the role of AI in Industry 4.0, Section 21.4 discusses the role of blockchain in Industry 4.0, Section 21.5 briefs on the manufacturing sustainability in Industry 4.0, Section 21.6 provides the discussion and future recommendation, and lastly Section 21.7 concludes the chapter.

21.2 BACKGROUND

21.2.1 Regarding Industry 4.0

The topic "Industry 4.0" was initially expressed in Germany in 2011 as a proposed plan for developing a new high-tech economic paradigm in Germany [7]. The idea has ushered in the fourth revolution of technology, which depends on thoughts and developments in online-offline systems, AI and block chain and depends on perpetual connection via Internet which enables interaction continuously and transfer of information not only between humans and machine, but also between the machines itself. This channel of communication has an effect on how knowledge management develops 4.0 [8].

21.2.2 Building 4.0 revolution with respect to Industry 4.0

Every industry has seen a revolution thanks to Industry 4.0 [8]. The conceptual framework known as "Building Revolution" was created to bring sustainability development objectives in the building sector. Any economy needs the construction sector, but it also has a big influence on the environment [9]. The biggest consumer of energy, materials, and water, as well as a significant polluter, is the building industry [10]. In light of these effects, it is necessary to use appropriate methods and take precautions to increase the sustainability of building activities. Increasing people's life and enabling them to stay in a safe environment with better economic, social, and environmental conditions is the goal of the sustainability idea [11]. Without doubt, the world's most intensive-resource industry is construction [9]. In order to construct smart building, proper monitoring and controlling of resources is necessary.

At the AIIB, the term "Intelligent Building" was first used in the United States in the 1980s. A collection of mechanical, technical, and electronic components working together to accomplish specified duties constitutes an intelligent building [12]. To increase the relaxation of living culture or environment, autonomous systems must be installed. Buildings will be automatically governed by these systems with regard to energy management, temperature, lighting, and other elements. Undoubtedly, one of the most intensive-resource industries in the world is the construction industry [13]. Through sensor networks, it may be able to perceive or regulate the environment. It may also identify objects by scanning their radio frequency identification (RFID) tags. Many industries, including healthcare, the smart grid, smart cities, and home automation, increasingly employ AI technology [14].

An intelligent technological system has the capacity to reason rationally and think like a person. With the appropriate approach, communities will be able to utilise human knowledge and experience alongside the Industry 4.0

concept at every stage of life, boosting both production and wellbeing[15]. People's need for intelligent devices for both indoor and outdoor living settings increases as technology advances. However, the traditional control system can only intelligently respond to the needs of human comfort by making mechanical modifications to living variables [16]. The automatic control of different sensors, devices, and appliances in the building is the present desire so that this system provides the comfort we live in (Figure 21.1).

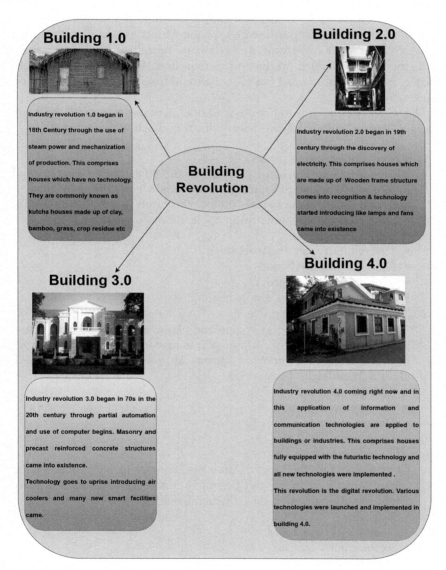

Figure 21.1 Building revolution.

21.3 ROLE OF AI IN INDUSTRY 4.0

AI is regarded as the primary building block of Industry 4.0, which aids in increasing the productivity and autonomy of industries. AI is a mix of various techniques which enables software and machines to comprehend, perceive, act, and learn based on self-learning or behaviours that are enhanced from those of humans. There are four main categories in which industry-related AI research may be divided: 1) maintenance and quality prediction; 2) design generative; 3) activities of supply chain; and 4) collaboration of human–robot. The main purpose of applying AI methods in manufacturing is to save cost, increase income, and eliminate mistakes. The blueprints for Industry 4.0's AI enabled production systems first. Additionally, it has been shown that application of AI is a growing trend for sustainable manufacturing systems [17].

The industry AI relates to the physical operations or system of an organisation. Various AI-based processes are illustrated below.

A. AI-based monitoring process
   There is a requirement of monitoring the efficiency of systems and finding ways to identify faults or situations which produce unacceptable outcomes. These models predict the future state of the systems. Various examples of monitoring application are:
   - Control of quality
   - Detection of fault and isolation
   - Maintenance prediction
   - Monitoring of inventory
   - Risk management of supply chain [18]
B. AI-based optimisation process
   AI-based decision support system and planning allows users to define a plan or path for getting a desired system state that optimise a target of business metrics. There are various AI tools which include genetic algorithm and neural networks:
   - Scheduling of job
   - Management of yield
   - Design of product
   - Location of facilities [19]
C. AI-based control process
   Control system is the heart of latest operations which is needed by the organisation to reap the full benefit of automation. There are many applications that benefit the AI which includes:
   - Robotics
   - Automation in factory
   - Autonomous vehicles
   - Automation of HVAC
   - Smart grids [20]

21.4 Role of blockchain technology in Industry 4.0

Industry 4.0's use of blockchain technology provides new tools for addressing sustainability challenges, system security, effectiveness, and resilience, as well as transparent and decentralised transaction platforms. Blockchain's traceability and transparency features improve the sustainability of production networks. Blockchain technology aids in cost reduction and real-time production transparency. Blockchain technology enhances the security and transparency of smart production, according to a survey of the dairy sector. Two different kinds of blockchain networks exist: those linked to the machine level to receive and collect data are 1) public and 2) private [17] (Figure 21.2).

A promising and often used technology is blockchain. It shows many applications in many different disciplines. Below is a description of a few blockchain industrial applications that offer benefits to industries based on their respective fields.

A. Financial industry: In the financial sector, blockchain is essential. In general, it offers interaction with third parties who carry out financial transactions between individuals and businesses. These third parties perform the following four tasks:
- Verifying the truth of the deals
- Avoiding the duplication of financial transactions
- Validating and registering financial activities
- Function of agents in support with clients or associates

Blockchain takes over two functions: it prevents double-entry transactions and validates and records financial activity. Blockchain makes it simple by avoiding numerous payments for the same thing [21].

B. Healthcare industry: Patient data is one of the most important elements in the healthcare sector. Typically, a patient's medical records are dispersed over several systems run and owned by one or more healthcare organisations. The capacity to convert patient information into an electronic medical record (EMR) has been made possible by the digital revolution. Due to many security and privacy concerns, sharing EMRs among several healthcare-related companies is fraught with difficulties. A secure EMR and other types of healthcare data may be shared across several providers securely using blockchain. A new start-up business by the name of Gem created a network based on the blockchain for creating healthcare applications and global healthcare data. Blockchains enable access easier and need fewer interactions with owners and third parties to approve and export the necessary data. Additionally, it saves a great deal of time and effort and lowers the cost of R&D [22].

C. Logistics industry: Applications for logistics management are computer programs that assist in controlling the transportation of goods, services, and raw materials between manufacturers and retailers and consumer destinations. These may operate inside a single organisation

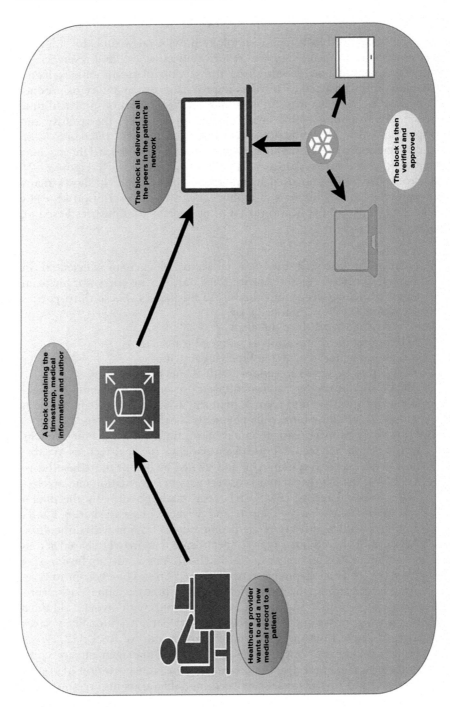

Figure 21.2 Blockchain-based patient data management.

or entity or span many. The blockchain can offer strong assistance, enabling these applications. The involvement of various firms in the operations is one of the challenges in management of logistics. This may involve a variety of cooperated sub-activities carried out by many businesses, such as storage facilities, manufacturers, transportation firms, and regularity agencies. Any management programme of logistics must include a collection of features that enable users to plan, organise, coordinate, monitor, and validate processes. Blockchain may effectively and securely enable such operations [23].

D. Manufacturing industry: The industry of manufacturing is moving forward quickly with autonomous operations and smart production, and blockchain can help in a variety of ways. Delays in time, management expense, and mistakes by human may all be decreased by using blockchain technology for industrial logistics management. Additionally, cloud manufacturing activities may be supported by blockchain technology. In order to transform manufacturing activities and resources into a collection of manufacturing services that can be intelligently connected and controlled, cloud manufacturing uses topics from cloud computing, IoT, virtualisation and service-oriented computing. A secure decentralised manufacturing architecture of cloud and privatise knowledge exchange for manufacturing of designs, such as developing and rebuilding injection moulding, are also made possible by blockchain technology [24].

E. Energy industry: In the energy sector, blockchain is preferred to microgrids. A micro-grid is a small network of connected electric power sources and loads that is designed to increase the reliability and efficiency of energy generation and consumption. Power buying and selling transactions in micro-grids may be facilitated, recorded, and validated using blockchain technology. Energy trade in smart grids can be made possible by blockchain on a bigger scale. Industry IoT energy trade is made possible by blockchain. Blockchain has the potential to lower energy costs and boost resilience in energy-related applications [25].

21.5 MANUFACTURING SUSTAINABILITY IN INDUSTRY 4.0

Instead of focusing on reducing carbon emissions and pollution, manufacturing sustainability focuses on the development and expansion of industries in a way that is socially just and environmentally responsible. The UN Sustainability 2030 Agenda, which encompasses low-impact industrialisation, smart manufacturing, and energy-efficient buildings, highlights sustainability as the foundation of corporate initiatives. According to the report, implementing sustainable policies in manufacturing procedures and utilising renewable energy sources both aid in achieving sustainability in Industry 4.0 activities. In Industry 4.0, developing countries have more potential for sustainability since their economies heavily depend on small- and medium-sized businesses (SMEs). The study used the ISM approach to assess the key function of sustainability for Industry 4.0 and discovered that innovation of business model and production efficiency are the two most important sustainability-related functions [26].

A. Economic sustainability: Only a few studies on the viability of the economy in Industry 4.0 have been published. Increased resource efficiency and higher taxes as a result of economic sustainability improvements have an impact on both social and environmental sustainability. Economic and political issues affect how Industry 4.0 develops industrial sustainability. Industry 4.0 techniques assist in lowering operating costs, advancing the circular economy, and increasing market share of goods that increase economic sustainability in industries [27].

B. Social sustainability: Another dilemma that is brought on by Industry 4.0 technology is the sustainability of the economy. Digitalisation and automation in production processes are the fundamental prerequisites of Industry 4.0. Less manual labour is required as Industry 4.0 becomes increasingly automated. The development of multidisciplinary skills that improve employee capacities and prepare them to handle the difficulties of Industry 4.0 should be the main emphasis of new training models in industries. Industries in the fourth revolution need to adopt fresh strategies to raise their skill levels: training skill development is followed by virtual training programmes, user-centred procedures, digital and soft skill development, and manufacturing system [28].

C. Environmental sustainability: One of the main pillars that contribute to sustainable business operations is environmental sustainability. To improve the quality of life, technology and innovation development is accelerating. The majority of environmental sustainability advances are concentrated on resource efficiency and climate challenges. Industry 4.0 technologies use a lot of energy since they are reliant on smart devices and sensors, which are not ideal for the environment. The disposal of hazardous waste and the production of greenhouse gases raise serious concerns on the sustainability of the environment. As a result, Industry 4.0 has to create new tools to address these problems [29].

21.6 DISCUSSION AND FUTURE RECOMMENDATION

The objective of Industry 4.0 is to attain sustainability. This section details the future recommendation in Industry 4.0.

- The majority of industrialised countries have knowledge of Industry 4.0 and prepared the ground for the adoption of different technologies in Industry 4.0. In order to successfully implement various technologies in Industry 4.0, sustainability must be taken into consideration. In the

future, it will be possible to determine the effect and relation between the aspects associated with sustainability and Industry 4.0 using multicriteria, decision-making techniques, uncertainty in decision-making and tools of statistics.
- The economic and operational benefits of blockchain have been applied across a variety of industrial disciplines. Resolving various issues will make blockchain more appealing and simpler to implement in commercial applications. To make blockchain more usable and effective, a lot of research studies have been done and a lot of features have been added. However, there are still certain unresolved problems that require additional investigation and analysis in order to enable its widespread practical and useful business applications, accomplishing the objectives. To solve these problems and fill in the gaps for more effective, scalable, and secure blockchain industrial applications, future studies in industry 4.0 need to be carried out.
- The use of intelligent security, parking, and safety is necessary for the advancement of smart buildings. AI technology is used to provide intelligent security. AI technology is capable of performing the analysis of acquired data as well as tracking authorised by user access. However, there is danger associated with making snap decisions which has to be overcome by conducting further research.

21.7 CONCLUSION

A new trend called "Industry 4.0" has emerged as a result of the rapid advancement of AI and blockchain in recent years and its integration into the Fourth Industrial Revolution. Blockchain technology has become a cutting-edge tool for decision-making and inclusion in many sectors. Because AI is being used, Industry 4.0 is more prepared for the future. This chapter provides the introduction of AI in Industry 4.0, including proactive process management, prescriptive optimisation, and passive monitoring. Manufacturing sustainability emphasises new SME trends and goes on to define dependence on the economy, environment, and society. Utilising the latest technology, Industry 4.0 has enabled monitoring, regulating, and optimising a variety of organisations. Further, the challenges faced by the technologies applied in Industry 4.0 have been explained and reviewed and the future recommendation has been provided accordingly.

REFERENCES

1. Jaramillo, L. G. Make cities and human settlements inclusive, safe, resilient and sustainable. In Cristian Parra, Brandon Lewis, & Saleem H. Ali (eds.), *Mining, Materials, and the Sustainable Development Goals (SDGs)* (pp. 99–116). CRC Press (2020).

2. Rahman, Z., Yi, X., & Khalil, I. Blockchain based AI-enabled industry 4.0 CPS protection against advanced persistent threat. *IEEE Internet of Things Journal* (2022). DOI:10.1109/JIOT.2022.3147186
3. Lee, J., Davari, H., Singh, J., & Pandhare, V. Industrial Artificial Intelligence for industry 4.0-based manufacturing systems. *Manufacturing Letters*, 18, 20–23 (2018).
4. Jan, Z., Ahamed, F., Mayer, W., Patel, N., Grossmann, G., Stumptner, M., & Kuusk, A. Artificial Intelligence for Industry 4.0: Systematic review of applications, challenges, and opportunities. *Expert Systems with Applications*, 216, 119456 (2022).
5. Attaran, M., & Gunasekaran, A. Blockchain-enabled technology: The emerging technology set to reshape and decentralise many industries. *International Journal of Applied Decision Sciences*, 12(4), 424–444 (2019).
6. Alladi, T., Chamola, V., Parizi, R. M., & Choo, K. K. R. Blockchain applications for industry 4.0 and industrial IoT: A review. *IEEE Access*, 7, 176935–176951 (2019).
7. Betül, G. Ü. R., & Franco Mosconi. The new European industrial policy: Global competitiveness and the manufacturing renaissance. *Industrial Policy*, 1(1), 33–37 (2015).
8. Roblek, V., Meško, M., & Krapež, A. A complex view of industry 4.0. *Sage Open*, 6(2), 2158244016653987 (2016).
9. Akadiri, P. O., Chinyio, E. A., & Olomolaiye, P. O. Design of a sustainable building: A conceptual framework for implementing sustainability in the building sector. *Buildings*, 2(2), 126–152 (2012).
10. Abidin, N. Z. Investigating the awareness and application of sustainable construction concept by Malaysian developers. *Habitat International*, 34(4), 421–426 (2010).
11. Ortiz, O., Castells, F., & Sonnemann, G. Sustainability in the construction industry: A review of recent developments based on LCA. *Construction and Building Materials*, 23(1), 28–39 (2009).
12. Novak, M., Belany, P., Bolf, A., & Roch, M. Design of a control system for an intelligent building. In *2018 ELEKTRO* (pp. 1–4). IEEE (2018).
13. Yan-Fang, L., Gui-Xian, Z., & Ai-Qin, L. "The design of intelligent building control system" in *Electric Information and Control Engineering (ICEICE)*, Wuhan, China (2011)
14. Rahman, A., Nasir, M. K., Rahman, Z., Mosavi, A., Shahab, S., & Minaei-Bidgoli, B. Dist Block Building: A distributed blockchain-based SDN-IoT network for smart building management. *IEEE Access*, 8, 140008–140018 (2020).
15. Karagöz, E., &Tecim, V. An integrated model and application for smart building systems with artificial intelligence. In Sezer Bozkuş Kahyaoğlu (ed.) *The Impact of Artificial Intelligence on Governance, Economics and Finance* (Vol. 2, pp. 15–40). Springer Nature Singapore (2022).
16. Graveto, V., Cruz, T., & Simões, P. Security of building automation and control systems: Survey and future research directions. *Computers & Security*, 112, 102527 (2022).
17. Jamwal, A., Agrawal, R., Sharma, M., & Giallanza, A. Industry 4.0 technologies for manufacturing sustainability: A systematic review and future research directions. *Applied Sciences*, 11(12), 5725 (2021).

18. Park, H. S., Phuong, D. X., & Kumar, S. AI based injection molding process for consistent product quality. *Procedia Manufacturing*, 28, 102–106 (2019).
19. Makkar, S., Devi, G. N. R., & Solanki, V. K. Applications of machine learning techniques in supply chain optimization. In *ICICCT 2019–System Reliability, Quality Control, Safety, Maintenance and Management: Applications to Electrical, Electronics and Computer Science and Engineering* (pp. 861–869). Springer Singapore (2020).
20. Bécue, A., Praça, I., & Gama, J. Artificial intelligence, cyber-threats and Industry 4.0: Challenges and opportunities. *Artificial Intelligence Review*, 54(5), 3849–3886 (2021).
21. Mainelli, M., & Von Gunten, C. Chain of a lifetime: How block chain technology might transform personal insurance. How Block chain Technology Might Transform Personal Insurance-Long Finance (2014).
22. Prisco, G. The block chain for healthcare: Gem launches gem health network with Philips block chain lab. *Bitcoin Magazine*, 26 (2016).
23. Hackius, N., & Petersen, M. Block chain in logistics and supply chain: trick or treat?. In *Digitalization in Supply Chain Management and Logistics: Smart and Digital Solutions for an Industry 4.0 Environment. Proceedings of the Hamburg International Conference of Logistics (HICL)* (Vol. 23, pp. 3–18). Berlin: epubli GmbH (2017).
24. Barenji, A. V., Li, Z., & Wang, W. M. Block chain cloud manufacturing: Shop floor and machine level. In *Smart Sys Tech 2018; European conference on smart objects, systems and technologies* (pp. 1–6). VDE(2018).
25. Al-Jaroodi, J., & Mohamed, N. Block chain in industries: A survey. *IEEE Access*, 7, 36500–36515 (2019).
26. Vinuesa, R., Azizpour, H., Leite, I., Balaam, M., Dignum, V., Domisch, S., & Fuso Nerini, F. The role of artificial intelligence in achieving the sustainable development goals. *Nature Communications*, 11(1), 233 (2020).
27. Braccini, A. M., & Margherita, E. G. Exploring organizational sustainability of industry 4.0 under the triple bottom line: The case of a manufacturing company. *Sustainability*, 11(1), 36 (2018).
28. Luthra, S., Kumar, A., Zavadskas, E. K., Mangla, S. K., & Garza-Reyes, J. A. Industry 4.0 as an enabler of sustainability diffusion in supply chain: An analysis of influential strength of drivers in an emerging economy. *International Journal of Production Research*, 58(5), 1505–1521 (2020).
29. Vrchota, J., Pech, M., Rolinek, L., & Bednář, J. Sustainability outcomes of green processes in relation to industry 4.0 in manufacturing: Systematic review. *Sustainability*, 12(15), 5968 (2020).

Index

Pages in *italics* refer to figures.

5G, 247, 249, 251, 252, 253

Air Pollution, 69, 72, 75
Antenna, 108, 110, 114, 118
Apache, 130
Artificial Intelligence, 137, 184, 191, 206, 271, 275, 300
Attacks, 38, 40, 46, 64, 201, 246
Authentication, 41, 201, 246, 251

Big Data, 18, 21, 23, 42, 214, 221
Blockchain, 18, 50, 52, 54, 64, 162, 273, 311
Blockchain Security, 63, 159
Building 4.0, 304

Classroom, 229, 234
Cloud Computing, 62, 230, 263
Constituents, 225
Convolutional Neural Network, 4, 16, 79
Cyber-Physical System, 128, 214
Cybersecurity, 38, 43, 78

Data Collection, 60, 71, 259
Deep Learning, 78, 88, 165
Diabetes, 31, 37
Discretization, 99

Education, 50, 63, 150
Evolution, 246

Hadoop, 130, 133
Healthcare 4.0, 18, 35
Heart Monitoring, 28

Identification, 29, 42
Importance, 80, 148, 232
Industry 4.0, 277, 303
Internet of Things, 18, 35, 38, 67, 107, 138, 195
IoT, 18, 29, 38, 65, 140, 196, 219

Land Pollution, 69

Machine Learning, 18, 77, 130, 196
Machine-To- Machine (M2m) Communication, 219
Medical Image Processing, 80, 91
MIMO, 247, 254
Monitoring, 257, 294

Network, 16, 26, 37, 39

Pollution Index, 69, 74

Resnet Neural Network, 86
Robotics, 214

Security, 19, 38, 48, 159, 185, 253, 275, 307
Smart Healthcare, 23, 199
Smart Networks, 24
Surgery, 27, 31
Sustainability, 36, 61

User Access, 41, 311

Water Pollution, 69, 75
Wireless Networks, 246, 255
Wvd-Based Methods, 97